Scott
Maple® for Environmental Sciences

Springer-Verlag Berlin Heidelberg GmbH

Bill Scott

Maple®
for Environmental Sciences

a Helping Hand

 Springer

Bill Scott

Murdoch University
School of Environmental Science
Perth, Western Australia, 6150
Australia
e-mail: scott@maple.murdoch.edu.au

Cip data applied for
Die Deutsche Bibliothek - CIP-Einheitsaufnahme

Scott, Bill:
Maple for environmental sciences : a helping hand / Bill Scott. - Berlin; Heidelberg; New York; Barcelona; Hong Kong;
London; Milan; Paris; Singapore; Tokyo: Springer, 2001
ISBN 978-3-540-65826-9   ISBN 978-3-642-56685-1 (eBook)
DOI 10.1007/978-3-642-56685-1

Mathematics Subject Classification (2000): 92F05, 68W30

ISBN 978-3-540-65826-9

Maple® is a registered trademark of Waterloo Maple Inc. This trademark is being used with kind permission of Maple Inc.

http://www.springer.de

© Springer-Verlag Berlin Heidelberg 2001
Originally published by Springer-Verlag Berlin Heidelberg New York in 2001

Cover design: *design&production,* Heidelberg
Typesetting by the author
Printed on acid-free paper   SPIN 10705563   40/3142ck-5 4 3 2 1 0

Preface

What is this book about? Please take this book as it is, a working document. It started as an idea that has grown. It will never be correct but should be self-correcting. In the limit, if there is one, the book should approach a 'correct' state. It is not the detail, and the numbers, that matter, but the structures and the order. These structures are inherently linked with the many minds that have made Maple, the minds of perhaps the best mathematicians, certainly some of the most useful.

Our environment is not separate from mathematics; mathematics is but one tool, of several, to help with understanding the environment. It is a harsh tool that requires numbers and symbolism; Maple handles the symbolism superbly; numbers need more consideration. We have included a substantial amount on reading and writing numbers, data, and dealing with floating point numbers.

It is the 'devil in the detail' that continually comes back to us in working with Mathematics and Maple. It becomes 'raw' and defined. Many of the things we do have rational and logical bases, but we don't know what they are. Often, in following the code and 'talking' with an input line to Maple, the detailed way of performing a task becomes clear. But not without frustration; the task is invariably simple, though.

Complexity comes from the simple, redone. Hopefully, this book reveals how one can work with the simplest of operations. This somehow grows, forming and defining the structure of the problem; the physics, boundaries and interconnections. And Maple does the grunge work, accurately and robustly.

This book starts at the basics and interleaves applications with an ever-growing level of mathematics. At the start, the material will be too simple; at the end, the material will seem too simple. In between, there will be difficulties, some that require a deal of thought and perhaps background reading. But all should be viewed as examples of what can be done. Some examples are full applications to ordinary environmental problems; many are simplified to reduce the number of pages required.

The Maple input to the computer needs to be tolerated. It is a little cryptic, but mostly English and logical. If you don't like the particulars, you can always define your own. Maple as a language is powerful; many of the raw commands can be viewed and even altered to suit your needs. You can use the maths mode that allows the input to *look* like mathematical symbolisms and this is agreeable to the eye. But, remember, you need brief, quick and unambiguous words to communicate with maths. The rather harsh Maple words may be a little too *massaged* in mathematical squiggles.

Presentation always comes into scientific endeavor; on paper, on screens, and over the Internet. There are many adequate ways of doing these things and Maple wouldn't be one of the best. But Maple does these things and may do admirably. Especially with graphical information, it seems *there are always better ways.* Importantly, Maple makes it possible to make and massage a graph to do 'what you want' the 'way you want'. The view in this book is that graphical techniques are essential, at the first and at the last of a problem and not just for presentation. The graph leads the scientist to a better view, or at least a different or broader view of what is happening.

Report writing is probably the last thing Maple should be asked to do. But Maple can. The reason for this is three-fold: One is the interlacing with La-TeX; the second is the ability to generate mathematical constructs to present graphical data; the third is the necessity, these days, for an 'active' worksheet. The worksheet can be prepared and converted into LaTeX or HTML for near-to-realtime viewing as a .dvi or .ps file or with Netscape. All of these document handling schemes are available through nearly free software and allow anyone to prepare an *active, professional document.*

Who is this book intended for? It is for anyone who needs a tool to help come to grips with complex environmental matters. Maple is good at doing difficult and tedious work.

Environmental sciences are many and varied. The examples and problems displayed here exemplify this variety. Therefore, whoever you are, the whole of this book is not for you – at least, not now. Some of it is for someone else now and perhaps for you later.

Because this is a book about Maple and how it is applied in the environmental sciences, basic principles of the sciences are not discussed; it is assumed that the reader will have acquired the necessary foundations elsewhere. This book can be used, at least initially, as a cookbook to do certain things, to allow you to absorb the science through rote learning, without knowledge as to why. The reality is that it will be necessary to read into and around the material of interest.

How should I read this book? The answer is: Don't. Take this book as a game to be played, or a reference book; or, better, a basis for learning, a helpful source of *working* information. Thumb through the pages, look for something of interest or need. Use the table of contents or index to find a topic or word of interest. Read as you want and when you want. The material is not particularly dependent on what went before, nor is it necessarily coherent. It should allow you to be adventurous, to *have a go*; Maple is that way. It doesn't respond as you expect but, through a *Trial and Error* process, you can cut and amend the examples from the help or out of the worksheet; then append, plot and *most of all* think about the problem, the data or the structure, or the way it is presented, or the way it massages your thoughts; the limit to a problem, the way it can be expanded or delayed in evaluation; the way mathematics and, at least, the Maple mathematicians think. These thoughts will usually be consistent, unlike the fuzzy way many thoughts seem to be.

or destroys them

The answer is: use the book as a tool to formulate a problem; Maple will do the solution, or at least prepare the information so that you are further along. Importantly, enjoy yourself and Maple.

Whatever is a symbol, anyway? It is usual to trip over the way strings, symbols and unevaluation are handled in recent versions of Maple. Unless converted or parsed into symbols or names, a string can not be evaluated and remains as a word or sentence to be used as, perhaps, a file name or a title on a graph; in recent versions of Maple strings are enclosed in double quotes " ". A symbol can be manipulated mathematically, multiplied and divided, and have another identity; it may be evaluated to become an expression or list or set of other symbols; you type its name but it can be more than simply a name. In Maple, most operations are done with symbols; specific designation of a symbol encloses the object in single backward quotes ` `. In contrast, Maple uses single forward quotes ' ' to delay evaluation or use the name one more time. Since evaluation usually occurs with execution of a statement, the forward quotes around an object are fragile and disappear after a single use.

How can I proceed? This is not a book to be read from cover-to-cover. First, get a copy of Maple and install it on your computer. Second, lay the book beside your computer, turn on your computer and stoke up Maple. Follow your nose from there, with the book's help. You will find that you fumble through the material and will need to backtrack. The first three chapters are mostly introductory and begin with the simplest of statements to promote interest; each chapter then grows into a comprehensive, sometimes tedious crescendo of detail in Maple. You may find starting with Chapter 3 valuable, or Chapter 4, or even Chapter 5 if you are mathematically inclined. Whatever approach you take, scan through what precedes: Chapter 1 has a listing of common errors and the end of the chapter contains valuable one-line references. Set up an appropriate initialization file for yourself; the last section of Chapter 1 should get you started. Should you need help, the latter part of Chapter 2 contains extensive self-help; you can go on to generate your own help files with Appendix AAArgh. Chapter 3 lists the details of plot options, operands to procedures and shows how to manipulate PLOT structures. Chapter 4 applies the methods of Maple to real data, display and analysis. More usual mathematical solutions appear in Chapter 5, particularly considering turbulence and kidney stones.

> Doing the exercises is a good idea, as a self assessment and further learning

Could this book make a course of study? Certainly, at four levels; considering roughly 12 weeks of study or a broken 80 hours of full-time study.

---

1 At primary school, secondary school or the 'complete idiot'; take the book in its given layout, leaving out sections 1.4–1.6, 2.3.3–2.4, 2.5.4–3.4, 3.6–3.7.2, 4.2–4.3, 4.4.2–4.4.7, 5.5–5.7.9, and the Exercises with ∗.

---

2 At first year university, cover sections 1.2–1.4, 2.2–2.4, 3.1–3.7, 4.1–4.4, 5.1.3–5.5, and 6.1–6.4. The Exercises with ∗ can be considered as projects.

---

3 Second and third year of university, cover 1.5, 2.2–2.4, 3.1–3.7, 4.2–4.4, 5.1–5.6, 6.1–6.5 . Complete two Exercises with a ∗.

4 Last year at university, postgraduates or computer-oriented individuals, cover 2.2–2.4, 3.1–3.9, 4.3–4.4, 5.3–5.7, 6.1–6.5, and appendix AAArgh, to Exercise A.2. In all, you should be able to generate your own, independent, self-driven course of study.

Is the material up to date? The book has been tested on Maple versions up to Maple V, Release 5.1; it is set up for use with that version. The main updates that will affect the users of Maple 6 are the use of | | for catenating strings and symbols, as well as the use of end do; end if; and end proc; in the place of od; fi; and end;   These syntax changes are noted in the marginal notes and workarounds are quick and transparent; see Maple Initialization, section 1.6, page 52. The material in this book is introductory and should be timeless and valid for most versions of Maple. The companion CD has been specifically updated to Maple 6; the material in many cases is different and offers a helpful compatibility with the material in the printed book. The real power of Maple 6 is in its preprocessing capability, interactions with other languages, faster and enhanced numerical capability, handling of large expressions, error checking features as well as compliance with other software in the exchange of figures and data. The material in this book has been tested on IBM compatible, Macintosh, and Unix (Linux) versions of Maple and has grown with teaching and tutorial experience since the earliest versions of Maple.

Who helped put this book together? This is a montage of material gathered through the years in teaching and research by the author. Some of the material is from the multitude of books available and the MapleTech Journal. If some part is not properly acknowledged, please accept my apologies.

The philosophy behind this book comes from the Symbolic Manipulation Group, in Waterloo, Ontario. A visit by the author, in 1992, started the thoughts rolling. Particularly important were the interactions with Stan Devitt, both in Waterloo and Saskatoon, where the idea of an active book was presented by Stan. A book intended to be replaced by the student, a book that continues to grow into more than ever seemed possible.

I hope this is that book.

Bill Scott 15Jun00

Acknowledgements

This book is dedicated to my wife, Pat, and my boys, Josh, Nathan and Evan. Suggestions and drawings from Pat, Nathan and Evan are included. Without Pat's continued support, particularly in foreign lands, this book would not be. Greg Fee and John Ogilvie were there at Simon Fraser University, at a turning point, and helped keep many ideas flowing. Stewart Greenhill and John Horgan have given unerring advice, written software and script files. Helen Cockerill and Leon Harris saved much of the material from oblivion and rebuilt various computer systems to make this book possible.

Contents

Chapter 1

Turning On: Hints

Turning On

Übersicht

Explains the use of help to get started. How Maple deals with
numbers and the need for a semicolon. Evaluate and roam within
the living morass of Maple. A quick look at the ditto operation,
solving equations and sets. A few hints and a listing of common
errors may keep you out of trouble.

Guidewords:

```
?     ;     :=     *     \     /     !     "     %
help( )  evalf( )  expand( )  ifactor( )  solve( )
```

Install Maple on
your machine

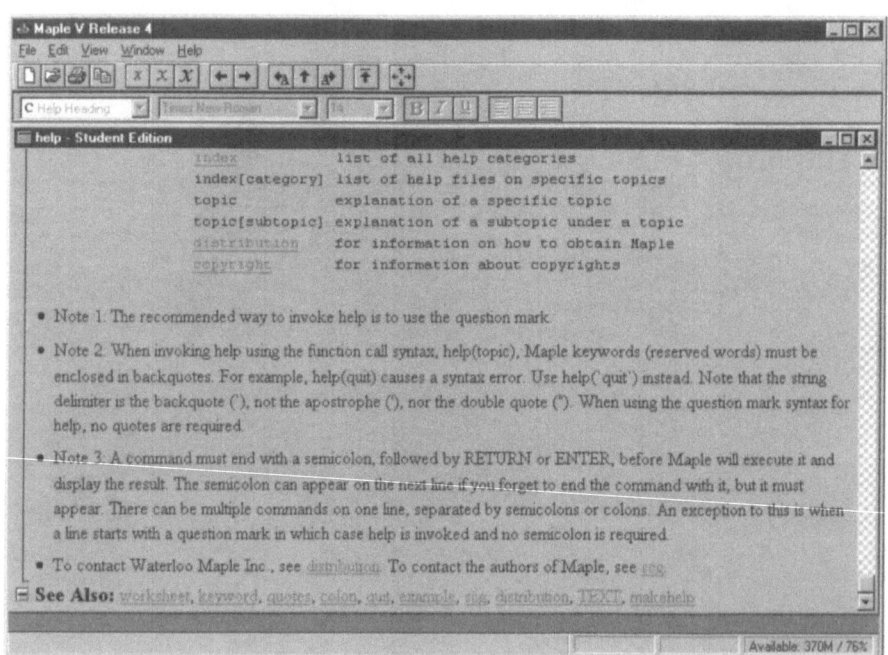

1.1   The Semicolon

Click on the
maple leaf icon

Turn on your machine and let it boot up. It may need an identification
number, password or whatever. This you should know or get from your tutor.
Double click using the mouse button until you find a 'Maple icon'; it should
run Maple on this machine. If there are any difficulties, see your local 'genius'.

Follow with Enter
or a Return CR

At the prompt, > type   ?   # you are asking for help

> ?

The help( ); calls are not very friendly. It is recommended that the ? be used as a first call for help. This can have several, different formats.

```
?topic or ?topic,subtopic or ?topic[subtopic]
```

Alternately, help is available through:

```
help(topic); or help(topic,subtopic); or help(topic[subtopic]);
```

But you will find it less frustrating to simply use the help on the menu bar at the top of the page.
Read the general help screen.
At the Maple prompt, try using the

;

This is a terminator of Maple operations. WITHOUT IT Maple WILL DO NOTHING! In all cases you are expected to type

Follow with an
Enter or a
Return CR

;

Enter or Return to get some activity.
At the prompt, type

```
> 3*5/51; # dividing 2 numbers by 51
```

$$\frac{5}{17}$$

This exemplifies several features of Maple. The expression is automatically evaluated if possible. Maple likes to work with integers and completes all calculations WITHOUT ERROR. Lastly, the result is presented in a 'pretty' form, 'echoed' to show that Maple understood. Note that, below, CR means Carriage Return, Return or Enter.

At the prompt, type

without a CR

```
> 1000050001*200000111;
```

Without CR, Enter or Return nothing happens. Now add a CR, an Enter or Return.

with a CR

```
> 1000050001*200000111;
```

200010111205550111

Use Shift+CR to
break Maple
input lines

Suppose you need a rather tedious number evaluated, say,

```
>  3*5+4*6^2-5^5*7+6*8^10+7*9+8^4*10+9*11\
>  + 3^15*8 - 10*44 + 88^3*4*5 + 11^4*7 - 10*22:
```

The colon
produces invisible
output

To keep a number or expression from becoming too long and untidy, \
is used to end the input or output at the end of the line. When the number
or expression is fully entered, the CR is still required to tell Maple to do
its thing.

When the
number is right,
use CR

```
>  3*5 + 4*6^2 - 5^5*7 + 6*8^10 + 7*9 + 8^4*10 + 9*11\
>  + 3^15*8 - 10*44 + 88^3*4*5 + 11^4*7 - 10*22;
```

$$6570992873$$

```
>  evalf(Pi,200); # Evaluates Pi to 200 decimal places
```

Note that the # generally is used to denote a comment. It can appear
anywhere on a line and makes the rest of the line invisible to Maple. This is
good for an odd comment to keep yourself in order.

The \ shows that
the number
continues

```
3.14159265358979323846264338327950288419716939937510582097494459235
  0781640628620899862803482534211706798214808651328230664709\
  8446095505822317253594081284811174502841027019385211055596435
  46229489549303820
```

Have a further look at what Maple has to offer.

At the prompt   >   type   ?intro # another intro

```
>  ?intro
```

doesn't need a ;

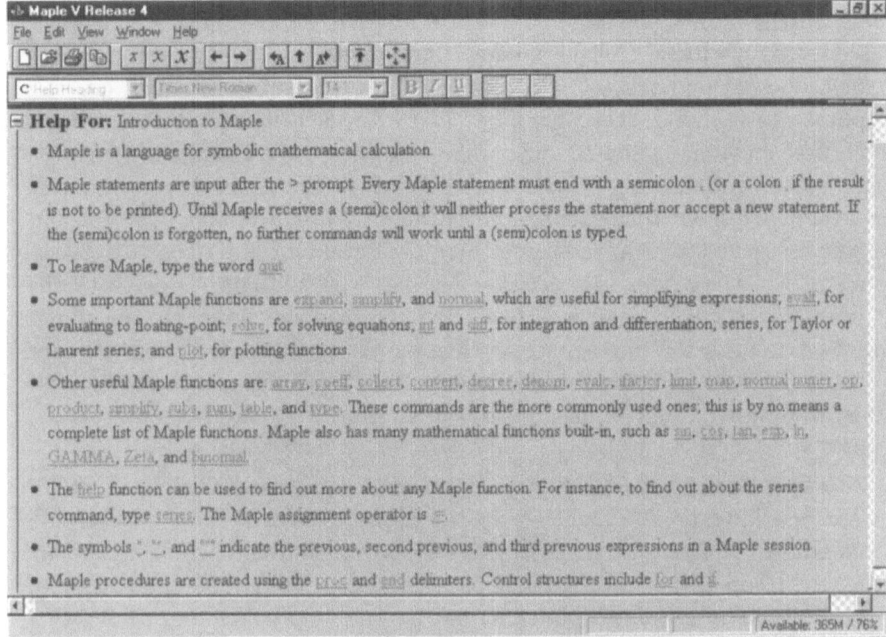

Quitting Maple

Maple is relative easy to quit; you may click on the upper corner, or type quit, done or stop, without the semicolon. Unfortunately, in starting with Maple, you will often have some sort of mess on the screen and Maple will object, repeatedly, to all sorts of swear words, or whatever, it is 'hearing'. Persist, type in a few semicolons, and, eventually, it will stop objecting and 'die'. Explore a few functions. Make a messy combination of x's, y's, numbers, letters, added or multiplied (with proper syntax, of course). Try expand( ); simplify( ); and normal( ); on these confusions. Do some numerical calculations with both integers and floating point (numbers without or with decimals). Add, multiply *, divide /, and subtract them; take them to powers, ** or ^. Use evalf( ); on them to convert them to decimals with any number of digits. Use solve( ); on a few algebraic or numerical problems. If your maths is up to it, consider dsolve( ); to solve a simple differential equation. Don't fear if many of your efforts go adrift.

Cleanup, as possible

In the Help, many of the items are active as Hyperlinks. Clicking on these creates an active connection that can be followed. It is recommended that you don't try to read everything about an operator or operation but jump directly to the examples at the end of the help files. Copy them directly out (cut and paste) and try them out, functioning, on your own worksheet.

Help button

Still, read the short introduction. Then, move the help window out of the way. It is not a good idea to 'simply click it away'. It is convenient to reduce it to an icon or use the upper drag bar to move the Help window to the right or left, off the screen, so that about 1/4 of the left side remains. Remember, you can copy a phrase from a window by highlighting (dragging the mouse cursor), followed by pasting. In most machines this is done by first copying the phrase with Ctrl-c, the control key and c key held down simultaneously; the phrase is now in the paste buffer ready for your use. Then, click with the mouse button (the left one, if there are several) at the position you want the phrase. Pasting is completed when you push Ctrl-v. Alternately, the cut and paste on the menu will do the same thing. Use the mouse button to activate the menu.   Drag and drop is also effective. You highlight: start with the left mouse button and drag it across the phrase, move the mouse away and come back to grab the phrase. It can then be placed (moved) to another spot or copied (push Ctrl) into the new space. Unix machines use the left mouse button to drag, the right mouse button to end and the middle mouse button to paste.

A Mac uses ⌘

?hotwin

drag and drop

good for a long expression

Do not do this now

You may want to save what you are doing for reuse or simply as a record of what you have done. Then, go to the appropriate directory or icon and open a editor (SimpleText is one); with an icon you simply click on some part of the icon to open the editor. The Editor window should allow full editing of a file, additions from your Maple window, as well as printing and saving to a floppy disk.

Alternatively, using the File menu, your Maple worksheet as a whole can be saved, with elaborate formatting as well as active and interactive structures.

Exact Numbers and Ratios

```
>   (2^6 + 5^8)/7^20;
```

$$\frac{390689}{797922662976612001}$$

```
>   100!;  # the factorial function 100*99*98*....1
```

933262154439441526816992388562667004907159682643816214685929638957\
21759999322991560894146397615651828625369792082722375825118\
5210916864000000000000000000000000

```
>   13051*896120193256*777 +
>   121*60*20*6666401*332;
```

$$9087541990170177912$$

```
>   evalf(log10(432),70);
```

2.63548374681491209274003928866331803467314611842052175973130676541\
  24120

Constants

Many absolute constants are defined within Maple; most have capitals as their first character; they are reserved names that can only be used for that purpose. There are reserved names for functions and operators as well, see section 1.5, page 38.

```
>  Pi;
```

$$\pi$$

The response shows nothing; Maple recognises Pi as an exact irrational number. Use evalf(Pi); to evaluate Pi as a floating point number, a decimal:

```
>  evalf(Pi);
```

$$3.141592654$$

Here we have Pi evaluated to 10 places. Suppose we wanted 1000:

```
>  evalf(Pi,1000);
```

3.1415926535897932384626433832795028841971693993751058209749445923\
0781640628620899862803482534211706798214808651328230664709\
8446095505822317253594081284811174502841027019385211055964\
4622948954930381964428810975665933446128475648233786783165\
7120190914564856692346034861045432664821339360726024914127\
7245870066063155881748815209209628292540917153643678925903\
0011330530548820466521384146951941511609433057270365759591\
5309218611738193261179310511854807446237996274956735188575\
7248912279381830119491298336733624406566430860213949463952\
4737190702179860943702770539217176293176752384674818467669\
0513200056812714526356082778577134275778960917363717872146\
4409012249534301465495853710507922796892589235420199561121\
9021960864034418159813629774771309960518707211349999998372\
7804995105973173281609631859502445945534690830264252230825\
3446850352619311881710100031378387528865875332083814206171\
7669147303598253490428755468731159562863882353787593751957\
8185778053217122680661300192787661119590921642019\9

try 10000   Do not try 100000! This doesn't mean Maple will be unable to do it; it only means that it will take a long time and we don't want to wait. Here is the catch with Maple. You can not ask it to do something that consumes the whole computer or it will either refuse to do anything or take forever. The 1000 shown is controlled by the reserved parameter Digits:

```
>  Digits := 100; evalf(Pi);
```

$$Digits := 100$$

3.1415926535897932384626433832795028841971693993751058209749445923\
0781640628620899862803482534211706\8

Assignment

The assignment above assigns the parameter Digits the value of 100. That is, it physically places 100 in the bin called Digits. The combination of

:

and

=

may be read as 'assigned equal to' or 'the same as'.

The ordinary equal sign = is used in equations to denote 'equality'. To Maple 'equality' could just as well be an 'inequality', using the symbols > , < , <= , >= , and <> This leads to a number of confusions as to what 'equals' means–one gets some understanding as to why Maple often finds it difficult to cope with 'equals', the sign =.

```
>  x = 2; 2*x;
```

$$x = 2$$

$$2x$$

Here we see no effect of the equals sign because Maple has no knowledge of the value of x. In fact the statement

```
>  x = 2;
```

is just a string of symbols with no real associated activity. If we do the assignment properly, Maple knows the value of

$$x$$

and does the calculation.

```
>  x := 2; 2*x;
```

$$x := 2$$

$$4$$

Solving Equations

Maple finds it easy to solve simple algebraic equations, as single equations or multiple equations, in as many variables as one desires:

```
>  y = 3*a * 7/(3*b) * c; solve(%,b);
```

$$y = 7\frac{ac}{b}$$

$$7\frac{ac}{y}$$

The first return is the result of Maple's acknowledgement that a relationship exists. Maple has, even here, simplified the mathematical statement.

The statements are chained in a line, separated by a semicolon   ;   The results are then stacked vertically in the output. Note that the first simplification is also reflected in the second, the answer to the operation solve. IMPORTANTLY, b has not taken on the given value, no assignment has been made. Solve has simply completed the solve operation and spit out the resulting algebraic expression. An assignment is necessary to make b equal to

```
>   7*a*c/y;
```

$$7\,\frac{a\,c}{y}$$

```
>   b:=%;
```

$$b := 7\,\frac{a\,c}{y}$$

```
>   b;
```

$$7\,\frac{a\,c}{y}$$

Ditto-take the last value

On two occasions, the last example used the repeat symbol   %   . It makes it easy to continue from the last value and collect the symbolic form. BUT IT SHOULD BE USED WITH CAUTION. Especially in worksheets *in any and all possible orders* when execution of a statement can follow any other execution, and all can be changed or even be made inactive.

```
>   5*3*10;
```

$$150$$

```
>   % * 2;
```

$$300$$

```
>   %%/2;
```

This operation can be repeated up to 3 levels:

$$75$$

```
>   %%%;
```

150

Sets

A set is denoted by { }. Note that a set can not contain two identical items and that the order of elements is unimportant and, probably is different with every presentation by Maple.

```
>  restart; # clears all variables
>{a,b,c,d};
```

The order is not
what you might
expect

$$\{c,\ a,\ b,\ d\}$$

```
>{b,b,a,c,d};
```

$$\{c,\ a,\ b,\ d\}$$

```
>{a,a,a,a,a,a,a,a,a,b,c,d,d};
```

$$\{c,\ a,\ b,\ d\}$$

```
>{d,d,d};
```

$$\{d\}$$

```
>{d,c,b,a};
```

$$\{c,\ a,\ b,\ d\}$$

Solving Simultaneous Equations

Here we just define the equations, individually, combine them as a set and complete the solution, with solve( ). Note that the arguments must be sets, lists or unambiguous, single elements.

```
>  e1 := x - y = 3;
```

$$e1 := x - y = 3$$

```
>  e2 := 4*x +10*y = 20;
```

$$e2 := 4\,x + 10\,y = 20$$

```
>  solve({e1,e2},{x,y});# solve the 2 equations
```

$$\{y = \frac{4}{7},\ x = \frac{25}{7}\}$$

```
>  solve({e1,e2},x,y);
```

Error, (in solve) invalid arguments

```
>  solve(e1,x);
```

$$y + 3$$

Units

Maple naturally substitutes units and cancels them as any other symbol. Consider Perth, on the coastal 'fringe' of ancient beaches, in Western Australia. A major underground aquifer , the Gnangara Mound, lies below the sand. Let us calculate the time required for groundwater to flow to the coast from the 'hip' of the Gnangara Mound. Uses Darcy's law and a porosity of 1/3.

```
>  restart:
>  vp := K*i/n; # the 'real' or pore velocity
```

$$vp := \frac{K\,i}{n}$$

Substituting
known data

```
>  subs({K=10*metres/day,i=10*metres/kilometre,n=1/3},%);
```

$$300\,\frac{metres^2}{day\,kilometre}$$

```
>  conversions:={kilometre=1000*metres,year=365*days};
```

$$conversions := \{kilometre = 1000\,metres,\ year = 365\,days\}$$

```
>  subs(conversions,%%);
```

$$\frac{3}{10} \frac{metres}{day}$$

> `velocity:=%;`

The units are
properly cared for

$$velocity := \frac{3}{10} \frac{metres}{day}$$

> `solve({velocity=distance/time},time);`

$$\{ time = \frac{10}{3} \frac{distance\ day}{metres} \}$$

> `subs(distance=10*kilometre,%);`

$$\{ time = \frac{100}{3} \frac{kilometre\ day}{metres} \}$$

> `subs(conversions,%);`

$$\{ time = \frac{100000}{3}\ day \}$$

Problem with
plurals

> `subs(day=days,%);`

$$\{ time = \frac{100000}{3}\ days \}$$

> `select(has,conversions,year);`

$$\{ year = 365\ days \}$$

> `solve(%,days);`

$$\{ days = \frac{1}{365}\ year \}$$

> `subs(%,%%%);`

$$\{time = \frac{20000}{219}\,year\}$$

```
>  evalf(%,3);
```

$$\{time = 91.3\,year\}$$

This also illustrates the nature of things. Our perception is better if we work with quantities of magnitude around 1. A large number is hard to comprehend, even 91.3 years!

```
>  restart;
```

Workspace

Extra Execution Groups are helpful

    restart; puts you into another mind and gives you a restart with Maple. It leaves the surface of your worksheet whole, with input and output, but removes all your assignments. That means that Maple's 'mind' has been reset, along with all your errors and confusions. This allows you to clean up, cut and paste your material, including input and output. You may want to add text material. If so, go to your worksheet and click on the [>, the Execution Group you want to affect. Using the T button, convert it to a [and type in whatever you want.

Alternatively, you may bring down another worksheet using the menu File/New or the special Worksheet Button. In any case you will find it advantageous to add a few execution groups to create space for working. This is easily done with the [> button or Insert/Execution Group. The keystrokes Ctrl-j or Ctrl-k will put an execution group after or before your selected statement.

After execution, after you press Return or CR, Maple goes to the next execution group. You choose your execution group by clicking on it. An execution group may be executed in any order, or repeated. Sequential execution, one execution group after the other, is easier to follow. However, repairs will usually be necessary and it may happen that a rather random structure is necessary. If so, give some guidance to your reader.

Exercises

Chapters contain various Exercises for the reader. Some Exercises follow, these are just to 'get started' in a thinking pattern. Others are more to 'train' or 'strain' the reader and full solutions may be the basis of a term project–these exercises are denoted with a *. Don't hesitate to skip them if you find them daunting or you don't have the time. It is expected that doing the exercises will bring you to a deep level of learning.

Exercise 1.1 Printing Help
Print the contents of help(help); and help(index,library);

Exercise 1.2 Factoring a Large Number
Calculate

```
>   7^23 + 2^53;
```

factor this. How many fives does it have?

Exercise 1.3 Large Addition
add

```
>   2/3 - 1/102 + 32*3*101/24;
```

Add up the series $1/2 - 1/3^2 + 1/4^3 - 1/5^4 + \ldots 1/10^9$;

Exercise 1.4 Ditto and Expand
```
>   (3^16 + 5^18)/7^20;
```

Consider

```
>   ifactor(%);
```

Try using %%; and %%%; expand(%);   What has Maple
                                                      done here?

Exercise 1.5 Pi and other Constants
Evaluate

```
>   gamma/Pi;
```

How close is Pi^2 to 10? What is gamma? What is the base of natural
logarithms?

Exercise 1.6 Factorials
```
>   100!;
>   10!  + 100;
>   %% + 333;
```

Exercise 1.7 Sets of Equations
Solve both equations for x.

```
>   e1 := 2*x + 3*y = 4;
>   e2 := 4*x + 1*y = 1;
```

Combine and solve for y.

Exercise 1.8 Different Solve
```
>   e1 := 2*x + 3*y = 5;
>   e2 := 4*x + 1*y = 0;
```

```
> ss := {e1,e2};
> solve(ss,{x,y});
```

Exercise 1.9 Floating Point Solve

```
> e = (((x^x)^x)^x)^x;   e := exp(1);
```

Use fsolve to solve for x. Substitute sqrt(15.25) and sqrt(15.26) for e. Solve for x in both cases. Maple may not be able to do the latter calculation.

Exercise 1.10 Newton's Law

The acceleration of gravity is roughly a constant 9.8 m/sec^2. Use this known acceleration to calculate the velocity of a rock 10 secs after being dropped (velocity=0) and the distance it has fallen.

```
> velocity := int(acceleration,t);
> distance := int(velocity,t);
> conditions :={t=10,acceleration=9.8};
> vel10 := subs(conditions,velocity);
> dis10 := subs(conditions,distance);
```

Exercise 1.11 Units

Put the appropriate units into the above calculations. Follow them through the integrations and substitutions to obtain an answer with the proper units.

1.2 Using Help

Again, start up a Maple session. Alternatively, you may bring down another worksheet using the menu File/New.

```
>  restart:  # cleans up, removes all assignments
>  ?help
>  |
```

Help on various topics and subtopics may be found using the syntax:

```
?intro  ?worksheet  ?index   ?mouse  ?syntax   ?semicolon  ?file
?statement   ?string   ?algebraic

or ?type,logical     ?worksheet,reference,hotwin  ?plot,options

or help(plots[sphereplot]);  ?help[INTERFACE_HELP];
help("quit"); help('convert,string');
```

Throw in a few Execution Groups

The Help system in Maple is less than friendly to the beginner. Help is
available as Balloon Help, the leading ? on the command line and menu Help.
Be sure you activate the Balloon Help, under the Help menu; this gives advice
on the different menu items, when you approach with the mouse. The ? on
the command line is straightforward and unencumbered with requesters. Most
seem to like to use the mouse, to click and activate the buttons on the menu
items. Locate the arrow tip on the item and activate the left mouse button, a
menu appears. Now, holding down the left mouse button, drag the arrow to
the selected menu item. Release the mouse button, and the item is activated.
Key strokes can be substituted for menu items, the underlined letters are
active.

Help is on the menu, far right

Otherwise Help is guided by ?topic,subtopic as

Activate Balloon Help

```
>  ?plot,options
```

In a search, you use the Help button. At the start, you may not know
any topics; so a search is best started with the Full Text Search option. You
click with the left mouse button on the Help, and, holding down the button,
come down to Full Text Search. A Full Text Search means typing in a word
or words you think are appropriate. Clicking on Search gives the results of
Maple's searching of its archives. Alternately, in Topic Search, with Auto-
Search on, if you type slowly and can spell (or can't spell), Maple may be able
to anticipate your word –a listing of possible topics appears in the subwindow.
An indication of the importance of each item is given by a number (0 to 1.00)
that gives the relative number of times the word appears in the particular
item. Double clicking on the item produces a help window in the background.
Closing the help requester leaves the help window, reading the help and using

tick Auto-Search

(clicking on) the active phrases or 'hot items' makes it possible for you to further focus your search.

Unfortunately, the requesters that appear make it difficult to see your original work; on Windows systems there is little room. With the mouse located on the upper bar of the window, you can move the requester to a convenient, off side position; it is better on the left to retain the right drag bar. An alternative is to activate the upper left Maple icon with the mouse and minimise the window to an icon. Still, you probably need to see and copy some of the help material, it may not be possible to move it aside or to the bottom; it may be easier for you not to try to interact with the help and the worksheet at one time. Rather, open another worksheet and, under the Edit menu, use Copy Examples. Copy them into the new worksheet; also you may copy small pieces by highlighting the parts you want. If you are copying into your working sheet, make a large hole with a number of execution groups [> to allow for errors of placement. If the copied material is text, convert the [> to a [to indicate ASCII text. This is done with the large T button to the left of the [> button on the menu bar. Once collected on a separate worksheet, the help can be shifted down or to the side more easily, and one can save the worksheet as a separate help worksheet, companion to the main worksheet.

The browser at the top of the Help window allows a general scan of the help topics and their structure. Remember that help links don't grow on trees and they don't have links between branches. You may need to exit the Help and do another, different search. Closing the window is completed by clicking with the mouse on the *lowest* set of square buttons, the far right one with a cross or X. Do not click on the square button in the far top right, Maple may close your worksheet! The window is made smaller or iconic by clicking on the *lowest* set of square buttons on the upper right. The history, under the help menu, can help with the search and allow you to recover your trail.

When doing a search within the requester, you click once on the topic you wish to read; then you click on Apply and a help page will open. The use of the Apply button allows that the dialog box remains on the screen and you can go to another topic. This can get around the problem of slightly difficult or different spellings, capitals and plurals; it allows you to have another go.

Maple Remembers

If you enter Maple in a regular way through the Maple program and Auto Save Settings (the default) is active on the file menu, it will put you at the same place you left in the worksheet. Maple will also remember to activate the Balloon Help and other settings. A listing of your files is located at the lowest level in the file menu, so you can go back at your pleasure. In Windows it is convenient to make a shortcut to the Maple program on your desktop; this is done by locating the program in a window or by locating the program in the explorer and clicking with the right mouse button.

Margin notes:

Close by clicking on the upper right button or activating the upper left Maple icon

left mouse down, drag, release

The File menu is on the upper left, after the Maple leaf icon

1.3 Helpful Hints

- Remember the ; Maple's # 1 rule

- Avoid assignments; Corless' 1st rule

- Look at the left side of an assignment; it should never contain a variable on the right

- Remove all Dittos, %; Fee's rule

- The right and left side of = are hard to deal with; use expressions

- Write comments in files, using #, or as paragraphs; use white space for clarity

- Retype as little as possible. Cut and Paste as much as possible

- Start with an example, and build on it

- Build the large from the understandable, and tested, small

- Maple is not always right

- If you are confused, so also will Maple be confused

- Don't hesitate to use Trial and Error

- Test and retest every result

- Try to keep a linear, top-down, thought pattern–but don't be addicted to it

- type all brackets together ( ), [ ], { }. Then fill in the space

1.4   Frequent Errors

- **Syntax**

These are errors in typing or thinking or simply not knowing Maple.

```
>  restart:
```
```
>  a = b; sub(%,ax+b);
```
$$a = b$$
$$\text{sub}(a = b,\ ax + b)$$

The operation was not performed. Usually, however, Maple can pick up these errors and can help.

```
>  suba=b,ax+b);
```

```
')' unexpected
```

A little clean up
and all is well

```
>  subs(a=b,ax+b);
```
$$ax + b$$
```
>  subs(a=b,a*x+b);
```
$$b\,x + b$$

The desired
result

The classic case of such errors was an O being substituted for a 0. This was a source of many ulcers since the earliest Fortran, it is an even bigger problem here. Of course, Maple can not know the subtle workings of the mind.

Maple 6 uses the
|| in place of .

```
>  'suds are everywhere'= s.u.d.s*a.re*every.where;
```
$$suds\ are\ everywhere = suds\ are\ everywhere$$

- **No semicolon ;**

This is something C programmers don't have problems with. The rest of us do. Look at some of the effects. You enter into a worksheet with this warning statement

```
>  x^2 + y^3 = 2
```

```
Warning, premature end of input
```
Unwittingly adding a useless ; after the warning compounds the problem.

```
>  ;
```
```
>  %;
```
$$suds\ are\ everywhere = suds\ are\ everywhere$$

In most cases, with modern versions of Maple, it tells you in a timely way that you have left out the ;

```
>  x + y + z = 2
```

`Warning, premature end of input`

Don't get flustered and retype the statement, that will only confuse things. Almost every time it will be a semicolon. Click on the statement, correct it and re-execute it with a CR.

```
>  (x11+10) + 10*exp(aaz+2)/phi - erf(Psi)
```

```
>  (x11+10)+10*exp(aaz+2)/phi-erf(Psi)
```

`Warning, premature end of input`

It is a waste of time retyping everything. The operation is cleaner and less prone to error if the statement is simply redone. Alternately, use cut and paste, with keys or the mouse.

• " or % Ditto

This is a bit of a problem with ordering of the mind. In fact, in the older, command line Maple the problem was minimal. Now, the last operation could be anything. Look at the execution groups below; groups one to three were executed serially; afterward, the second execution group was activated.

The newest Maple uses %; the " is used to indicate a string

```
>  eq:= exp(3*sin(5*t));
```

$$eq := e^{(3\sin(5t))}$$

```
>  int(%,t);
```

`Error, (in int) wrong number (or type) of arguments`

```
>  subs(t=x^2,%);
```

$$\int e^{(3\sin(5x^2))}\,dx^2$$

It is clear that if we can flex our mind in all directions, much nonsense can be produced. Still, this back substitution can be very productive and useful. It is frowned upon, however, because reading it requires a road map. Confusion is minimised when one statement follows another.

• Equals =

The difference between = and assignment := is not clear in most minds. The first is a statement, best considered to relate two algebraic or other complex quantities; the second is an actual 'name' or tag for the item, or, if you like, the label for the box in which it is kept. Following from above,

```
>  x^2 + y^3 = 2;
```

$$x^2 + y^3 = 2$$

this is simply the statement of 'equality'. It can also have a 'name' or identity

```
>   eq1 := x^2 + y^3 = 2;
```

$$eq1 := x^2 + y^3 = 2$$

which is different than the left hand 'expression'

```
>   exp1 := lhs(%);
```

$$exp1 := x^2 + y^3$$

Still, we might type

```
>   x = sqrt(y^3 - 2);
```

$$x = \sqrt{y^3 - 2}$$

thinking that x has now taken on the value given by the square root. In fact, this is far from the case.

```
>   x;
```

$$x$$

x is still unassigned. Carrying on with a number of other operations can be very misleading.

- **Right Single Quote ' '**

Look at them, they slant up to the right. They are ordinary apostrophes or 'right single quotes'. They are neither character strings nor symbols in Maple; they refer to a level of evaluation.

```
>   f(x) := 1/(x+a);
```

$$f(x) := \frac{1}{x + a}$$

```
>   ''int(f(x),x)'';
```

$$\,' \int f(x)\,dx\,'$$

```
>   %;
```

Principle of last evaluation

$$\int f(x)\,dx$$

```
>   %;
```

$$\ln(x + a)$$

We see that, at each level of evaluation, the ' ' are removed. This could have been done by eval( ) or eval(,2). Each evaluation (or ; and CR) does the same thing.

```
>   eval(''int(f(x),x)'');
```

$$\int f(x)\, dx$$

```
>  eval(%);
```

$$\ln(x + a)$$

Don't use these right single quotes for a symbol or a string.

```
>  read 'results.m'; # should be a string or symbol
```

should be a
string or symbol

```
Error, read must have a name
```

But they can be used effectively to cause unassignment.

```
>  x := 'x';
```

$$x := x$$

this simply declares the x is a pure symbol, a name ready for use.

- **Left Single Quote ' ' or Full Quotes " "**

Left Single Quotes bend to the left and are used to define symbols or
character strings that can be evaluated.

After Maple V5r5
strings are in
double quotes "

```
>  'these are' := 'real symbols';
```

$$these\ are := real\ symbols$$

```
>  read 'results.m';
```

Maple has read the file name. A file with .m, in a binary Maple format, is
available and read into the Maple session.

```
>  read 'nofile'; # no such file exists in Maple's path, libname
```

```
Error, could not open 'nofile' for reading
```

Maple knows the
file name but
could not find it

This idea can be very confused if one is making help files with many lines
of symbols, so as to be easily associated with Help:

```
>  'help/world/money' := TEXT(
>  'Cree Indian Saying:  Only when the last tree has died',
>  ' and the last river has been poisoned',
>  '  and the last fish been caught',
>  '   will we realise we cannot eat money');
```

$$help/world/money := \text{TEXT}(Cree\ Indian\ Saying: \ Only\ when\ the\ last\ tree\ has\ died,$$
$$and\ the\ last\ river\ has\ been\ poisoned,\ and\ the\ last\ fish\ been\ caught,$$
$$will\ we\ realise\ we\ cannot\ eat\ money)$$

This is fine. All the variables are, more or less, symbols, but look what
happens if we forget and carry on, without a second ' somewhere.

```
> 'The fact that we have not closed the symbol with
> a second left quote can make the entire workscreen useless.
> This is particularly difficult when Maple has also put in a
> carriage
> control.    See what happens when I type in a proper statement,
> thinking all is well
> x := 3;
```

```
Syntax error, missing operator or ';'
```

Note that the colour of the active screen is red. The lack of a prompt is a clue. Maple is confused. The solution is to repair the situation and add a last ` and a ; to the execution group. The string enclosures " " can have the same effect.

• end; in Procedure

In early versions of Maple, one could carry on for a time with nothing happening. Maple was waiting for you to continue on with the procedure. The use of end; and several ;; might be required to carry on in the command line or worksheet. Now, provided one doesn't enter into procedures strangely, Maple gives due warning. This is not a big problem.

```
> PP := proc(x); exp(x); end;
```

$$PP := \mathbf{proc}(x) \exp(x) \, \mathbf{end}$$

```
> GG := proc(x); exf(x);
```

```
Warning, incomplete statement or missing semicolon
```

When you realise you are in a procedure, you simply add end; to the execution group

• Wrong Assignment

This is a little subtle; in some ways it is the 'right' versus 'left' side of an operation. Consider the examples.

```
> f := 3;
```

$$f := 3$$

```
> f(x) := (x+5)^2;
```

```
Error, invalid left hand side in assignment
```

We have tried to define 3 as an expression, an algebraic quantity.

```
> x^2 + 2 := 4;
```

```
Syntax error, missing operator or ';'
```

Again, an illogical operation. An expression can not be assigned the value 4–it would not be an expression.

- **Recursive Definition**

This comes back to the use of programming. Many people are used to having Fortran or Basic make a reassignment to make a counter, by one unit.

```
>  y := y + 1;
```

`Warning, recursive definition of name`
$$y := y + 1$$

Maple allows this but has a brain confusion, goes away for a long time doing, and redoing, a serial, never-ending problem when we ask for more to be done to y. Hopefully, after a minute or so, Maple doesn't simply crash the system.

```
>  sin(y);
```

`Error, too many levels of recursion`

```
>  y;
```

The solution is to unassign y ASAP!
$$y$$

```
>  y := 'y';
```

$$y := y$$

```
>  sin(y);
```

$$\sin(y)$$

Now we are OK. The above seems simple; with complex symbolism this comes up from the back like a burglar with a cudgel!

- **Operation Precedence**

This seems obvious but the order of many operations is unclear. Maple reads from left to right and takes that as an order of precedence, with exponentiation first, then multiplication and division and then addition and subtraction.

```
>  2 + 3*5;
```

$$17$$

```
>  3*5 + 2;
```

$$17$$

```
>  4*5^2 - 1;
```

$$99$$

```
>  4*5 - 3 + 6^2;
```

53

Use ( ) whenever it is unclear.

```
>   4*(5 - 3) * 6^2;
```

288

• Something Simple Missing

Something is not completely typed and names and operations are confused or not done. Look carefully at how Maple interprets the following statements.

```
>   e1*rr^42-(e+1)*x*z*(x-2)*exp(y);
```

$$e1 \; rr^{42} - (e+1)\,x\,z\,(x-2)\,e^y$$

```
>   e1*rr^42-(e+1)*x**(x-2)*exp(y);
```

$$e1 \; rr^{42} - (e+1)\,x^{(x-2)}\,e^y$$

```
>   e1rr^42-(e+1)*x*z*(x-2)*exp(y);
```

$$e1rr^{42} - (e+1)\,x\,z\,(x-2)\,e^y$$

Multiple errors of this sort can lead the calculation totally astray. It is recommended that the standard conventions be followed; keep variables roughly in lexigraphic order, with a preference for simple alphabetic characters and the use of space during operations, particularly with additions and subtractions.

• Part of an Assignment :=

The making of an assignment is not an easy operation and can cause much cursing

The typing of an assignment operation := requires three operations: pushing the shift key and the colon key simultaneously followed by finding and pushing the equals on the upper right of the keyboard (with a shift as well on European keyboards). It is easy to err. With the simplest of errors, disasters occur. There are at least three variations that happen when using the assignment operation. If the equals component is simply left out, all is well but the statement essentially ends there, the : is the same as a ; and there is no assignment.

```
>   restart;
```

restart is used here, and elsewhere, to 'tip the bin' or clean up the worksheet

```
>   f(x) :  (x+3)^2;
```

$$(x + 3)^2$$

```
>   f(x);
```

$$f(x)$$

When the : is left out, there is also no real activity. We see only the equality.

```
>   f(x) = (x+3)^2;
```

$$f(x) = (x + 3)^2$$

With neither of the two parts of the assignment operation, or a space,

```
>   f(x)    (x+3)^2;
```

$$f(x)(x + 3)^2$$

```
>   f(x)  :   = (x+3)^2;
```

```
Syntax error, '=' unexpected
```

the operation is corrupted and you may be missled. The form f(x)(x+3)^2 could be a complex operation.

```
>   whattype(f(x)(x+3)^2);
```

- **\ instead of /**

This problem is aligned with the different way DOS systems handle directories; the \ is usually difficult to find and so should only plague DOS or LaTeX users. The \ is used by Maple to join long numbers through line breaks or carriage returns. If in the middle of a number, it may simply join the number or letters together

```
>   2\3; a\3;
```

$$23$$
$$a3$$

- **# the hash sign or number sign**

The hash or number sign has been used by Maple since the earliest days as a comment line. That is, Maple ignores as a comment anything from where # begins to the end of the line. It is recommended in documenting within an active sheet or on the end of a line. It may also deactivate a working line:

```
>   sqrt(x^2 + y^2); # the diagonal of a triangle
>   or the length of a vector
```

$$\sqrt{x^2 + y^2}$$

In typing this statement, Maple has rolled the line around into the next line. Without a Shift/CR, the last part of the line is part of the original line. Later, it is necessary to use the result of this calculation. Going back to the second line, an assignment statement is added.

```
>   sqrt(x^2 + y^2); # the diagonal of the triangle
>   result := %:  or the length of the vector
```

Nothing happens, as expected. But the name result is not assigned because the line is deactivated by the # sign.

- **Maple Errors**

Maple can get it wrong. Consider the integral

```
>  int(1/(1-x+x^4),x=-infinity..infinity);
```
$$0$$

```
>  plot(1/(1-x+x^4),x);
```

The function is everywhere positive, yet it has an integral or enclosed area of zero! This is an error or bug that Maple hasn't fixed. There are ways of tackling this to get the right answer, but the value of the integral is wrong. One way to 'see' the answer is to look at the limit as the size of the integral gets larger and larger. Here we use the limit to a sequence; the numerical answer is is clearly nonzero.

Maple 6 works; evalf( ) includes a small imaginary part

The numerical result is a positive number, as given below

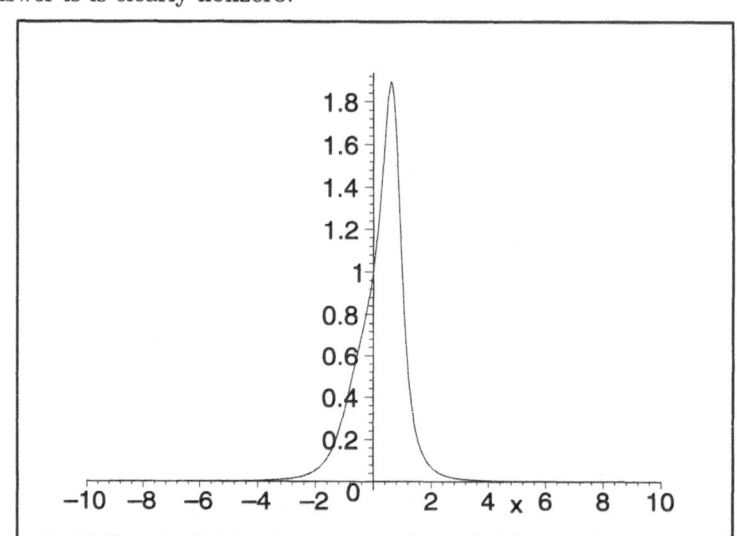

```
>  ff := proc(x); 1/(1-x+x^4); end;
```
$$ \mathit{ff} := \mathbf{proc}(x)\, 1/(1-x+x^4)\,\mathbf{end} $$

```
>  Digits:=8;
>  seq(evalf(int(ff(x),x=-i..i)),i='1,10,20,50,100');
```
$$ \mathit{Digits} := 8 $$
$$ 2.1330584,\ 2.6828816,\ 2.6834649,\ 2.6835429,\ 2.6835476 $$

Another one. This has to do with the way evalf is mapped onto the powers.

```
>  NN := 4/5*10^(2/3)*5^(1/3);
```

$$NN := \frac{4}{5} \, 10^{(2/3)} \, 5^{(1/3)}$$

```
>  evalf(NN);
```

$$6.3496040$$

```
>  evalf(NN,1), evalf(NN,2), evalf(NN,4);
```

$$8., \, 6.2, \, 6.350$$

```
>  evalf(NN);evalf(%,1);
```

$$6.349604210$$
$$6.$$

8 is not a one
digit estimate of
6.35

Maple has mapped cvalf( ) on the integer combination, not rounded an accurate floating representation of the number. If a certain accuracy is required, follow the pattern above.

A last error, where Maple does not see the missing multiplier $*$.

```
>  exp(-1/2(y/sigma));
```

$$e^{(-1/2)}$$

This is a part of Maple's composing of functions. Let us hope you are not affected by it.

- **Can not evaluate boolean**

The use of other than logical (true or false) arguments is common. It usually means the type of variables can not be compared.

```
>  x:= []; # this is a list
```

$$x := []$$

```
>  while 1<x do; print(help) od;
```

```
Error, cannot evaluate boolean
```

Getting your thoughts together, you will see the logic of only comparing like things. There is some confusion about types and conversions sometimes will be necessary. This is particularly true with symbols and strings, especially when the symbols can be evaluated to form a complex structure.

For more help with errors use the Maple Help and the many Maple books. See section 1.5.

1.5   Reference Material: types and commands

Guidewords:

> books, packages, share library, archives, Maple Users Group, MapleTech, operators, types, reserved words, commands

Review the Maple Worksheet

Don't become
confused

Review what you have been doing. Look at what the Maple Worksheet is, what it does, and how it works.

```
>?worksheet
>?glossary
>?update
```

Next, you may want to look at a few books, either in the library or, you may like to buy a book or two. Since the start of Maple, perhaps 20 years ago, a large number of books have appeared. In fact, some of the earliest have a reasonable summary of types, commands and other operations, in the command line interface version of Maple.

as compared to
the GUI versions,
Xmaple and
Wmaple

Some Useful Books

In the references at the end of this book, page 354, books on Maple cover Maple basics to high levels of programming. In addition, Maple material is available in the various packages, the share library, the Maple Archives, the journal MapleTech, and the Maple Users Group, MUG.
Of these books

Ellis et al's *The Maple V Flight Manual, Tutorials for Calculus, Linear Algebra and Differential Equations*, 183 pgs (1992)

is a pleasing, coloured introduction, filled with marginal notes and the structure of 'getting into Maple'. It is a bit dated, however. The Maple Worksheet, the organisation of Maple, the generation of documents in Maple are better covered in the various printings of

Heal et al *Maple V Learning Guide*, 274 pgs (1996) and
Heal et al *Maple 6 Learning Guide*, 314 pgs (2000)

which are superb compliations from Waterloo Maple to guide you up the learning curve. But, by far the most comprehensive, single book is

Andre' Heck's *Introduction to Maple*, 2nd Edition, 699 pgs (1996)

It is a little old, but a complete text with exercises, much detail and reference material, including details of formatting, plot options, Maple programming, input/output and many examples.

Markus Eikelberg's *Einführung in die Arbeit mit Maple V, Eine Anleitung für die praktische Anwendung*, 359 pgs (1998)

is an more up-to-date source, with a CD-ROM, in German. It deals with the GUI interface, buttons, spreadsheets and smartplots.

Monagan et al, *Maple V Release 5, Progamming Guide*, 379 pgs (1998) and Monagan et al, *Maple 6, Progamming Guide*, 586 pgs (2000)

are comprehensive, well-written books from Waterloo Maple inc. They reveal programming methods and particulars even to non-programming minds. There is no doubt that Maple is much more than simply a mathematics tool and Michael Monagan points out many of the subtle features of Maple with a critical and constructive eye. Programming tools in Maple, even 'Hackware Tools', become accessible and assessable without trama.

Rob Corless, *Essential Maple, An Introduction for Scientific Programmers*, 218 pgs (1995)

Rob Corless has put together a reflective book for programmers that reveals the most common constructs of Maple. It is now dated but has the property that, though it 'is not necessarily right', it reveals the minds and processes that have made Maple. Maple is a phenomenon that is evolving and dynamic; as one expects in today's world. Maple must also have its critics.
A last suggested reference is

Michael Kofler's *Maple V Release 4, Einführung und Leifaden für den Praktiker*, 582 pgs (1998)

This German book also comes with a CDrom with active worksheets as well as programs to massage plots and convert data from Maple to Mathematica; the book is also upgraded to Release 5 on the Addison-Wesley web site.

Packages

A listing of packages is available from the help files.

Setup problems? consider ?initialize and section 1.6

```
>?package
># or
>?index,packages
>?updates,R5,packages
```

The Share Library

You may need to change the file extension to .tgz

The share library is either supplied with Maple or it can be obtained from the Waterloo Maple Inc. web site.

http://www.maplesoft.ca

It is easiest to download in its entirety as 'crunched' file share.tar.Z When expanded in the Maple V Release 5 directory, the share library becomes accessible to Maple. The expansion is done with winzip or pkunzip, also available on the web.

```
>?shareusage
```

Next you must include the name of the share library location in your library path.

```
>libname;
>libname:=libname,"C:\\PROGRAM FILES\\MAPLE V RELEASE 5/share";
>libname := "C:\\PROGRAM FILES\\MAPLE V RELEASE 5.1/update",
 "C:\\PROGRAM FILES\\MAPLE V RELEASE 5.1/lib",
 "C:\\PROGRAM FILES\\MAPLE V RELEASE 5/share"
```

Once installed, the library becomes part of the help system and the commands can be accessed through the shortened names. It is a good idea to include this library path in your .ini file as well. See section 1.6

```
>with(share);
```

An alphabetic index is also available

```
>?share,contents
># or
>?share[contents]
># or
>?share/contents
>?entry_name
>?share[entry_name]
>with(entry_name);
```

where entry_name is the share library entry. Using with( ) in this way allows the use of the share library procedure names directly.

Maple Archives – march

The march archive is available through help in Maple 6

The Maple archive is available as a separate executable, once you decide on the package and/or the commands of interest. The easiest way to get through to march is to use the DOS-Prompt. Here it is presumed that the Maple files and you are on the same hard drive. The Unix commands are similar, with different directories. Unix usage is explained in help. The -l option lists the files. The | pipes the data through sort into the file libcontents.txt The DOS editor edit allows viewing of the contents of the Maple archive.

```
>?march
```

```
cd "\Program Files\Maple V Release 5.1\bin.wnt"
march -l ..\lib | sort >libcontents.txt
edit libcontents.txt
```

The various options in march allow

```
adding a file to an existing archive             -a
creation of a new archive in a given directory   -c
deleting an existing file from the archive       -d
listing all files and contents                   -l
packing the library archive                      -p
reindexing the library archive                   -r
updating an existing archive                     -u
extracting files without removal                 -x
```

march is the only mechanism for direct manipulation of the library archives, for a reason: A corruption of the libraries or the library path makes Maple unworkable. Hence it is recommended that you only look at the archives and only allow Maple to change them.

Maple Users Group–MUG

Information on the Maple Users Group MUG is available through majordomo@daisy.uwaterloo.ca
You may join the group by simply subscribing. See

```
>?maplegrp
```

MapleTech

This journal has usable Maple codes and procedures for the solutions of countless problems in Biology, Engineering, Physics or Mathematics. These are at both the lowest and the highest levels of sophistication, given the philosophy of the journal and the 'open book' philosophy of the creators of Maple. The articles include many hints on the use of Maple as well as details on new share libraries.

As a source of 'hands-on' information in research and teaching, the journal couldn't be more effective, to the credit of the editor, Tony Scott. However, at the end of 1998 publication ceased. Still, the material is timeless, workable and

scott@
uni-paderborn.de

usable material the reader can 'take on board' and do. Highly commended.
Some issues may still be available from

MapleTech@birkhauser.com

Abstracts of some articles are available as a part of the Maple help system.
Some examples in this book were derived from articles in MapleTech.

```
>?mtn
># or
>?MapleTech
```

Internet Sites

Ross Butterworth has prepared an active list of internet sites that can
be useful. This list is in bookmarks.htm; the email addresses are listed in
bookmarks.txt and appear below. The author takes no responsibility for the
internet linkages; on 15 Sep 00 the addresses were active.

```
http://www.maplesoft.com/
http://s3.haifa.ac.il/math/maple.html
http://help.unr.edu/maple.htm
http://www.chem.brown.edu/maple/
http://web.mit.edu/afs/athena/software/maple/www/
http://www.rpi.edu/Computing/Consulting/Software/Maple/Hints/maple.html
http://courses.ncsu.edu/classes-a/maple_info/www/
http://rhodium.chem.utk.edu/~chem483/maple.html
http://www.math.utah.edu/lab/ms/maple/maple.html
http://www.cc.utexas.edu/math/Maple/
http://www.chemie.uni-regensburg.de/maple_software.html
http://www.rz.uni-karlsruhe.de/~Maple/index.html
http://www.math.vanderbilt.edu/cgi-bin/mapleFE.pl
http://www.df.uba.ar/~jakubi/maple2.html
http://www.chem.brown.edu/maple/menus.html
http://www.astro.queensu.ca/~grtensor/maple2.html
http://www.tu-harburg.de/rzt/tuinfo/software/comp_alg/kurse/web/maple.html
http://daisy.uwaterloo.ca/SCG/MUG.html
http://www.mcs.dundee.ac.uk:8080/~dfg/MapleExternal.html
```

Frequently Asked Questions FAQs

These are responses by Maple experts to questions asked by Maple users.
Some are published regularly by Maple Waterloo Inc. as PR in the Maple
Reporter. These and others are at their web site
http://www.maplesoft.com/support/Faqs/index.html
Others are printed in Nancy Blachman's book [5].

Syntax and operators

The syntax of Maple allows for many operators and operations. All are somewhat dependent on the Maple type.

a.b is a||b in Maple 6, see upgrades page 38

```
>?syntax
>

>?operator
```

;	command separator
:	silent command separator
:=	assignment
%	Ditto, what just went before
%%	DittoDitto, what went before what went before
%%%	DittoDittoDitto, what went before what went before what went before
&*	matrix multiplication
.	non-communicative multiply in Maple 6
[1,2,3]	list of the three numbers, ordered
{3,4,1}	set of the three numbers, unordered and with no duplicates
x*y	x times y
x*(y+3)	x times the result of adding y to 3
x^y or x**y	x to the power of y
x = y	x is equal to y
x <> y	x is not equal to y
x < y	x is less than y
x > y	x is greater than y
x <= y	x is less than or equal to y
x >= y	x is greater than or equal to y
(sin@cos)(x);	sin of the cos of x, the composition operator
(sin@@2)(x);	sin(x)^2, the repeated composition operator
R and S	boolean and
R or S	boolean or
not R	boolean negation
f(x)	function call
{A} union {B}	joining of sets
{A} minus {B}	removal of the elements of B from A
{A} intersection {B}	common elements of sets A and B
p::list	an association of a type or property
a.b	catenation (not Maple 6)
a||b	catencation in Maple 6
"a" "b"	is catenated string "ab" in Maple 6
'a||b'	unevaluated catenation in Maple 6
1..6	range of 1 to 6, inclusive
$1..5	evaluation operator, this becomes 1,2,3,4,5
O(x)	Order of function
Order	of the order of, truncation order in expansion
procedure	proc(x) x^3 end;
'x + y'	unevaluate, delay the evaluation of the sum
"Murdoch"	string, can not be evaluated
'gun jel'	symbol or name, can be evaluated
!dir	in DOS Prompt ! escapes to DOS
!cat File	in Unix command line maple ! escapes to the host system
?\	display the last help window

Maple types

It is rather difficult to come to grips with Maple types. There are many and they are not exclusive. Heck [13], page 82, has a good explanation. The basis may be explored with commands that search the nested and structured types.

Whattype–responds to surface type

type–tests for nested and structured types

hastype–tests the existance of subexpressions of a given type

typematch–matches a structured type

Other than this general advice, types are individual features that make up the identity of an element in Maple. Their properties are important when they are sorted or used in commands. It is possible to define your own types, see Michael Monagan's book [23]. Types are declared at the start of a procedure with :: ;when this is done special procedures are activated.

Here we simply list types for reference. Some types you know include

```
>whattype(x); whattype(x^3); whattype(f(x); whattype(x+y);
>whattype(x<2); whattype(33); whattype(1/2); whattype({B});
>whattype(1..9); whattype([A]); whattype(''x+y'');
>
>?type,defn
```

The full type list valid for your version of Maple should be found from your help files. Here we list types known to be valid on Maple 6 as of Jun00. This list is not all inclusive.

```
>?type
```

Maple 6 replaces
. with ||
. is for non-
communicative
multiplication

'='	'<>'	'<'	'<='	'||'	'.'	'..'	'!'	'**'	'*'	'+'	'^'
'and'	'intersect'	'minus'	'not'	'or'		'union'	RootOf	TEXT
algebraic	algext	algfun	algnum	algnumext	anyfunc	anything	arctrig
array	atomic	boolean	complex	complexcons	constant	cubic	dependent
disjcyc	equation	even	evenfunc	expanded	exprseq	facint	float
fraction	freeof	function	hfarray	identical	indexable	indexed	indexedfun
infinity	integer	intersect	laurent	linear	list	listlist	literal
logical	mathfunc	matrix	minus	monomial	'module'	name	negative
negint	nonnegative	nonnegint	nonposint	nothing	numeric	odd	oddfunc
operator	point	polynom	posint	positive	prime	procedure	protected
quadratic	quartic	radalgfun	radalgnum	radext	radfun	radfunext	radical
radnum	radnumext	range	rational	ratpoly	realcons	relation	rgf_seq
rtable	scalar	sequential	series	set	specfunc	sqrt	square
suffixed	string	symbol	symmfunc	table	tabular	taylor	trig
type	uneval	union	vector				

The surface and nested types may be checked with the Maple command 'type(something,sometype)' which checks the top level. Nested types are when an expression tree is involved. Most types are surface types encoded only at the top node of the expression tree.

Surface types

Most of the types in the list above are Surface types; Surface types include set and algebraic.

```
>?type[surface]
>
>type({x1,x2,x3,x4}, set);
>
>type(a + b, algebraic);
```

Nested types

In contrast, 'constant', completely scans an expression for any non-constant parts; it is a Nested type. Nested types include

algfun	algnum	constant	cubic	expanded	polynom
linear	quadratic	quartic	radnum	radfun	ratpoly

Structured types

Structure types care for combinations of types that are algebraic or logical.

```
>?type[structured]
```

gives the combining rules. For instance, {f(r),g(s),h(t)} corresponds to a set of functions.

```
>type({f(r),g(s),h(t)},set(function));
```

Structured types include

```
name^integer,name^posint), algebraic^integer, '&+'(name,radical,integer), exp(name),
'&*'(algebraic, algebraic,algebraic), TEXT(string,string), procedure(integer,string)
specfunc(string,TEXT), anyfunc(string), anyfunc(string,string), [name,integer],
list(integer), list(integer,integer), list({name,integer}), name(TEXT(string))
```

Skip ahead, now, and preview section 2.5.5, page 109, consider the detail near the heart of Maple. Nested types and other types may be found by exploring the way types are nested and how types are formed. Use eval( ) and lprint( ) on the symbols that define types, a Directory Structure. See 2.5.5, page 103.

```
>   eval('type/cubic');
>   lprint(%);
```

$$\mathbf{proc}(f, x) \ldots \mathbf{end\ proc}$$

```
proc (f, x) option 'Copyright (c) 1991 by the University of Waterloo.
All rights reserved.'; if nargs = 1 then type(f,polynom) and degree(f)
= 3 elif type(x,list) then type(f,cubic({op(x)})) else
type(f,polynom(anything,x)) and degree(f,x) = 3 end if end proc
```

Maple 6 has Aggregate Types sequential, tabular and rtable

We see that the symbolism is 'type/cubic' and that it is a procedure. Within the procedure lies several type check statements, including a recursive checking of 'type/cubic' itself.

Reserved Words and initially known names

```
>?ininames
```

Maple 6 uses end
do; end if; and
end proc;

Many names in Maple are reserved, protected or initially given values.

```
and  by  do done elif else end fi for from if in intersect
local minus mod not od option options or proc quit read save
stop then to union while constant false gamma infinity true
Pi Float Integer Fraction Complex factorial restart tracelast
catch error exports finally module return try use
```

```
FAIL -- unknown truth for 3 level logic
Digits -- default is 10
```

There are also initially known functions, including more than the standard functions.

```
>?inifcns
```

Upgrades to Maple 6

Maple V Release 5.1 is the recommended companion software to this book. Otherwise the book material should be usable with most versions of Maple with minor changes. Maple should detect most errors and remind you of necessary changes. Some changes to upgrade from earlier versions of Maple are
–The change of the ditto " from the full quote or double quote to percent % with Revision 5. The " is used for strings.
–The use of :: for declarations in the arguments to procedures.

The main upgrades to Maple 6 are
–The catenating operator is the || and the . is used for non communicative products. Provided you are not using . for vector dot products or other order dependent products, the continued use of . is effected by running a workaround macro at startup in your maple.ini file. In Windows, this .ini file is located in the Bin.wnt directory, the users directory or the current directory. See section 1.6, page 52.

In unix the
initialization file
is .mapleini

```
>macro('.'=cat):
```

–The replacement of end; in procedures and the ends of if . . fi conditionals and the ends of loops for . . do . . od with end proc; end if; and end do; These changes comply with major languages like Fortran, Pascal and C and allow the use of alternative conditional structures including try . . catch . . end try; An option in interface allows that the older useage be retained.

```
> interface(longdelim=false);):
```

Again, this may be placed in your .ini file (section 1.6, page 52).

Maple Commands

Following is a list of selected commands; they are not all the commands in Maple. Not included are many recent changes, including commands for the algcurves and Grobner packages. In any case, more complete explanations are available by typing

```
>?help,command
>#or
>help(command);
```

where command is one of those listed below, or a name that is known to Maple. Note that there may be no direct help file; the command may be a component of a package. Also some commands may not be accessible without evoking readlib( ) and/or with( ) to acknowledge the shortened names.

library reading is automatic in Maple 6

An index is also available through help.

```
>?index[category]
>#or
>help(index, category);
```

specifically, try

```
?index(expression)-  operators for forming expressions
?index(function)  -  list of functions
?index(misc)      -  miscellaneous
?index(packages)  -  library packages
?index(procedure) -  procedures and programming
?index(statement) -  Maple statements
```

Convenient Commands: one line summaries

These commands are listed so that you will be aware of their existence. They are not complete; it is expected that further information will be obtained from the help system, the help button ?whatever or help(whatever); or reference books. See page 30. These entries are available in a soft form as goodies and goodies.tex on the companion CD.

```
abs(x); # absolute value of x
add(x,y,z); # simple addition
algsubs(a+b=c,exp(a+b+d)); # substitute for a+b
alias(B=binominal); # a new name for binominal
alias(alpha=RootOf(x^2-2)); # alias to a RootOf
allvalues(expr); # evaluate RootOf's
anames(); # sequence of assigned names
anames(environment); # sequence of environmental variables
 unames(); # sequence of unassigned names
appendto(file); # output is appended to the file and does not go to the screen
 writeto(file); # all future output is stored in the file
applyop(expr,2,p); # apply operator p to operand 2 of the expression
applyrule([r1,r2],expr); # powerful substitution
array(0..2,2..4); # specialized table with indices with an integer range
 table(); # create a table
 table([chicken,corn]); # table of two entries
 table([(2)=b,(26)=z]); # a look-up table
 table(symmetric,[(b,c)=x]); # symmetric so that T[b,c]=T[c,b]
 # arrays and tables can be symmetric,antisymmetric,sparse,diagonal,identity
 indices(table); # indices of a table or array
 entries(table); # entries of a table or array
assemble - assemble a sequence of addresses into an object
 disassemble,assemble,pointo,addressof; # Maple 'hackware' package
ASSERT(a>1,cat('a is small: ',a)); # with kernelopts
assign({x=1,y=3}); # assigns the values
 unassign('a'); # unassigns the unevaluated name a
assigned(a); # checks a for assignment, boolean response
assume(x>0); # x is between 0 and infinity
 tassume(a<b,b<c,c=d,d<f); # totorder assume
 tis(u<f); forget(u); forget(everything); ordering(); init();
asubs( x^2 + 3 =k, (x^2 + 3*x + 3 ) ^ 3 + x ); #substitute for a sum
asympt(x/(1-x-x^2),x); # asymptotic expansion of the expression
attributes(a); # find the attributes attached to a
 setattribute(a,blue); # set blue as an attribute of a
branches(f); # plot the branches of a multi-valued function
bspline(d,x,k); # B-spline segment polynominal of degree d
C(expr,optimized); # convert expression to C, optimized code
cat(a,b,c); # concatenate the 3 symbols or strings
 "a"."b"."c"; # concatenate the 3 symbols or strings
ceil(x); # smallest integer greater than or equal to x
 frac(x); # fractional part of x
 floor(x); # greatest integer less than or equal to x
 round(x); # round x to the nearest integer
 trunc(x); # truncate x to the next nearest integer towards 0
charfcn[A](expr); # the characteristic function of A in expr
charstrip(PDE,f(x,y,z)); # find the characterisic strip of the PDE
 splitstrip(PDE,f(x,y,t)); # divides solution into ODE subsets
```

```
pdesolve(PDE); # find an analytical solution to the PDE
pdetest(sol,PDE); # returns 0 if sol matches the PDE
chrem(list,mlist); # the chinese remainder for modulo list mlist
close(filenm); # close filenm; write data to the disk
coeff(polynom,x,2); # the coefficient of $x^2$ in polynom
 lcoeff(polynom,x); # the leading coefficient of x in polynom
 tcoeff(polynom,x); # the trailing coefficient of x in polynom
 coeftayl(expr,x=a,5) # 5th coeff of the Taylor expansion about a
coeffs(s,x,'t'); # coeffs of powers of x; t is the sequence of powers
col(A,4..6); # selects columns of matrix A in a sequence of lists
collect(expr,x); # collect powers of x
 collect(expr,x,factor); # collect powers and factor the coefficients
combine(Int(x,x=a..b)-Int(x^2,x=a..b)); # combine the two integrals
comparray(A,B); # compare elements of the two arrays
compiletable([equations relating entries]); # make look-up table
 tablelook(expr,table); # look-up expression from table
 insertpattern(expr,pattree); # insert pattern in table
 patmatch(expr,pattern,'s'); # boolean pattern match with match in s
compoly(r,x); # returns a composited polynomial that makes r
conjugate(expr); # complex conjugate of expr
content(a,x,'pp'); # content of a multivariate polynominal
 primpart(a,x,'co'); # prime part of a multivariate polynominal
context[function](args); # context package--GUI and dynamical menus
convergs(listA,listB,n); # convergences of continuing fraction
convert(string,symbol); # convert to a symbol
 convert(listlist,matrix); # convert to a matrix
 convert(expr,confrac); # convert to a continuing fraction form
 convert(9, binary); # convert to 1001
 convert(1.23456,fraction); # convert to 3858/3125
 convert(x-2x^3+O(x^5),polynom); # drop the order term
 convert([1,2,x],'+'); # robust, list conversion to a sum
copy(a); # make a duplicate of a
cos,cosh,cot,coth,csc,csch; # trig and hyperbolic functions
curry(f,1,2); # M6 function adds arguments 1,2 before other arguments
curry(Psi,2); # M6 function is a one argument 2nd polygamma function
 recurry(f,1,2); # M6 function adds arguments 1,2 after other arguments
cost(a); # operations cost in terms of multiplications, etc.
csgn(a); # sign function for real and complex arguments
currentdir("c:/user/work"); # find and set current directory
currentdir(".."); # step back one directory
D[1,2](f(x,y)); # differential operator wrt y, then x
dawson(x); # Dawson's integral exp(-x^2) * int(exp(t^2), t=0..x)
dchange({TR},PDE,known=psi); # transform the arguments of the PDE
?debugger- Maple DEBUG - invoked by stopat(), stopwhen(), stoperror()
 # at the prompt DBG> the following commands are active:
 # cont next step into outfrom return quit where[] showstat[][]
 # showstop list stopat unstopat stopwhen unstopwhen stoperror
declare(y::numeric,g); # declare y numeric in procedure g
 dontreturn(a,f); # don't return a from procedure f
 makeglobal(b,f); # make b global in procedure f
 makeparam(x,f); # make x a parameter of procedure f
 makevoid(f); # do not return any values from procedure f
 packlocals(f,[w,t],A); # pack locals of procedure into array A
define(f,linear); # define the characteristics of an operator
```

```
definemore(f, f(2)=t); # add a characteristic to an operator
degree(poly,x); # degree of x in the polynomimal
 ldegree(poly,x); # low degree of x in the polynominal
denom(expr); # denominator of an expression
 numer(expr); # numerator of an expression
depends(int(f(x),x=a..b),x); # boolean check of dependence on x
diff(f(x),x$2); # second derivative of f(x) wrt x
difforder(PDE,x); # find the differential of the PDE wrt x
dilog(x); # the dilog function int(ln(t)/(1-t), t=1..x)
disassemble- assemble a sequence of addresses into an object
disassemble,assemble,pointo,addressof; # Maple 'hackware' package
discont(expr,x); # find the discontinuities
discrim(poly,x); # the discriminant of a polynominal
 resultant(poly1,poly2,x); # the resultant of two polynominals
display(P1,P1); # overlays PLOT structures
divide(x^3-y^3,x-y,'t'); # boolean indication of exact divide
dismantle(T); # displays large tables and arrays with blank entries as.....
dismantle(op(f)); # display the Maple Data Structure of procedure f
dontreturn(a,f); # don't return a from procedure f
dsolve(deqn,f(x)); # solve the differential equation
echo("hello"); # print to the screen, depends on interface(echo=0 to 4)
eliminate({eqns},{x,y}); # eliminate x and y from the equation set
entries(table); # entries of a table or array
 indices(table); # indices of a table or array
eqn(expr,filename); # troff/eqn output for printing
erf(x),erfc(x); # error function and complimentary error function
erfi(x); # the imaginary error function call
escape, !; # escape to the host environment
 system("xedit"); # runs program xedit in host system, no return
 ssystem("capture",3); # runs capture outside Maple with two element return
eulermac(1/x,x); # asymptotic approximations to sum(1/x,x)
eval(symbol,2); # evaluate to level 2
evala(expr); # evaluate RootOf as algebraic number or function,
'evalapply/F'(f,g)([a,b,c]); # for function calls of lists or arrays
evalb(x>2); # evaluate as a boolean
evalc(x); # valuate as a complex number
evalf(x,5); # evaluate as a floating point number to 5 digits
evalf(Int(sin(ln(x)),x=0..1)); # numerically integrate
evalhf(x,10); # evaluate using hardware floating point
evalm(A); # evaluate the matrix
evaln(x); # evaluate as a name
evalr(sin(INTERVAL(2..5))); # evaluate using range arithmetic
 shake(Pi,5); # compute a bounding interval with 5 digits
evalrC(expr); # evaluate using complex range arithmetic
exp(x); # the exponential function call
expand(expr,haltingexprs); # distributes products over sums
expand(expr,sin,cos,tan); # prevents trig expansions
 frontend(expand,[expr]); # prevents all functions from expansion
'expand/f'; # defines particular expansion for f
expandoff(funct); # suppresses expansion of function, uses option remember
expandon(funct); # unsuppresses expansion of function
factor(expr); # factors the expression
 factor(x^3+5,RootOf(x^3-5)); # factors using a RootOf
fdiscont(1/sin(x),x=0..Pi); # numerically find the discontinuities
```

```
feof(File); # boolean test for end of file
 filepos(File,infinity); # shows the file position, then goes to the end
 iostatus(); # determines the I/O status of all files, streams of pipes
fflush(fd); # flush the output of a given file
FFT(m,RArray,IArray); # complex fast Fourier transform of length 2^m
 iFFT(m,RArray,IArray); # inverse complex fast Fourier transform
Float(25,-7); # the floating point number 2.5E-6
floor(x); # greatest integer less than or equal to x
 ceil(x); # smallest integer greater than or equal to x
 frac(x); # fractional part of x
 round(x); # round x to the nearest integer
 trunc(x); # truncate x to the next nearest integer towards 0
fnormal(e, 4); # floating point normalization to 4 places
foldl(f,id,a,b,c); # fold operator, produces f(f(f(id,a),b),c)
fopen("file.m",APPEND); # open file.m for buffered appending
 open("file.m",READ); # open file for unbuffered READ
forget(proc,a,1,x); # forget this argument sequence to proc
forget(proc); # removes whole memory table from proc
fortran(s=x^2,precision=double); # convert to double precision fortran
fprintf(fd,"x=%d, y=%g",x,y); # print to file in formatted string
fprint(ssystem("dir")[2]); # properly formatted system directory
freeze(x+y); # replace x+y by a name
 thaw(z); # remove the freeze on the variable, allow evaluation
fremove(file); # remove the file
frontend(expand,[expr]); # keeps functions from expansion
frontend(degree,[abs(x)+x^3]); # doesn't fail because of suppression
frontend(indets,[e],[{'^','*'},{x-1}]); # types '^','*' and x-1 unfrozen
fscanf(1,"%a%a%d%a%a"); # parse from file 1 following the format
fscanf(fl,"%[^\n]c"); # parse characters until a CR is found
 scanf("%a"); # collect a variable from the screen, the default
 sscanf("25 1/x^4","%d%a"); # uses format to parse the strin
fsolve(eqn,x); # solve using floating point arithmetic
 fsolve(poly,x,-1..1); # search for roots on the closed interval
 fsolve(sin(x),x=3.1); # with a specified starting value
 fsolve(eqn,x,complex); # find at least one complex root
 fsolve(poly,x,maxsols=3); # find the 3 least roots
 fsolve(sin(x),x,avoid={x=0,x=Pi},0..10); # avoiding some values
 fsolve(eqn,fulldigits); # doesn't decrease Digits during calculation
GAMMA(a,z); # the gamma function int( exp(-t)*t^(a-1),t=z..infinity)
gc(); # garbage collection, delete data without references
gcd(poly1,poly2); # greatest common denominator of polynominals
 lcd(poly1,poly2); # least common multiples of polynominals
getenv("HOME"); # shows environment
getenv("PATH"); # get the full PATH used by Maple
GRADIENT(F); # generate the gradient of procedure F
harmonic(n)=sum(1/i,i=1..n )= Psi(n+1) + gamma; # harmonic function
has(expr,x); # boolean check for the presence of x
hasfun(expr,exp,y); # boolean check to see if expr has exp(y)
hasoption(Options,style,'s','otherOpts'); # boolean, s becomes style
hastype(expr,integer^fraction ); # boolean check for 2^(4/5)
 type(1+y,'+'); # boolean type check
 whattype(x,y,z); # gives top level type
 typematch(x^2,p::(b::anything^n::integer),'s'); # boolean type match
heap[new](lexorder,greg,tony,bruno); # priority Queque data structure
```

```
queue[enqueue](Q,1); # place 1 at the end of the Queque data structure
stack[push](Ian,s); # pushes Ian to the top of the Stack data structure
help(topic,subtopic); # opens help
hfarray(1..3,1..3); # array of hardware floating point data
history(); # initiate a history session; on starts, off quits
 timing(expr)); # time to compute the expression, appears in normal output
identity - argument to a table or array
ifactor(expr); # factors integer expressions
implicitdiff(eqn,y,x); # implicit differentiation, y wrt x
implicitplot(x^2+y^2=1,x=-1..1,y=-1..1); # plots a unit circle
implicitplot({x^2-y^2=1,y=exp(x)},x=-Pi..Pi,y=-Pi..Pi); # multiple
implicitplot3d(r^2+z^2=9,theta=-Pi..Pi,r=0..3,z=-3..3,coords=cylindrical);
implicitplot3d((x-1)^2+(y+1)^2+z^2=4,x=-1..3,y=-3..1,z=-2..2,color=blue);
indets(expr); # finds all the indeterminates
?index - index of help descriptions
indices(table); # indices of a table or array
 entries(table); # entries of a table or array
infolevel[a1]:=3; # set outside procedure to gain access
 userinfo(3,a1,'entered with',x,y); # set in procedure to give information
 # 1: required information  2,3: general information  4,5: more detail
?inifcns - initially known functions - see above
?ininame - initially known names - see above
initialcondition[f](Diff(f(x),x,x)=sin(x)); # initial condition for f(x)
?initialize - initializing Maple - see Section 2.3
int(exp(4*x^5),x); # integrate expression indefinitely wrt x
int(ln(x),x=1..e); # integrate expression definitely wrt x
 evalf(Int(1/(1+x^2),x=0..infinity)); # numerically evaluate the integral
 Int(x^3/(1-x^4),x); # unevaluated integral -- use value()
intat(f(a),a=y); # integral evaluated at a point
interface(echo=2,prompt="$ "); # sets echo and prompt
interp([0,1,2,3],[0,3,1,3],x); # the interpolation polynominal for the x's and y's
invfunc[f] - inverse function table, eg, replaces sin@@(-1) with arcsin
iostatus(); # determines the I/O status of all files, streams of pipes
 feof(File); # boolean test for end of file
 filepos(File,infinity); # shows the file position, then goes to the end
iperfpow(125,'p'); # determine if the integer is a perfect power, p is the power
 issqr(n); # boolean test if an integer is a perfect square
iquo(63,3,'r');# the integer quotient of 63/3, r is the remainder
irem(21,4,'q'); # the integer remainder of 21/4, q is the quotient
iroot(10,2); # integer approximation to the sqrt(10)
irreduc(x^3+5,5^(1/3)); # boolean check for irreducibility wrt 5^(1/3)
 factor(x^3+5,5^(1/3)); # factors over the given number field
is(x*y<50); # boolean check on the product
iscont(1/x,x=-1..1); # boolean check for continuity
isdifferentiable(ftn,x,2,'la'); # check at class c^2, discontinuities in la
ispoly(poly,linear,x,'a0','a1'); # boolean check with coefficients in a0 and a1
  issqr(n); # boolean test if an integer is a perfect square
isolate(4*x*sin(x)=3,sin(x)); # isolates sin(x) on the left
isolve(expr); # solve for integers
joinprocs([f,g]); # join the body of two procedures together
latex(expr); # convert the expression to LaTeX
lcm(poly1,poly2); # least common multiple of polynomials
 gcm(poly1,poly2); # greatest common divisor of polynominals
lcoeff(polynom,x); # the leading coefficient of x in polynom
```

```
 tcoeff(polynom,x); # the trailing coefficient of x in polynom
 coeff(polynom,x,2); # the coefficient of $x^2$ in polynom
 series('leadterm'(x^x),x=0,3);
length("Perth"); # length of string or symbol
lexorder(string1,string2); # boolean check of lexigraphical order
lhs(eqn); # left hand side of an equation
 rhs(eqn); # right hand side of an equation
linsolve(A,B); # matrix solution of AX = B
limit(sin(x)/x,x=0); # limiting value of sin(x)/x
lnGAMMA(1.234+2.345*I); # logarithm of the gamma function of the imaginary number
 GAMMA(a,z); # the gamma function int( exp(-t)*t^(a-1),t=z..infinity)
ln(x); # logarithm to the base e
 log10(x); # logarithm to the base 10
 log[2](x); # general logarithm to the base 2
 ilog10(x); # integer approximation to logarithm
lprint(expr); # linear printing of the expression
 print(x,y,z); # print as ordinary output
 printf("x=%+6.2f y=%+5.2f z=%a\n",x,y,z); # formatted print
 fprintf(fd,"x=%d, y=%g",x,y); # print to file in formatted string
makeglobal(b,f); # make b global in procedure f
 makeparam(x,f); # make x a parameter of procedure f
 makeproc(f,[x,y]); # make a procedure from a formula
 makevoid(f); # do not return any values from procedure f
map(x->x^2,x + y); # apply the procedure to each operand of function or procedure
map2(op,2,[a+b,c+d,e+f]); # passes the 2 as a first parameter to op
map2(op,(1..2),[[a,b,c],[3,4,8,9]]); # lists first two operands of lists
map2(map,convert,listlist,string); # converts all of listlist to strings
map2(map2,op,(2..3),listlist); # lines up arguments 2 and 3 of listlist
map2(subsop,4=NULL,listlist); # peels off the 4th argument of listlist
mapde(PDE1,psi,canop); # map the PDE to make it easier to solve
maple2intrep(f); # converts a Maple procedure to an abstract tree
match(ln(k)^(1/2)=A*ln(k)^P,k,'s'); # boolean match, parameters in s
 patmatch(sqrt(3.)*x-Pi,a::float*x+b::constant,'s'); # boolean match
 typematch(x^2,p::(b::anything^n::integer),'s'); # boolean, s defines b,b,n
matrix(2,3,[x,y,z,a,b,c]); # a 2 row, 3 column array beginning at 1
 # a matrix is a dimensioned array with columns and rows beginning at 1
max(1,3,10,4.3); # maximum of the given values
maximize(x^2+y^2,{x,y},{x=-10..10,y=10..20}); # maximum value in the range
 minimize(x^x,x,{x=0..2}); # the minimum of the function in the range
maxnorm(poly); # polynominal coefficient with the maximum absolute value
member(x*y, [x*y, w+u, y]); # boolean check on presence
 member(w, [x, y, w, u], 'k'); # k becomes the position in the list
min(a,b,c,d); # minimum of the given values
minpoly(1.234,3); # minimum polynominal with the approximate root of degree 3
mint - special executable program to check Maple syntax
minus - {a,b,c} minus {a,c};
 union - {a,b,c} union {x,y}
 interection - {a,b,c} intersection {c,d}
modp(12,7); # the remainder of 12/7 or
 12 mod 7; # 12 modulo 7
 mods(12,7); # symmetric representation (negative) modulo 7
mtaylor(x*exp(y^2),[x,y],8,[2,1]); #  8th order multivariate expansion, weights 2,1
mul([2,5,6,1,0,2][i],i=2..4); # multiply those three elements together
multiapply[fn](listdata); extension of zip, with multiple lists (stats)
```

```
zip((x,y)->x*y,[0,14,8],[2,6,12]); # multiply two lists together
nextprime(n); # the next largest prime number
 prevprime(n); # the next smallest prime number
nops(list); # number of operands of the list
norm(array([1,-1,2]),3); # cubic root of the cube of magnitudes added (linalg)
norm(array([1,-1,2]),2); # sqrt of the sum of the squares (vector magnitude)
normal(quotient); # simplifies polynominal quotients
not(a<0); # boolean not
nprintf("\n%a\n",x = .5); # formatted print to a name
 print("x = ",x); # print as ordinary output
numboccur(expr,sin(x)); # number of times sin(x) occurs in the expression
numer(expr); # numerator of an expression
 denom(expr); # denominator of an expression
odetest(sol,ODE); # returns 0 if the solution to the ODE is satisfactory
op(3,list); # finds the third operand of the list
 op([3,1,1],expr); # repeatedly uses op on the 1st,1st,3rd operands
open(File,WRITE); # opens the file for unbuffered writing
 fopen("file.m"); # open file.m
optimize(expr); # optimise the code generation in fortran and C (codegen)
order(series); # the truncation order of a series
packargs(f,[x,y,z],A); # pack arguments of procedure into array A
packlocals(f,[w,t],A); # pack locals of procedure into array A
packparams(f,[x,y,z],A); # pack arguments of procedure into array A
parse("x+2:"): # parse the string as a Maple statement
 parse(); # indicates true if the last parse ended in a colon
 parse('cos(1.)'); # like evaln
 parse('sin(3.)',statement); # parse with option statement evaluates
 parse(readline(),statement); # like readstat
patmatch(sqrt(3.)*x-Pi,a::float*x+b::constant,'s'); # boolean match
 match(5*x^2-3*x+z*x+y=a*(x+b)^2+c,x,'s'); # boolean match, parameters in s
 typematch(x^2,p::(b::anything^n::integer),'s'); # boolean, s defines b,b,n
pclose(pipe); # closes a pipe to a process started with popen
 popen(pipe); # opens a pipe
pdesolve(PDE); # find an analytical solution to the PDE
pdetest(sol,PDE); # returns 0 if sol matches the PDE
 charstrip(PDE,f(x,y,z)); # find the characterisic strip of the PDE
 splitstrip(PDE,f(x,y,t)); # characteristic uncoupled system divided into subsets
piecewise(x<0,-1,x<1,0,1); # uses combination condition, function values, otherwise
plot(sin,0..1); # plot the sin with arguments between 0 and 1
plot([sin(3*t),cos(t),t=1..3*Pi]); # parametric plot
plots[display](seq(plot(i*sin),i=1..20),insequence=true); # animation
plots[setoptions}(title='pilgarlic',axes=BOXED,coods=polar); # global plot settings
plot3d(sin(x*y),x=-Pi..Pi,y=-x..x); # 3D plot with variable endpoints
plotsetup(ps,plotoutput='plot.ps'); # B/W postscript output, probably rotated
plotsetup(ps,plotoptions='colour=rgb,landscape,noborder'); # landscape .ps output
plotsetup(ps,plotoptions='colour=cmyk,width=4in,height=3in'); # size set
plotsetup(cps,plotoutput="plot.ps",plotoptions="noborder"); # colour .ps option
plotsetup(default); # the default plotsetup for output(inline)
pochhammer(x,n); # pochhammer function x*(x+1)*...*(x+n-1)
pointto - obtain the expression pointed to by an address
 disassemble,assemble,pointo,addressof; # Maple 'hackware' package
poisson(f,[x,y],3,[1,2]); # 3rd order mtaylor expansion, combined trig coefficients
polar(3+4*I); # polar representation of the complex number
polylog(a,z); # the polylog function sum(z^n/n^a,n=1..infinity)
```

```
prep2trans(f); # with(codegen) prepares procedure f for translation
 split(h); # prepare procedure h for automatic differentiation
prevprime(n); # the next smallest prime number
 nextprime(n); # the next largest prime number
 content(a,x,'pp'); # content of a multivariate polynominal
primpart(a,x,'co'); # prime part of a multivariate polynominal
print(red,rouge,rot); # print as ordinary output
printf("%A",%); # prints leftflush without ' ' or " "
printf("x=%+6.2f y=%+0*.*f z=%a\n",x,5,2,y,z); # print formatted to default
printf("%-2.5s:%0.5s:%10.5s",M,Map,MapleV); # print formatted to default
printf("op %-3d %-15A has % 3d terms\n",i,x,n); # print formatted to default
printf("The listing of ranges is %a"); # prints whatever as a part of the sentence
 fprintf(fd, "x = %d, y = %g",x,y); # print to a file or pipe
 sprintf("x =%d, y= %g",x,y); # formatted print to string
 nprintf("%-5.5s:%9.5s","Go","MapleV"); # formatted print to a name
 fscanf(1,"%[^,]%[^,]"); # open file, read 1st two common separated fields as text
proc(x) local a,b statement; statements end; # procedure
procbody(b); # 'neutralized form' of a Maple procedure
procmake(%); # make the 'neutralized form' into a procedure
 unapply(expr,x); # returns an operator from an expression and arguments
 type(op(4,eval(pr)),table); # checks for presence of remember table
product(a[k],k=0..4 ); # indefinite product
profile([procedures],[sortflag]); # runtime information tables
 profile(proc); # starts the profiling of proc
 showprofile(proc); # shows the table of usage
 showprofile(time); # shows the data on timewise usage
 unprofile(); # terminates the profiling
proot(poly,n); # nth root of the polynominal
 psqrt(poly); # square root to the polynominal
protect(x); # make x protected from assignment
 unprotect(Pi); # make it possible to change the value of Pi (silly)
quo(x^3+x+1,x^2+x+1,x); # quotient of the polynominal
queue[enqueue](Q,1); # place 1 at the end of the Queque data structure
 stack[push](Ian,s); # pushes Ian to the top of the Stack data structure
 heap[new](lexorder,greg,tony,bruno); # priority Queque data structure
radnormal(3*3^(1/3))^(1/3)+3^(1/3)); # normalize the radical numbers
radsimp(1/1 + 2^(1/2)),'ratdenom'); # simplify the radical denominator
rand(); # 12 digit random number generator
rand(1..6); # procedure the digits 1..6 randomly    die()
randomize(Seed); # reset the seed for the random number generators
randpoly([x,y],terms=20); # random polynominal generator
rationalize(expr); # rationalise the multivariate radical denominator
readbytes(file); # read bytes from file, default=binary
 writebytes(file); # write bytes from file, default= binary
readdata(file,integer,3); # read 3 columns of integers from the file
readlib(linalg); # read library file linalg.m to define linalg
readline(file); # reads one line from a file or pipe
readstat(); # reads the next statement from the input stream
 x:=readstat("x will ="); # after prompt, the following input is assigned to x
 readstat(prompt,ditto3,ditto2,ditto1); # reassigns %,%%,%%% as well
realroot(poly,1/1000); # isolating intervals for the real roots
recipoly(poly,x,'p'); # determine if the poly is self-reciprocal
recurry(f,1,2); # new function adds arguments 1,2 after other arguments
 curry(BesselJ,1); # new function is a one argument first order BesselJ
```

```
rem(x^3+x+1,x^2+x+1,x,'q'); # remainder of the polynominal, q is the quotient
remove(type,list,name); # removes the list elements that are names
remove(has,[a,c,40,b,1],[1,40]); # removes 40 and 1 from the list
rename, unrename, forget; # for renaming inside an object (tools package)
residue(f(x),x=a); # coefficient of 1/(x-a) of the Laurent expansion
rhs(eqn); # right hand side of an equation
 lhs(eqn); # left hand side of an equation
round(x); # round x to the nearest integer
 frac(x); # fractional part of x
 trunc(x); # truncate x to the next nearest integer towards 0
 floor(x); # greatest integer less than or equal to x
 ceil(x); # smallest integer greater then or equal to x
row(B,1..4); # sequence of lists of the first 4 rows of matrix B
rsolve(f(n)=3*f(n/2)+5*n,f(n)); # recursion relation solver
rsolve({y(n)*y(n-1)+y(n)-y(n-1)=0,y(0)=a},y);
rsolve({F(n)=F(n-1)+F(n-2),F(1..2)=1},F,'genfunc'(x));
save a, b, c, "file.m"; # save a,b,c in file.m (maple format)
save x file; # save x in ordinary text format
savelib('a','b',"file.m"); # saves a and b in libname[3]/file.m
savelibname:=libname[3]; # set save library to the 3rd libname
scanf("%a"); # collect a variable from the screen, the default
 sscanf("25 1/x^4","%d%a"); # uses format to parse the string
 fscanf(1,"%a%a%d%a%a"); # parse from file 1 following the format
searchtext(Vwx,rstuvwxyz); # case insensitive pattern search
SearchText(pattern, string, range); # search for the given string
 substring("wink",2..3); # select "in" from "wink"
select(has,set,x); # selects the set elements with x
select(isprime,list); # selects the prime numbers in the list
seq(x.i,i=1..12); # a sequence of 12 x names
series(expr,x=a,n); # Taylor, Laurent or general series expansion
setattribute(a,red); # set red as an attribute of a
 attributes(a); # find the attributes attached to a
setoptions(labeldirections=[HORIZONTAL,VERTICAL]); # use with(plots)
shake(e,3); # compute a bounding interval for e with 3 digits
showtime(); # time and space statistics for commands; on starts, off quits
 time(); # total CPU time used since start of the Maple session
 time(x); # time to evaluate x
 history(); # maintain a history of all values computed
 timing(expr)); # time to compute the expression, appears in normal output
showstat(pr); # shows statement numbers in given procedure
sign(x); # sign of x
sign(poly,[y,x],'a'); # sign of the leading coefficient a
signum(x); #  x/abs(x) or the sign of x
signum(1,x); # derivative of signum(1,x), 0 for all non-zero real numbers
signum(0,x,1); # temporarily sets _Envsignum0 to 1, allowing evaluation
simplify(expr); # apply simplification rules to the expression
simplify(expr,name); # name is one of
   # Ei,GAMMA,RootOf,atsign,hypergeom,ln,polar,power,radical,sqrt,trig
simplify(g,assume=real); simplify(g,assume=positive);
simplify(a,f); # invokes `simplify/f`(a) if `simplify/f` is defined
sin,cos,tan,arcsin,arccos,arctan; # standard functions and inverses
singular(x*y+1/(x*y),x); # finds the singularities
sinh,sech,sinh,sech,arcsinh,arccosh,arctanh; # hyperbolic trig calls
sinterp(f,[x,y],10,101); # sparse modular polynomial interpolation
```

```
thiele([1,2,a],[3,4,5],z); # Thiele's continued fraction formulation
 spline([0,1,2,3],[0,1,4,3],x,cubic); # natural spline from the x,y values
smartplot(sin(2*x)); # drag and drop from Maple output, right mouse button
smartplot3d(sin(x^2+y^2)); # form, rotate, rerange, resize, remove
solve({eqns},{x,y}); # solves for x and y in the set of equations
solvefor[x,y,z]({eqns}); # isolates (solves for) variables
sort(expr,x); # sort wrt x
 sort([c,a,d],lexorder); # use lexigraphic ordering on list
 sort(expr,[x,y],plex); # use lexigraphic and dominant powers
sparse - table(sparse,[A=a]): or array(sparse,1..2): all other elements 0
spline([0,1,2,3],[0,1,4,3],x,cubic); # natural spline from the x,y values
split(h); # prepare procedure h for automatic differentiation
splits(poly); # complete factorization of a polynominal
sprintf("x =%d, y= %g",x,y); # formatted print to string
sqrfree(poly); # square free polynominal factorization function
sqrt(x); # square root of x
sscanf("123.456E7 123.456E7","%g%d.%d %[Ee] %d"); # format parses the string
 fscanf(1,"%g%a%d%a%a"); # parse from file 1 following the format
 scanf("%a"); # collect a variable from the default (screen)
ssystem("edit"); # 2 element list; return code, string from system
 system(date); # similar to system, simple return
 !ls; # operator, list from host directory in command line maple
stack[pop](s); # takes from the top of the Stack data structure
 heap[new](lexorder,greg,tony,bruno); # priority Queque data structure
 queue[enqueue](Q,1); # place 1 at the end of the Queque data structure
sturm(poly,x,a,b); # real roots of the polynomial in interval a,b
sturmseq(poly,x); # Sturm sequence of the polynomial
subs(x=a,expr); # substitutes equations or sets or lists of equations
subsop(2=y,expr); # substitute for operand 2 in the expression
subsop(3=NULL,expr); # removes operand 3 in the expression
substring(string,range); # extract a substring or subsymbol
substring(st,4..length(st)); # last part of string
substring(st,-2..-1); # last 2 characters of string
sum(x^n,n=1..4); # sum the 4 powers of x
surd(8,3); # the 3th root of 8 with (complex) argument is closest to 8
symmdiff(a,b); # set symmetric difference (a union b) minus (a intersect b)
symmetric - symmetric indexing function, array(symmetric,1..5,1..5);
system(date); # collects the date from the system
 ssystem("capture"); # runs capture outside Maple with two element return
 !del junk; # deletes file junk on the host system (DOS) -- command Maple
 printf(ssystem("dir/w")[2]); # wide listing of files
swapargs(f,1=2); # swap arguments of a procedure
table(); # create a table
table([chicken,corn]); # table of two entries
table([(2)=b,(26)=z]); # a look-up dable
table(symmetric,[(b,c)=x]); # symmetric so that T[b,c]=T[c,b]
 array(0..2,2..4); # specialized table with indices with an integer range
 # tables and arrays can be symmetric,antisymmetric,sparse,diagonal,identity
 indices(table); # indices of a table or array
 entries(table); # entries of a table or array
 compiletable([equations relating entries]); # make look-up table
 tablelook(expr,table); # look-up expression from table
 insertpattern(expr,pattree); # insert pattern in table
tan,arctan,sin,arcsin,cos,arccos,sec,arcsec,csc,arccsc; trignometric
```

```
   # standard function calls that use radians
tanh,arctanh,sinh,arcsinh,cosh,arccosh,sech,arcsech,csch,arccsch; # hyperbolic
tassume(a<b,b<c,c=d,d<f); # totorder assume
 tis(u<f); forget(u); forget(everything); ordering(); init();
testeq(poly1=poly2); # boolean random polynomial-time equivalence tester
testfloat(1.23,1.25,2); # quasi-boolean test, floating point equivalence, 2 places
thaw(a); # allow evaluation of symbol a
 freeze(a); # disallow evaluation of a
thiele([1,2,a],[3,4,5],z); # Thiele's continued fraction formulation
 sinterp(f,x,t,p); # sparse multivariate modular polynomial interpolation
time(); # total CPU time used since start of the Maple session
time(x); # time to evaluate x
timelimit(0.5,f()); # limits the CPU computation time of f() to .5 secs
trace(pr); # detailed trace and printout of the inner workings or a procedure
 untrace(pr); # removes the trace on procedure pr
translate(poly,x,1); # linear translation of poly by one unit right
traperror(f(x)); # traps the error in calculating f(x) in variable lasterror
 ERROR("out ot lunch ",x); # exit procedure with given message
 lasterror; # contains the last error message, possibly 'division by zero'
 WARNING(errMSG); # write a warning message to the current output stream
trigsubs(expr); # returns a table of identities equal to the expr
trigsubs(eqn); # returns 'found' if the table has the identity
trigsubs(eqn,expr); # applies the identity eqn to the expr
trunc(x); # truncate x to the next nearest integer towards 0
 round(x); # round x to the nearest integer
 frac(x); # fractional part of x
 floor(x); # greatest integer less than or equal to x
 ceil(x); # smallest integer greater then or equal to x
type(x+y,polynom); type(x+y,'+'); type(x^(-2),algebraic^integer);
type([x,1],list({name,integer})); type(exp(x),exp(name));
type(op(4,eval(pr)),table); # checks for presence of remember table
 # boolean check of types, surface and nested types, structured types
typematch(x^2,p::(b::anything^n::integer),'s'); # boolean type match
 hastype(expr,integer^fraction ); # boolean check for 2^(4/5)
 patmatch(sqrt(3.)*x-Pi,a::float*x+b::constant,'s'); # boolean match
 match(ln(k)^(1/2)=A*ln(k)^P,k,'s'); # boolean match, parameters in s
unames(); # expression sequence of active names which are 'unassigned'
unapply(expr,[x,y]); # unapply expression wrt x,y to make a procedure
 makeproc(f,[x,y]); # make a procedure from formula f
 proc(x) local a,b statement; statements end; # procedure
 procmake(%); # make the 'neutralized form' into a procedure
unassign('b','ss'); # unassigns b and ss
union - {a,b,c} union {x,y}
 minus - {a,b,c} minus {a,c}
 interection - {a,b,c} intersection {c,d}
unload(f); # removes the remember table entry for f
untrace(pr); # removes the trace on procedure pr
userinfo(3,a1,'entered with',x,y); # set in procedure to give information
 infolevel[a1]:=3; # set outside procedure to gain access
 # 1: required information  2,3: general information  4,5: more detail
value(expr); # evaluate the Inert functions (with capitals) in expr
vector(4,[1,x,x^2,x^3]); # 1D array starting at 1     with(linalg)
verify(L1,L2); # checks the sameness of the two lists
verify('int(sin(x),x)');; # boolean verify of the Maple integration
```

```
verify('solve({x+y=1,2*x+y=3},{x,y})');
verify('msolve({3*x-4*y=1,7*x+y=2},19)');
whattype(anything); # gives the basic Maple type (see above list)
 hastype(expr,integer^fraction ); boolean check for 2^(4/5)
 type(1+y,'+'); # boolean type check
with(linalg); # allows the package procedures to be simple commands
writebytes(file); # write bytes from file, default= binary
 readbytes(file); # read bytes from file, default=binary
 readdata(file,integer,3); # read 3 columns of integers from the file
writedata(terminal,A,float); # Print all data in the table as floats
writedata(terminal,A,[string,integer,float,integer,float]);
writedata(terminal,A,integer,proc(f,x) fprintf(f,'%a',x) end); # complex
writeline(default,"We are now","Maple command experts"); # to a file or pipe
 readline(file); # reads one line from a file or pipe
writestat(1,"let us ",a+b," writestat",next); # writes to file or pipe 1
 readstat(); # reads the next statement from the input stream
 x:=readstat("x will ="); # after prompt, the following input is assigned to x
writeto(file); # all future output is stored in the file, not on the screen
 appendto(file); # output is appended to the file
zip(gcd,[0,14,8],[2,6,12]); # zip together two lists
zip((x,y)->x+y,[1,2,3],[4,5],0); # 0 is a default value for values not present
 [0,14,8]+[2,6,12]; # same as zip, above
 multiapply[fn](listdata); extension of zip, with multiple lists (stats)
ztrans(sin(Pi/2*t),t,z); # z transform, sum(f(n)/z^n,n=0..infinity)
 invztrans(a*exp(3/z),z,n); # inverse z transform wrt n
                WDS 15Aug00, Maple V Release 5.1 and Maple 6
```

use
?index,function
for an up-to-date
listing

libname := "c:/Program files/Maple V release 5/lib", "c:/mplbook/Mylib"

1.6   Your own Environment: .ini and interface

Guidewords:

>   libname, setoptions( ), initialization, mapleini file, interface( ),
>   plotsetup, echo, prompt, % for substitution, worksheet style

libname; # finds
the library path

For convenience and ease of operation, it is possible to operate Maple from
your own environment. On startup, Maple runs a special file to set its parameters. This is called an initialization file and is called maplev5.ini or maple.ini
on DOS machines and .mapleini on Unix machines. Using a simplest editor
and saving in ASCII, you can create a file with any Maple commands, definitions or other setup, as you like. When placed in your currentdirectory, the
users directory or the bin (Bin.wnt) directory, it is be applied automatically
to your session on startup. For a good feeling, a message can appear so that
you will know that the initialization setup has been completed.

Maple Initialization

At the top of the page is an echo to the command used to begin this session. It simply establishes a path for Maple to use in finding libraries. The
mapleini file that produced this statement is below. The macro( ) and inter-
face( ) actions set up a workaround environment that uses the older   .   and
end; rather than the || and end proc; introduced with Maple 6; these state-
ments are not needed with earlier versions of Maple. Maple 6 uses the period
or full stop   .   for non-communicative multiplications instead of &* . Using the workaround removes the capability of using   .   as a multiplier and
requires using &* for non-communicative matrix multiplications. Noteworthy, however, is the smooth way Maple can generate a binary operator using
the neutral operator &

Next, from section 2.5, the AAArgh command is a comfort. The plotsetup(
) line sets the inline option for plotting in colour, with portrait orientation and
without borders. These default options suggest how to influence printing to a
file.

"It isn't
reasonable to
expect the first
attempt to solve
a problem will be
error-free"

The echo default follows Rob Corless' suggestion that Maple should echo
input with output when reading Maple statements from an ASCII text file.
This particular maple.ini file is followed by a friendly statement to put Bill
Scott into the right mood.

```
>   libname:="c:/Program files/Maple V release 5.1/lib":
>   libname:=%,"c:/mplbook/Mylib";
```

```
>   # macro('.'=cat); # workaround catenation in Maple 6
>   # interface(longdelim=false); # sets defaults end;  do;  fi;
>   #   All three delimiters can be simply end; in Maple 6
>   #   This default gives the shorter output

>   macro(AAArgh=ssystem("edit readme",3)[2]);

>   plotsetup(inline,plotoptions='colour=rgb,portrait,noborder');

>   # interface(echo=2); # Corless' Recommendation

>   "Hello Bill.   May the world treat you well.";
```

libname := "c:/Program files/Maple V release 5/lib", "c:/mplbook/Mylib"

"Hello Bill. May the world treat you well."

With plotsetup, the default option is inline but variations of this setup
allow one to put the plot in a separate window, (X11 on Unix), make a printout
on the terminal screen in ascii, or form a postscript file.

setoptions

The plots package contains the command plots[setoptions]( ) that sets
global variables for plotting. These variables include the titles, labels, legends
and axes strings as well as specific fonts and font sizes; linestyles and thick-
nesses, symbols and symbol sizes; colour, coordinate system, scaling, view,
and the possibility that the lower part of the curve is filled with colour. See
the detail under plot Options, section 3.6, page 143.

```
>   ?setoptions
>   ?setoptions3d
```


plotdevice

This must be the first argument to plotsetup. They may be inline,window,
ps, X11 and default. postscript or ps produces an encapsulated postscript
file and X11 displays the plot on a X, Unix window. With postscript, the
file name must be specified by setting the variable plotoutput. Note that
with inline, and export to latex, .eps files are automatically produced with
sequential numbering of file names, by adding 01.eps, 02.eps, 03.ps etc. to the
original worksheet name.

With each different platform, and different versions of Maple, there are
variations to the setup. Suggestions are contained in the one line commands
of the last section and on the next page. Some details in the use of postscript
are presented in section 3.6, page 149. Most of Chapter 3 deals with plotting
and section 2.4.6, page 93 has further information on producing plot files.

```
>   ?plotsetup
>   ?plot,device
>   ?plot,device,ps
>   plotsetup(ps,plotoutput="fred.ps",plotoptions=noborder);
>   plotsetup(ps,plotoptions=`colour=cmyk,
>   width=4in,height=3in`);
```

Echo

if quiet=true,
there is no echo

The level of echo can also be annoying. This is changed through the interface command. First it is good to find out what the default setup is. This is valuable for every session. Echo has the following alternatives.

- echo=0, no echo in any circumstances

- echo=1, echo only when input/output from terminal interactions are involved–the default

- echo=2, echo only when input/output is not from the terminal

- echo=3, echo only with a read statement

- echo=4, echo always

Interface

The Maple interface( ) command adjusts many of the front-line parameters of Maple. Interface communicates with the Maple 'Iris' and sets and reveals behind-the-scene information that does not affect the computations.

```
>   ?interface
>   interface(echo);
```

$$2$$

```
>   interface(quiet);
```

$$false$$

```
>   interface(errorbreak);
```

$$1$$

```
>   interface(version);
```

Maple Worksheet Interface, Release 5.1, IBM INTEL NT, Nov 5 1998

```
>   interface(echo=2, prompt="$ ");
$   interface(plotdevice);
```

inline

To generate postscript output in a file

```
$   interface(plotdevice=postscript,
$   plotoutput="fuzzy.ps",plotoptions="landscape");
```

*The plot device
you are using*

To reset the plot output to a window

```
$   interface(plotdevice=window);
```

To look at the Maple code in library routines

```
$   interface(verboseproc=2);
```

To store the current value of quiet and set it to true

```
$   oldquiet:=interface(quiet,quiet=true);
```

*This works the
same as
plotsetup,
page 53*

oldquiet := false

Go back to the default prompt and echo

```
>   interface(echo=1, prompt="> ");
```

An Example Initialization file

An edited version of an initialization file from John Borwein shows the variation that can be included in the startup of a Maple session. These commands and variable settings can be altered to suit your needs. Note, however, that not all platforms allow extensive interactions with the system. For instance, it is difficult to remove or delete files and change directories from inside Maple in DOS systems.

Note that warnlevel suppresses all warnings at the start; it is reset to 3 at the end. readlib( ) and with( ) are companion commands; readlib( ) reads the package.m from one of the libraries; with( ) takes the table structures package[command] and scrapes off the package to leave the command. Also, with( ) looks for either a package[init] or a 'package/init' directory structure; this is read and activated first and is usually a procedure that is a vehicle for definitions of global variables, variable types or the making of help information, perhaps through makehelp.

```
#                         John Borwein's .mapleini file                        #
                            (Edited by Bill Scott)

#2345678901234567890123456789012345678901234567890123456789012345678901234567890123456789#

 interface(warnlevel=0): # suppress all warnings (for this .ini file)

#                    System Specifics, Global Variables                         #

 kernelopts(printbytes=false):
 interface(prompt='*:'):
 nodes:=25: Digits:=20:

#                                    macros                                      #

 macro(ls=system('ls -xF')): # use ls (list or directory)  inside Maple

#                         Friendly Read statements                              #

 rd:=proc(file)  read(file); end: # simple, abbreviated read
 vi := proc(file) local bbb; bbb:='vi '.file;system(bbb); read(file); end:# vi, read
 readq:=proc(filnme) interface(verboseproc=0): read(filnme): end:  # no reply
 readn:=proc(filnme) interface(verboseproc=1): read(filnme): end:  # some echo
 xfred:=proc(myfile) system('xedit '.myfile); read myfile end: # xedit read

#              Read in Libraries and Packages, including Share Library           #

 readlib(FFT): readlib(ellipsoid); # FF Transforms, Ellipsiod  Areas
 with(plots): # display, implicitplot, animate, fieldplot, odeplot
 with(share): readshare(gfun,analysis):  with(gfun);
 read('/home/jborwein/Maple/fred'):

#                                Plot Specifics                                  #

 #plotsetup(tek,vt100);
 #interface(plotdevice=vt100,verboseproc=2);
 interface(plotdevice=x11);

#                                Full Procedures                                 #

 incr:=proc(g,f) local k,s,t;s:=op(1,f);for k to nops(s)
  do t[k]:=[op([1,k,1],f),simplify(g(op([1,k,2],f))),
    op([1,k,3],f)];od; # nested op structure, i.e., op(1,op(k,op(1,f)))
 cf(convert(t,list),op(2,f));end;

#                                    ----                                        #
 comp:=proc() local f,e,x,l,t,k,s,y;
 f:=args[1];e:=args[2];if nargs>2 then x:=args[3];else
 x:=op(2,convert(e,cf));fi;
 l:=unapply(e,x);s:=op(1,f);y:=op(2,f);
 if D(l)>0 then for k to nops(s)
 do t[k]:=[l(s[k][1]),subs(y=e,s[k][2]),l(s[k][3])];od;
 else for k to nops(s) do t[k]:=[l(s[nops(s)+1-k][3]),
 subs(y=e,s[nops(s)+1-k][2]),l(s[nops(s)+1-k][1])];
```

```
od; fi; cf(convert(t,list),x); end;
```

```
#                      Information to the User                    #
```

```
interface(warnlevel=3): # print library, kernel, and parser warnings (the default)
print('The current graphics interface is X11 with Digits',Digits);
```

```
#23456789012345678901234567890123456789012345678901234567890123456789#
```

Exercise 1.12 Analyse Borwein's .mapleini

Go through the above lines and add comments as to what they may be doing. Look into the share libraries. There are a large number of interesting bits.

The assume tilde

The presentation of tilde before special assumed variables can be annoying. Lopez suggests that, if you don't want it , turn it off. Here and now, it is better not to use the assume command; it has too many complex implications. If you use it on a variable within a session and you need to change it, it is often better and easier to simply restart .

Do not use
assume

```
>  restart:
>  interface(showassumed); # the default setting
```

$$1$$

```
>  interface(showassumed=1);
>  assume(aa>0);
>  about(aa);
```

Originally aa, renamed aa~:

  is assumed to be: RealRange(Open(0),infinity)

```
>  sumap:=sqrt(sum(aa^j,j=1..10));
```

$$sumap := \sqrt{aa^{\sim} + aa^{\sim 2} + aa^{\sim 3} + aa^{\sim 4} + aa^{\sim 5} + aa^{\sim 6} + aa^{\sim 7} + aa^{\sim 8} + aa^{\sim 9} + aa^{\sim 10}}$$

```
>  interface(showassumed=0);
>  sumap;
```

$$\sqrt{aa + aa^2 + aa^3 + aa^4 + aa^5 + aa^6 + aa^7 + aa^8 + aa^9 + aa^{10}}$$

Automatic Substitution

If you are in other than *Editable Math* output, Maple often uses the % sign to stand for a common subexpression within a large expression. This can be annoying and somewhat destructive to structure. Blachman suggests changing the interface variable labeling. It is possible to have the automatic substitution happen only with large subexpressions by using, say, labelwidth=10.

```
> restart:
```

Convert the output to other than *Edible Math* (or *Standard Math Notation*), or the effect will not appear. This is under Options, Output Display.

The output is
now in Typeset
Notation

```
> interface(labeling);
```

$$true$$

```
> interface(labelwidth=5);
```

This is made small to show the effect of labeling.

```
> oldlabel:=interface(labelwidth,labelwidth=10);
```

$$oldlabel := 5$$

```
> interface(labelwidth);
```

$$10$$

The following, from Blachman, is an awkward expression that Maple wants to simplify by substitution.

```
> ans:=solve(x^3-x^2-x-1,x);
```

$$ans := \frac{1}{3}\,\%2 + \frac{4}{3}\,\%1 + \frac{1}{3},\ -\frac{1}{6}\,\%2 - \frac{2}{3}\,\%1 + \frac{1}{3} + \frac{1}{2}\,I\,\sqrt{3}\,(\frac{1}{3}\,\%2 - \frac{4}{3}\,\%1),$$
$$-\frac{1}{6}\,\%2 - \frac{2}{3}\,\%1 + \frac{1}{3} - \frac{1}{2}\,I\,\sqrt{3}\,(\frac{1}{3}\,\%2 - \frac{4}{3}\,\%1)$$
$$\%1 := \frac{1}{(19 + 3\,\sqrt{33})^{(1/3)}}$$
$$\%2 := (19 + 3\,\sqrt{33})^{(1/3)}$$

```
> interface(labelwidth=28);
```

```
> ans;
```

$$\frac{1}{3}\,(19+3\,\sqrt{33})^{(1/3)} + \frac{4}{3}\,\frac{1}{(19+3\,\sqrt{33})^{(1/3)}} + \frac{1}{3}, -\frac{1}{6}\,(19+3\,\sqrt{33})^{(1/3)} - \frac{2}{3}\,\frac{1}{(19+3\,\sqrt{33})^{(1/3)}}$$

$$+ \frac{1}{3} + \frac{1}{2}\,I\,\sqrt{3}\,(\frac{1}{3}\,(19+3\,\sqrt{33})^{(1/3)} - \frac{4}{3}\,\frac{1}{(19+3\,\sqrt{33})^{(1/3)}}), -\frac{1}{6}\,(19+3\,\sqrt{33})^{(1/3)}$$

$$- \frac{2}{3}\,\frac{1}{(19+3\,\sqrt{33})^{(1/3)}} + \frac{1}{3} - \frac{1}{2}\,I\,\sqrt{3}\,(\frac{1}{3}\,(19+3\,\sqrt{33})^{(1/3)} - \frac{4}{3}\,\frac{1}{(19+3\,\sqrt{33})^{(1/3)}})$$

If this is of concern, it is advised to simply use the *Editable Math* in Output Display, the default, in the Options menu. Otherwise, you might want to change the interface variable labelwidth or set labeling=false.

Keeping a Stable Style

Your environment is also set up to make documents. Maple is a powerful mathematics word processing environment, with active worksheets and capable of manufacturing HTML files that can be viewed on the web. Here we should try to establish a convenient or appropriate style for working with documents and making them look good.

section 2.4.6, on page 94 takes HTML further

The Save Automatic Settings on the File menu or the Save as Default on the Format/Styles menu can retain errors you commit, and make a mess of your worksheet. You are freed from the problem if the first time you open Maple you save the style sheet: In an appropriate library, Mylib, lodge the file as Original.mws. Using the Format menu, select Styles. Click on Save as Default in the Style Management dialog box that appears. Type in Original.mws in the file name holder; make sure your folder is correct and click OK. The Styles Management window reappears; click Done; this saves the style Original.mws in your folder.

Reinstalling Maple may be necessary

You can now proceed to make your own style files, companions to this one. This is done by a return to the Format menu and Styles. In the Styles Management window you adjust the styles to your liking; that includes text, Maple Input, and 2D Output. With each of these, you select from the Styles panel and click modify; various possibilities are selected. Finally, you click OK, and, as before, Save as Default in the Styles Management window, and, finally, Done. This, in a nutshell, should allow that you have some choice in the style.

A journey through a maze of semicolons, dittos and assignments into solutions and units. Through errors into success with the reward of a growing sense of awareness of symbols and structure. Your statements, your commands to Maple are getting somewhere. Get your ini file right, remember the listings of commands and note some of the different types. Comment your statements, keep records and backups and take care; the Maple command line can be friendly.

From here you venture into pulling Maple apart with some calculations of lake areas and trends in carbon dioxide levels. The Maple structure is part of the solution.

Acknowledgements for Chapter 1

Most of the introductory material has links to Maple Help files for Versions 5r3, 5r4 and 5r5. The recursive definition is from Corless [9], pg 26. The error in the integral comes from Maple V Release 4, by Michael Kofler [20], pg 109. Ross Butterworth of Murdoch 'couldn't stop himself' from looking up the Maple web sites, bookmarks.htm   The example on labeling came from Blachman and Mossinghoff [5], pgs 19 and 425. The initializing file is from John Bornstein of CECM, Simon Fraser University. The advice on styles comes from Lopez, MapleTech 3.3, pg 46.

Chapter 2

Structures and Analysis: Help

Finding and Changing

Environmental Measurements

The Worksheet and Execution Groups [>

Self Help

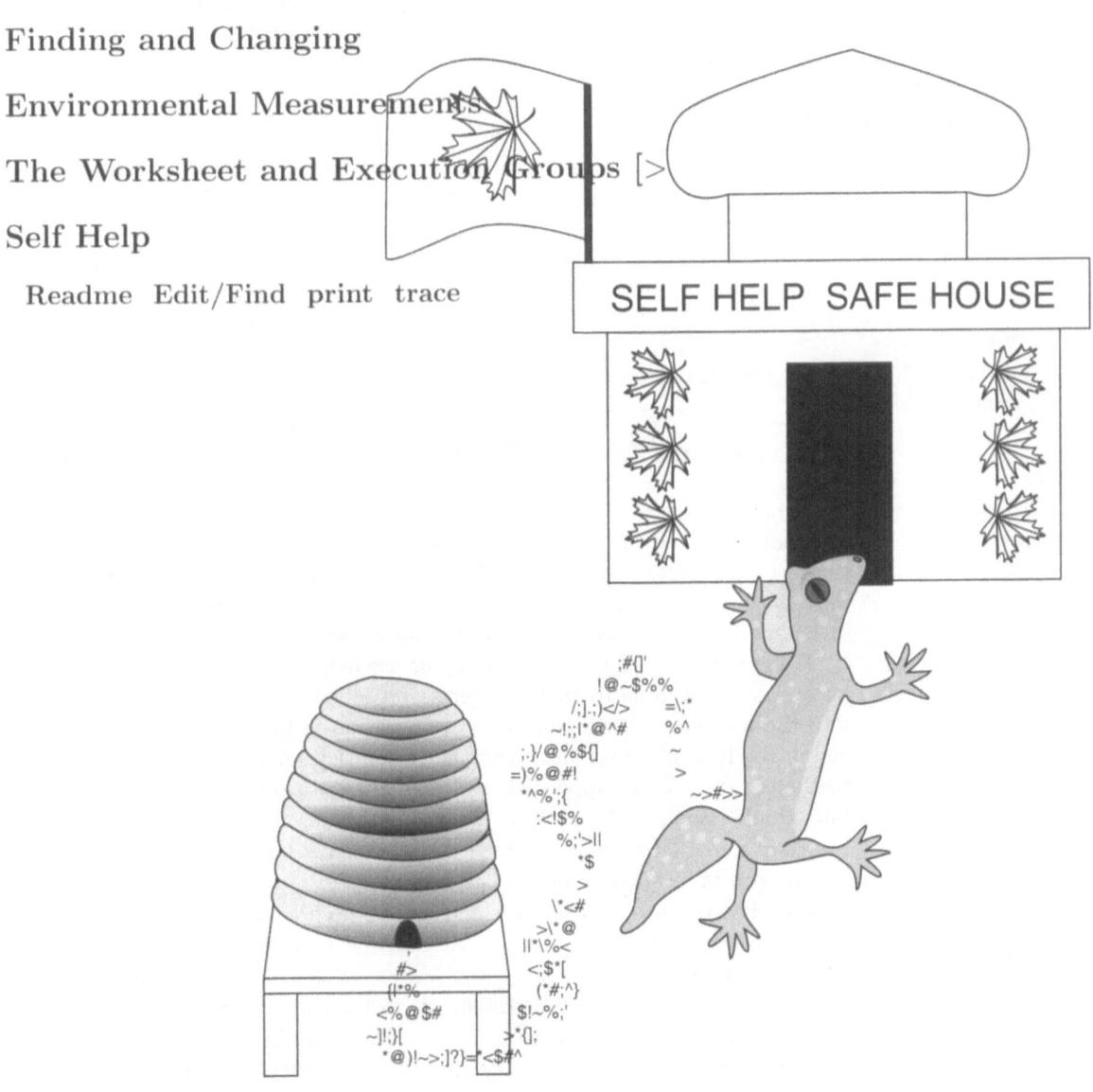

SELF HELP SAFE HOUSE

Finding and Changing

Übersicht

Understand the 'bones' of Maple structures; arguments, indexes and ordering. Analysis of units and and a greenhouse chamber. The details of execution groups and producing documents. Set up a self-help system, using readme files and help from Maple. Combine initialization and help to make an environment you enjoy.

for .ini files see section 1.6, page 52

Guidewords:

op( ), nops( ), convert( ), has( ), whattype( ), member( ), seq( ), sum( ), " or % , . , union, minus, intersection, symbol, sequence, list, array; units, data, calibration, precision; hotwin, ASCII, LaTeX, HTML, worksheet, execution group; command line Maple

see upgrades page 38 and Maple 6

```
>  restart;
```

2.1 Pulling Maple Apart

Turn on the machine. Expect to have trouble with one thing or another. One problem relates to the security system. If you make too many errors, it may refuse to acknowledge you exist. Try to keep the Caps Lock key in the off mode and use only small letters; Maple is sensitive to whether you use upper or lower case. If you have a choice, use lower case.

under Windows

Bring down the Help menu above, or use help( ); The menu is convenient because it allows one to click on items and follow a tree to the manual built into the machine. A simple mode of operation is to cut examples from the end of the help files and paste together a solution. An alternative and secure method is to have a floppy disk and read the data (commands, etc) directly into Maple. Such a file, with a .mws extension can be opened directly by Maple as a Mapletext file; provided the statements have a prompt > before the lines to be executed and all the comments are preceded by a # it will convert directly into a .mws file and can be used as a .mws file thereafter.

A backup on a floppy is a good idea

If the same file is kept as an ASCII raw text file, it will probably remain timeless, independent of the Maple version or platform you use. It is convenient to record the statements in the same format as one types them into Maple, without the prompt > Such files have historically been saved as .mpl files. Files without the >   copy directly with the mouse, drag and drop, Edit/Copy and Paste, Ctrl-c and Ctrl-v. They can be read directly into Maple with the read command. When read in and echo=2, (see page 53) the statements appear as they are read in, along with the output.

```
>  interface(echo=2); # echos statements as they are read in
>  read("myfile.mpl");    # reads your file
```

A scheme is to make a raw ASCII file of your statements in an editor. Then the statements can be copied with 'cut and paste' into any part of your workspace. The 'raw' text is used for development of your thoughts. Later, when it is convenient, your workspace may be saved as a .mws file.

A separate editor in a separate screen/window can usually be used to copy bits into your worksheet. A favorite editor can be used for this purpose, but simple editors are better, like Wordpad, SimpleText and Xedit. A variant of this is to use another worksheet in Maple; set up the worksheet as text with the T button; you can 'cut and paste' freely into it as if it were an ordinary editor.

Do remember, though, to make up your mind as to what you are pasting and why. It is easiest to bring up several execution groups into which you paste the material. Try not to use CR (Enter or Return) in the middle of your copy; Maple will prematurely execute the statements or you may have a repair job. Maple mistakes CR, without Shift, as a sign to execute the statement. With a Shift, a proper, typed carriage return CR is entered into the Maple code.

look up help

With a direct read, the operations are immediately executed. Also, with read and save, be sure to use a ' or " around the file name or Maple will misread it. Saving as a .m file is for saving the full maths background, in a consolidated, binary form. On reopening a .mws file, the input will need to be 'replayed' or reexecuted, as some results will be needed.

The .mws files only save the 'face' of the workings

However you do it, you should have a good record of your Maple session. Save often so you won't lose your work. But, before you save permanently, remove the rubbish; also, clean up your worksheet now and then, so you won't confuse yourself. Depending on your version, menu items allow collection of the whole session or the output only, as desired. If there are several worksheets, Maple saves only the one that is active.

2.1.1 Functions and Assignment

Consider the function call

```
>  f(x,y);    # f is a function of x and y
```

$$f(x, y)$$

Don't forget to put in a ; as well as an enter, return CR

```
>  whattype(%);
```

function

```
>  nops(%%); # number of operands in the above
```
In Maple, this is two execution groups back.

$$2$$

```
>  %%%; # 3 operations back
```
Make sure you know what ditto " means here

In Maple 5r5 this
is a %

$$f(x, y)$$

```
>  %%%;
```
After 3 levels,
you lose
    Maple has 'spit out' the statement as in bad taste. It is not a good idea
to extend the ditto operation too far.

```
missing operator or ';'
>  op(1,%);      # operand 1
```

$$x$$

```
>  op(2,%%);
```

$$y$$

```
>  op(0,%%%);
```
Maple defines operand 0 as the operator (function) itself

$$f$$

    The operation nops( ) gives the number of operands, op( ) gives the sequence of operands or it may be used to select an individual operand. The full function call(above) has three parts and can be separated as required.

```
>  nops(f);
```

$$1$$

```
>  nops(f(x,y));
```

The symbol f is within f(x,y)

$$1$$

```
>  f := 10; # f is assigned the value 10
```

A bad idea

$$f := 10$$

```
>  f(x,y);
```

$$10$$

```
>  op(f(x,y));
```

$$10$$

```
>  f(x,y) := x^3 - 2*x^5;
```

```
Error, invalid left hand side in assignment
>  whattype(%);
```

$$integer$$

```
>  op(%%);
```

$$10$$

The confusion is the result of considering f to be two things; a name or symbol and the number 10. Clearly if f has the value 10 it cannot be a function. The illogic is compounded because both f and f(x,y) now take the value 10. It is not possible for the integer 10 to take an algebraic value.

An assignment of a variable to a version of itself can create an infinite loop that crashes Maple when another operation is performed. A simple solution is to unassign the variable. This may be done with unassign but is simpler using the unevaluate quotes ' '.

Maple will usually warn you

```
>  f := 'f';
>  # f again becomes the name f with no assigned value
```

$$f := f$$

This operation leaves the function f with the simple name 'f'; it is left unassigned. This is necessary in order to use f again in another calculation. An alternative is to restart but you will lose all your calculated values.

In some versions this is Revert from the File menu

Another alternative is simply to quit Maple and start again, a drastic solution.

2.1.2   Lists, Sets and Sequences

** and ^ are the same

A list is an ordered set, enclosed by [ ] with elements separated by commas. The list A can be easily explored:

```
>  A := [x,x^2,x**3,x^4]; # [a sequence of variables]
```

$$A := [x,\, x^2,\, x^3,\, x^4]$$

```
>  nops(A);
```

$$4$$

```
>  op(A);
```

Just the sequence, inside

$$x,\, x^2,\, x^3,\, x^4$$

```
>  op(1,A);
```

$$x$$

```
>  op(2,A);
```

$$x^2$$

```
>  op(3,A);
```

$$x^3$$

```
>  op(4,A);
```

A has been assigned so there are no problems with ditto %

$$x^4$$

```
>  op(1..3,A);
```

$$x, \, x^2, \, x^3$$

A set is an unordered list. It may and probably does have a different order every time you see it in Maple. Remember, ordering can be consuming of time and, often all we want to know is contained in the elements themselves. The Maple people made the decision, early on, that they would not order anything, unless necessary.

```
>{x,y,z};
```

$$\{x, \, y, \, z\}$$

```
>{x,x,y,z}, {x,y,z,x}, {z,y,x}, {x,y,y,z};
```

$$\{x, \, y, \, z\}, \{x, \, y, \, z\}, \{x, \, y, \, z\}, \{x, \, y, \, z\}$$

The commas, between sets, present an ordered sequence

```
>{10,y,x^2,D2,q,a};
```

$$\{10, \, y, \, a, \, D2, \, q, \, x^2\}$$

We can make list A into a set in several ways:

```
>   As := {op(A)};
```

$$As := \{x, \, x^3, \, x^2, \, x^4\}$$

```
>   convert(A,set);
```

$$\{x, \, x^3, \, x^2, \, x^4\}$$

A set doesn't preserve order

```
>{A[1],A[2],A[3],A[4]};
```

$$\{x, \, x^3, \, x^2, \, x^4\}$$

```
>{x,x,y,y+4,y+5,y+4};
```

Duplicates overlay one another in a set

$$\{x,\, y,\, y+4,\, y+5\}$$

```
>  seq(x.i,i=1..10);
```

a robust way to
make a sequence

$x1,\, x2,\, x3,\, x4,\, x5,\, x6,\, x7,\, x8,\, x9,\, x10$

```
>  x.(1..10);
```

The catenation
operator is || in
Maple 6

$$x1,\, x2,\, x3,\, x4,\, x5,\, x6,\, x7,\, x8,\, x9,\, x10$$

2.1.3 Convert

The convert( ) command will often convert from one type to another.

```
>  whattype(%);
```

$$exprseq$$

```
>  [%%]; whattype(%);
```

$$[x1,\, x2,\, x3,\, x4,\, x5,\, x6,\, x7,\, x8,\, x9,\, x10]$$

$$list$$

```
>  convert(%%,set);
```

$$\{x1,\, x2,\, x3,\, x4,\, x5,\, x6,\, x7,\, x8,\, x9,\, x10\}$$

```
>  convert(%,array);
```

The argument
has to be a list in
this instance

```
Error, (in convert/array) cannot convert to an array
```
An array is ordered and has organised sequencing of rows and columns

```
>  [x.(1..8)]; whattype(%);
```

$$[x1,\, x2,\, x3,\, x4,\, x5,\, x6,\, x7,\, x8]$$

The x. is x|| in
Maple 6

list

A list can be converted into an array, a table indexed with a range of numbers

```
> convert(%%,array); whattype(%);
```

$$[x1,\, x2,\, x3,\, x4,\, x5,\, x6,\, x7,\, x8]$$

array

```
> op(0,%%);
```

array

```
> op(%%%);
```

$$1..8,\, [1 = x1,\, 2 = x2,\, 3 = x3,\, 4 = x4,\, 5 = x5,\, 6 = x6,\, 7 = x7,\, 8 = x8]$$

This structure shows the correspondences in the array; the right hand side of each element is the index; the left side, the entry.

```
> convert([x1,x2,x3,x4,x5],array);
```

$$[x1,\, x2,\, x3,\, x4,\, x5]$$

```
> convert([[x1],[x2],[x3],[x4],[x5]],array);
```

$$\begin{bmatrix} x1 \\ x2 \\ x3 \\ x4 \\ x5 \end{bmatrix}$$

```
> convert({x1,x2,x3,x4,x5},list);
```

$$[x4,\, x3,\, x2,\, x1,\, x5]$$

Converting a set to a list or array is ambiguous because an array without order has little significance. Starting with a sequence, however, preserves order.

```
>{y.(1..4)};
```

$$\{y1,\ y2,\ y3,\ y4\}$$

```
>   op(%);
```

$$y1,\ y2,\ y3,\ y4$$

```
>   [%]; whattype(%);
```

$$[y1,\ y2,\ y3,\ y4]$$

list

Other conversions may also be necessary but are not ethical mathematically. The following statement puts a '+' between every operand.

This is rather
unscrupulous

```
>   convert(sin(x)*exp(y)=z,'+'); # puts + between operands
```

$$\sin(x)\,e^y + z$$

```
>   convert(%,'*');
```

$$\sin(x)\,e^y\,z$$

2.1.4   Set Operations and Types

The operations of sets use logic for additions and subtractions.

```
>   Bs := {y,y^2,3+y};
```

$$Bs := \{y,\ 3+y,\ y^2\}$$

```
>   Cs := As union Bs;
```

$$Cs := \{x,\, y,\, 3+y,\, x^3,\, x^2,\, x^4,\, y^2\}$$

```
>   Ds := Bs union {x+2,x+3,x+4,x,y};
```

$$Ds := \{x,\, y,\, 3+y,\, x+2,\, x+3,\, x+4,\, y^2\}$$

```
>   As minus Ds;
```

$$\{x^3,\, x^2,\, x^4\}$$

```
>   As intersect Ds;
```

$$\{x\}$$

```
>   member(x+2,Ds);
```

$$true$$

```
>   member(x+2,%%);
```

$$false$$

But some general exploration is also possible:
```
>   A;
```

$$[x,\, x^2,\, x^3,\, x^4]$$

```
>   has(A,x^5);
```

$$false$$

```
>   has(A,x);
```

$$true$$

Recapping, there are many different types of variables in Maple:
```
>   whattype(x^2+y^3+4);
```

$$+$$

```
>  whattype(A);
```

$$list$$

```
>  whattype(As);
```

$$set$$

```
>  whattype(x-y=2+3*z*q -4^z);
```

$$=$$

A listing of different types appears in section 1.5 on page 36

2.1.5   Catenation

The operator .   has already been used to make the names of variables. It comes in various forms. An alternative is the cat( ) command.

Maple 6 uses ||
see page 38

```
>  x.y;
```

$$xy$$

```
>  x.symbol;
```

$$xsymbol$$

```
>  x.symbol.y;
```

$$xsymboly$$

```
>  whattype(%);
```

$$symbol$$

```
>  (1..4);
```

$$1..4$$

```
>  x.%;
```

$$x1,\ x2,\ x3,\ x4$$

```
>  (a,b,c,d);
```

$$a,\ b,\ c,\ d$$

```
>  y.%;
```

$$ya,\ yb,\ yc,\ yd$$

```
>  $ (1..5);      # complete the evaluation
```

$$1,\ 2,\ 3,\ 4,\ 5$$

```
>  g[$ (1..5)];   # multiple subscripts
```

$$g_{1,2,3,4,5}$$

```
>  cat('Create a ',new,' Dir/topic');
```

Create a new directory Dir/topic

Exercise 2.1 Using op

Find the different parts of the following:

```
>  f(cat,dog,rat,pig);
>  A := [[2,3,4],[0,0,0],[10,-1,10],[a,b,c]];

>  height[cow] := 150; height[mouse] := 2;

>  height[cow];
>  f(x,y,z);
>  f(x,y,z(t));
>  D[1,2,1,2](f)(x,y);
>  a + b + c + f;
>  -(1 + x)*y*(g + h^3) + exp(x)*sin(y)*(x^10-y^10);
```

hint: use
op(op())
look at op(0,%)

look at 2-3 op
levels

Exercise 2.2 Whattype

Whattype are these? What are the indeterminates? (Try ?indets)

Exercise 2.3 Two Lists

Use eval, op, rhs and lhs to make the table into 2 separate lists with ages in one and age classification in the other. Using map makes the operation quick.

```
>  t := table([old = 80, young = 20, ordinary = 45]);
```

Exercise 2.4 Indices and Entries

Use these commands to separate the two parts of the table. See page 40 or use ?indices

Exercise 2.5 Double-Sided Lookup table

Reverse the entries above to make a proper data base that is double sided. Consider zip( ).

Exercise 2.6 Extended op Command

Try the two equivalent forms of the op command on the table t, above.

```
>  op(1,op(1,op(2,eval(t))));
>  op([2,1,1],eval(t));
```

Exercise 2.7 Expand and Select

Have a look at this. How many terms contain ln(x)?

```
>   diff(x^x,x$10):  expand(%):
```

Exercise 2.8 Sum the series

```
>   sum(1/k^3,k=1..10); sum(1/k,k=1..n);
>   sum(1/exp(k),k=1..infinity);
```

Exercise 2.9 Use catenation
Produce the following without retyping the whole.

```
>   x11,x12,x13,x21,x22,x23,x31,x32,x33;
>   [[x11,x12,x13],[x21,x22,x23],[x31,x32,x33]];
>   whattype(%);
```

Remove the list [x21,x22,x23] from the above without retyping.

Exercise 2.10 Redundant Data
Consider the following sequence of data. Remove the redundant values by converting it to a set. Separate the floating point values, the integers and the symbols. Consider the commands select( ), remove( ), and hastype( ).

```
>   data :=
>   6.66,1.25,c,c,3.12,6.66,q,6.66,6.66,15,2,4.51,a,b;
```

Exercise* 2.11 Lake Data
The following data are from a traverse around WestLake, see section 6.2, page 316. With conversion to numbers using digitisation, there are numerous errors; one is the 'stutter' of the cursor or mouse, which creates multiple values in a sequence. Work out how to automatically remove the redundant values and keep the original order.

[179, 204], [179, 204], [176, 204], [176, 204], [174, 204], [174, 204], [174, 204], [173, 204], [175, 207], [175, 207], [175, 207], [175, 207], [179, 212], [184, 216], [188, 218], [188, 218], [192, 219], [192, 219], [192, 219], [194, 218], [194, 216], [192, 214], [192, 214], [189, 212], [186, 209], [186, 209], [182, 207], [179, 204]

A difficult problem if done for the general case. Careful conversion between sets and lists may work. Consider numbering the entries and sorting. Tables, indices( ) and entries( ), can make a two-way data base. Procedures (page 116 and section 3.4, page 129) can easily deal with the general case.

2.2   Area of a Lake

Calculating the area of a lake covered by water is a bit awkward, especially if there are peninsulas and islands. The lake perimeter is presented as a list of lists or a listlist in Maple. This is a set of coordinates of each point passing around the lake in one direction. A rather clever surveying technique, from Ned Magnus, calculates contorted areas as a series of 'signed trapezoids'. A planar view of the lake and trapezoidal areas is shown here; this may be done by hand or using Maple (see chapter 3, page 134).

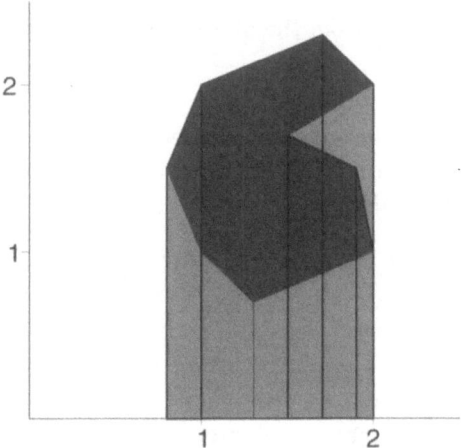

The horizontal axis and each pair of points forms a trapezoid; the average of the two vertical values multiplied by the signed difference of the horizontal values calculates a signed trapezoidal area. Adding these areas gives the area of the lake; it allows for peninsulas and odd shapes. Moving in a positive x direction produces trapezoids with a positive area; these add as one continues around the lake; but, a reversal in direction along the x axis produces a negative area which subtracts from the positive ones. In all, a complete following of the lake perimeter in one direction results in a differencing of trapezoidal areas, leaving the lake area.

The given data are in kilometers and represent a lake with a rough peninsula on the eastern side. Note that the individual coordinates are, respectively, the distance East of a reference point and the distance North of the reference point. The double indices prise out the data using the idea of a lookup (page 192); each row in the execution group is a trapezoid; they are all simply added. Also note that this calculation is much easier using procedures,

Chapter 3, page 116. Since the lines in the second execution group are so similar, they can be copied with Ctrl-c and Ctrl-v or Drag and Drop with Ctrl to copy; hand editing completes the job.

```
>  d:=[[.8,  1.5],[1,1],[1.3,.7],[2,1],[1.9,
>  1.5],[1.5,1.7],[2,2],[1.7,2.3],[1,2]]:
>  (d[1][2]+d[2][2])/2*(d[2][1]-d[1][1])+
>  (d[2][2]+d[3][2])/2*(d[3][1]-d[2][1])+
>  (d[3][2]+d[4][2])/2*(d[4][1]-d[3][1])+
>  (d[4][2]+d[5][2])/2*(d[5][1]-d[4][1])+
>  (d[5][2]+d[6][2])/2*(d[6][1]-d[5][1])+
>  (d[6][2]+d[7][2])/2*(d[7][1]-d[6][1])+
>  (d[7][2]+d[8][2])/2*(d[8][1]-d[7][1])+
>  (d[8][2]+d[9][2])/2*(d[9][1]-d[8][1])+
>  (d[9][2]+d[1][2])/2*(d[1][1]-d[9][1]);
```

$$-1.240000000$$

The lake has a negative area simply because of our conventions; we passed anti-clockwise around the lake. The area is $1.24km^2$.

Exercise 2.12 Area with Commands

Use Maple operations to add the first point into the listlist, as the last point. A direct add, with a running index, finishes the calculation.

```
>  x:=[op(d),d[1]];
>  add((x[i][2]+x[i+1][2])*(x[i][1]-x[i+1][1]),i=1..9)/2;
```

Why is the area now positive?

Exercise* 2.13 Triangular Areas

Check the calculation of the lake area using Heron's formula for the area of a triangle. It uses the parameter $s = (a + b + c)/2$, with the $area = \sqrt{s(a-s)(b-s)(c-s)}$; a, b and c are the lengths of the three sides of the triangle. Use a 'focus' for all the triangles; the 'innermost point', [1.5,1.7]; click on the graph, the x, y position appears in the box in the lower left menu position. Remember that the distance between any two points is the square root of the sum of the squares of the differences in positions.

```
> restart;
```

2.3   Environmental Numbers

Consider some real scenes, when an environmental scientist is faced with units conversion, calibration and mass balance. It is clear that the maths is well done by Maple. The individual must evolve the correct structures to 'produce a correct answer' and understand the trends in the data.

2.3.1   Units Conversion

Science is invariably about the manipulation of numbers and modern computational software, such as Maple, offer possibilities undreamed of by previous generations. The sheer speed of number crunching generates results, almost as fast as keyboard skills allow. Indeed, the very nature of the beast promotes fast and efficient computation without the hassle of acquiring programming skills or a deep understanding of the 'engine under the hood'. Once the basics are mastered, it's very easy to get blasé and develop calculations 'on the fly'. But some words of warning! Maple, like any other computer program, is subject to the same basic law of computation; **GIGO**, or garbage in, garbage out; an expression emphasising the importance of entering accurate data and valid equations.

Data is usually generated through measurements with commonly agreed standards and the quantities measured are very carefully defined. There are a variety of standards with which the environmental scientist must comply that describe and quantify the physical, chemical and biological properties of matter. In addition to standards, measurements are made up of a combination of a dimension, a unit and a precision; each is a property that describes the measurement. Thus 1.25 seconds is made up of the dimension of time, the unit of seconds and the precision or accuracy inferred by the decimal places. The accuracy of a measurement is defined as how close the result comes to the true value and is usually dependent on the calibration of the method with a known certified standard. Precision is the reproducibility of multiple measurements and is described by statistical tests such as the standard deviation, standard error or confidence interval. Each of these properties should be taken into account when formulating calculations or presenting results. The analyses of dimensions or units are especially useful in checking the validity of equations and ensuring that unlike quantities are not compared. This may be undertaken using a conventional scratch-pad, but the analysis of dimensions or units using Maple is simplicity itself since it treats non-numerical input by the rules of algebra, presenting the results in dimensionally correct units. Remember to include brackets and multiplication signs when appropriate and ensure you use identical symbols to represent a particular unit.

Non-commutative multiply . in Maple 6, can work with units

Consider the working equation to convert a measurement of 400 parts per million by volume (ppmV) of nitrogen dioxide at Standard Temperature and Pressure STP into a concentration of grams per cubic metre at a 20C and 95 kPa.

The equation is quite straightforward, although sometimes students make the mistake of inverting the temperature and pressure corrections. When working from first principles, remember that gases subjected to a decrease in pressure undergo a decrease in concentration, while a decrease in temperature results in an increase in concentration.

```
>   400/10^6*46*gram/(22.4*litre)*273.2*K/((273.2+20)*K)
>   *95*kPa/(101.3*kPa)*1000*litre/metre^3;
```

$$.7177954377 \frac{gram}{metre^3}$$

Notice that Maple performs the rules of algebra on the units and cancels out like terms in the equation, ensuring that the results are presented with correct dimensions. The initial ratio is in volume units or litres of NO_2 per litre of air; identification of the litres with NO_2 could also be included for preciseness but it tends to make the calculation a little verbose. The 46 is the Molecular Weight MW in grams per gram mole which is $14 + 2*16$ for the nitrogen and oxygen atoms in the NO_2; it is known is that a gram mole of NO_2 has a volume of 22.4 litres at STP or 46 grams of NO_2 is equivalent (in volume) to 22.4 litres. Next the absolute temperature ratio, followed by the pressure ratio flow on to convert from litres to cubic metres.

Finally, when presenting a result, the precision of the calculation should be taken into account. Maple has automatically converted all the integers to floating point values and completed the calculation to 10 digits. The result infers that the precision of this calculation is 0.1 ng, though the molar volume, molecular weight and pressure are only approximated. A general rule for presenting results is not to exceed the precision of the least precisely known parameter. Better to limit the number of digits to 3, or at most 4 digits.

```
>   evalf(%,3);
```

$$.718 \frac{gram}{metre^3}$$

Try re-entering higher precision values of atomic weights and standard volumes; recalculate and compare results.

Consider that we want to calculate the amount of carbon dioxide produced from the combustion of 1000 cubic metres of landfill gas at STP, assuming that landfill gas consists of pure methane. If the gas burns completely, it follows the reaction:

$$16g \ CH_4 + 2 \ O_2 => 44g \ CO_2 + 2 \ H_2O$$

where we have purposely put the weights involved in grams before the carbon species of interest; these, of course, are the molecular weights. Thus 1 mole of methane produces 1 mole of carbon dioxide and since 1 mole of a gas occupies 22.414 litres, 1000 cubic metres of gas contains:

```
>  1000*metre^3*1000*litre/metre^3*mole/(22.414*litre);
```
$$44614.97278 \ mole$$

1 mole of CO_2 contains 44g, thus the weight of CO2 produced is:

```
>  %*44*gram/mole*kilogram/(1000*gram);
```

$$1963.058802 \ kilogram$$

Don't forget to express your answer in terms of a realistic mathematical precision.

```
>  evalf(%,3);
```

$$1960. \ kilogram$$

2.3.2 Calibration of Permeation Tubes

Permeation tubes are a useful method of producing a known gas concentration for the calibration of some analytical instruments. The rate of permeation through a PTFE membrane is dependent on the membrane thickness, the pressure that a gas exerts on the membrane surface, the concentration gradient across the membrane and its temperature. The saturated vapour pressure exerted by a liquid/gaseous phase in an enclosed vessel is dependent on the temperature; as long as some liquid remains in equilibrium with the vapour phase, the internal pressure remains constant. A small tube held at a constant temperature and containing a suitable liquid/vapour with a PTFE window allows the gas to permeate the tube and enter an enveloping flow at a constant rate. The fresh air flow is steady to maintain a constant concentration gradient across the permeation membrane. This reasonably simple method can produce accurate concentrations of a variety of gases when the 'target gas' has a saturated vapour pressure below 2 MPa at ambient temperatures.

Permeation tubes are usually placed in a small constant temperature oven at a slightly elevated temperature, normally $35C$. A preheated air stream of about 200 ml/min is passed through the oven and across the tube and then mixed with a variable air flow to dilute the gas to a suitable concentration. By carefully weighing the tube at suitable intervals, the rate of emission can be easily calculated and is generally expressed in nanograms per minute. Permeation tubes are usually supplied to produce a concentration range of several ppm at a flow rate of one litre per minute. The concentration of the diluted gas is easily calculated, as demonstrated below.

A sulphur dioxide analyser is located in a constant temperature room at 25C with a ambient air pressure of 100.20 kPa. A permeation tube with an emission of 1024 nanograms of sulphur dioxide per minute is placed in a calibrator with a combined outlet flowrate of 2 litres per minute measured with a flowmeter calibrated at 20 degrees Celsius. What is the concentration in ppb (by mass) if the flow meter is at ambient temperature?

Firstly, correct the flow rate f (ml/min), for temperature and pressure:

```
>  f:=2000*100.20/101.32*(273.2+20)/(273.2+25);
```
$$f := 1944.727981$$

Standard pressure is 101.32 kPa; this lowers the volumetric flow. The temperature is lower in the room and this decreases the volumetric flow. The concentration of the calibration gas in ppm generated by a permeation tube is:

emission rate*molar volume/flowrate

```
>  (1024*24.05/64)/f;
```
$$.1978682899$$

Here 24.05 is the molar volume in litres at 25C and 64 is the molecular weight of SO_2. Hence, the units of this number are

```
>  ng[SO[2]]/min*litre[SO[2]]/gm[SO[2]]/(ml[air]/min);
```

$$\frac{ng_{SO_2} \, litre_{SO_2}}{gm_{SO_2} \, ml_{air}}$$

The level of detail in units specification makes it clear that we have not used the correct conversion of volume units. The use of the molar volume is an incorrect substitution. The substitution should be the molar volume of air; 24 litres of air per 29 gms of air (only to 2 digits).

```
>  (1024*24/29)/f;
```
$$.4357669989$$

with units

```
>  ng[SO[2]]/min*litre[air]/gm[air]/(ml[air]/min);
```

$$\frac{ng_{SO_2} \, litre_{air}}{gm_{air} \, ml_{air}}$$

which is more appropriate when we adjust for litres

```
>  %%*subs(litre[air]=1000*ml[air],%);
```

$$435.7669989 \, \frac{ng_{SO_2}}{gm_{air}}$$

or

```
>  round(op(1,%))*ppb;
```
$$436 \, ppb$$

better yet, since we have only kept 2 decimal places

```
>  evalf(%,2);
```

$$440.\,ppb$$

Of course, this is in ppb by mass. Notice that we would be more than 100% larger numerically had we used the wrong conversion

```
>  round((1024*1000*24.05/64)/f);
```

$$198$$

But, indeed, the 'normal' units are usually in ppmV, on a volume basis. Returning to our original calculation and units and remembering that in one gram there are 10^9 ng,

These execution
groups are joined,
using the Edit
menu

```
>  rr:=(1024*24.05/64)/f;
>  'has units of';
>  ng[SO[2]]/min*litre[SO[2]]/gm[SO[2]]/(ml[air]/min);
>  'and, combined';
```

$$rr := .1978682899$$

$$has\ units\ of$$

$$\frac{ng_{SO_2}\ litre_{SO_2}}{gm_{SO_2}\ ml_{air}}$$

$$and,\ combined$$

```
>  rr*%%;
```

$$.1978682899\,\frac{ng_{SO_2}\ litre_{SO_2}}{gm_{SO_2}\ ml_{air}}$$

```
>  %*gm[SO[2]]/(10^9*ng[SO[2]]);
```

$$.1978682899\,10^{-9}\,\frac{litre_{SO_2}}{ml_{air}}$$

```
>  %*1000*ml[SO[2]]/litre[SO[2]];
```

$$.1978682899\,10^{-6}\,\frac{ml_{SO_2}}{ml_{air}}$$

```
>  round(op(1,%)*10^9)*ppbV;
```

$$198\,ppb\,V$$

Often a multipoint calibration is required to satisfy standard calibration methods. Rather than repeating the calculation by substituting a number of values for f, Maple allows us to produce a calibration curve. By activating the graph by clicking on it, the cursor can then be placed on the plot line and clicked again to obtain x and y values in the upper left box. More detailed plotting is considered in Chapter 3 and active plotting is reconsidered in Chapter 6, section 6.1, page 312.

```
>  restart;
>  plot((1024*1000*24.05/64)/f,f=500..5000,axes=BOXED,
>  title='SO2 Concentration',titlefont=[TIMES,BOLD,14],
>  labels=["flowrate ml/min","SO2 in ppb   "]);
```

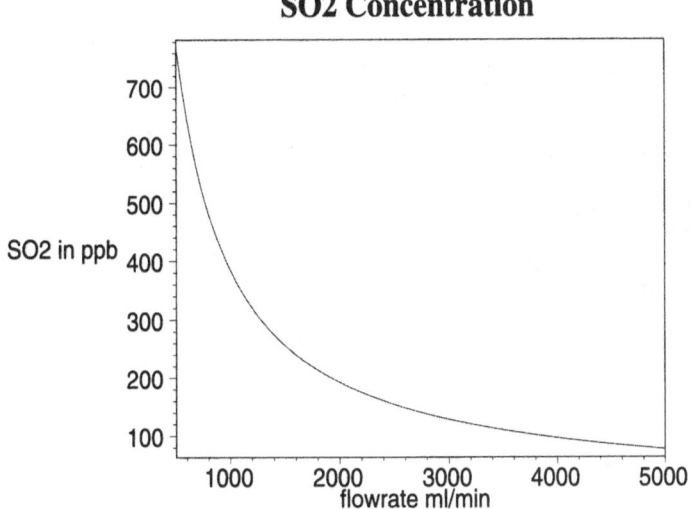

2.3.3 Greenhouse Atmospheres

It is now common practice in the horticultural industry to heat greenhouses during the winter months to promote plant growth. However, intense market competition within the industry has resulted in the introduction of reclaimed oil, a cheaper alternative fuel, bringing a substantial saving compared with the relatively high cost of natural gas or refined oil. Other recent innovations include the enrichment of greenhouse atmospheres with carbon dioxide, a growth stimulant, which is readily obtainable by diverting some of the stack gases into the greenhouse atmosphere. Specially made reclaimed oil burners are capable of combustion efficiencies of 99.95% or better, making such a scheme theoretically possible. Using data obtained from a horticultural farm and emissions from a reclaimed oil burner stack, we use Maple to explore the possibilities.

```
> restart; with(plots):  with(plottools):
> ttl := TEXT([0,5],"Greenhouse Chamber",FONT(TIMES,ROMAN,16)):
> opts := axes=NONE, scaling=constrained:
> cr := colour=red:  cg := colour=green:
> afnt:= ALIGNABOVE,FONT(TIMES,BOLD,12):
> a1 := arrow([-4,1], [-2,1], .2, .4, .1, cr):
> a2 := arrow([-3.3,2], [-2.3,2], .2, .4, .1, cr):
> a3 := arrow([2,1.5], [5,1.5], .2, .4, .1, cr):
> t1 := TEXT([-3.1,1.3],f1,afnt):
> t2 := TEXT([-2.8,2.3],f2,afnt):
> t3 := TEXT([3.3,1.8],f3,afnt):
> d := ellipticArc([0,0],3,4, 0..Pi,filled=true,cg):
> plots[display](a.(1..3),t.(1..3),ttl,d,opts);
```

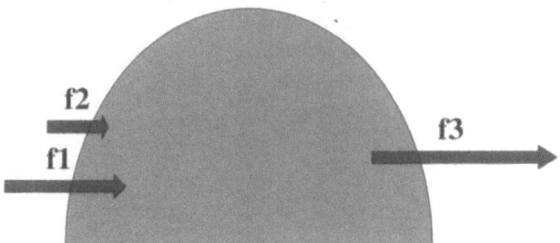

Greenhouse Chamber

Greenhouse Data

The volume of the greenhouse is 2640 cubic metres with 2 air changes per hour and a internal air circulation of 5 metres per sec allows good mixing throughout the growing area. The target carbon dioxide level for optimum plant growth is 1200 ppm.

corrected for STP

Stack Gas Data   Flowrate – 680 m^3 per hour

Carbon Dioxide concentration – 8.0% by volume

Sulphur dioxide – 0.1 g/m^3

Nitrogen dioxide 1 – 7 ppmV

Note that the average CO_2 at ground level in ambient air is currently 370 ppm; in rural areas the average daytime CO_2 may fall to 290-310 ppm due to carbon fixation. In heavy industrialised areas daytime CO_2 levels may not fall below 400ppm. We presume a rural area here.

```
>  conds :=
c1 = 8/100*44/22.414*1000,# grams CO2/m^3 of stack gas
c2 = 300*44/22.414*1000/1000000, # grams CO2/m^3 of ambient air
c3 = 1200*44/22.414*1000/1000000, # target grams of CO2/m^3 air
V = 2640, # volume of the grow chamber in m^3
f2= 2*2640; # Ventilation rate of two chambers of air per hour
```

$$conds := c1 = 157.0447042, \ c2 = .5889176408, \ c3 = 2.355670563, \ V = 2640, \ f2 = 5280$$

The scene is one in which the concentration in the chamber is c(t), a function of time. The concentrations of the two inflow streams are, respectively, c1 and c2; with a well mixed chamber, the concentration of the outflow is c(t). The flow of stack gas is f1 m^3 per hour; the flow of ambient air is f2 = 2*V m^3 per hour.

Mass Balance

The rate of accumulation of CO_2 is the difference between the inflow of CO_2 and the outflow of CO_2. To proceed, analyse the units of the matter. The air flows have units of

```
>  m^3/hr;
```

$$\frac{m^3}{hr}$$

The flow of CO_2 is the product of the concentration and the airflow

```
>  %*g/m^3;
```

$$\frac{g}{hr}$$

Hence the inflow and outflows, the advection or movement of CO_2, is a product of the appropriate concentration and the containing airflow.

```
>  'inflow of CO2':=c1*f1+c2*f2;
```
$$inflow \ of \ CO2 := c1 \ f1 + c2 \ f2$$

```
>  'outflow of CO2':=c(t)*f3;
```
$$outflow \ of \ CO2 := c(t) \ f3$$

With no change in the volume or pressure in the chamber, the outflow of air is the sum of the two inflows

```
>  f3:=f1+f2;
```
$$f3 := f1 + f2$$

When the chamber has reached a 'steady-state', there is no accumulation or change in the concentration with time; the inflow of CO_2 and outflow of CO_2 are the same. This allows that we can specify the 'target concentration' c3 and adjust f1 to allow that, after some time, the concentration c3 is reached. The accumulation is zero when the target concentration is reached.

```
>  'rate of accumulation of CO2':='inflow of CO2'-'outflow of CO2';
```
$$rate \ of \ accumulation \ of \ CO2 := c1 \ f1 + c2 \ f2 - c(t) \ (f1 + f2)$$

The steady solution, when the rate of accumulation is zero, predicts the steady stack flowrate f1; then the outflow concentration is equal to the target concentration c3.

```
>   subs(c(t)=c3,%)=0;
```

$$c1\,f1 + c2\,f2 - c3\,(f1 + f2) = 0$$

```
>   solve(%,f1);
```

$$-\frac{f2\,(c2 - c3)}{c1 - c3}$$

Bad idea produces a set of unknown order

```
>   #vars:=indets(%);
>   vars:=f2, c1, c2, c3:
>   f1:=unapply(%,vars);
```

$$f1 := (f2,\ c1,\ c2,\ c3) \to -\frac{f2\,(c2 - c3)}{c1 - c3}$$

This last operation has made f1 a function of the four arguments f2,c1,c2,and c3. Because they are all known constant values, we could simply calculate the flow rate of stack gas as a number. The command unapply( ) introduces the idea of a procedure or function, the generalised symbolism that calculates f1 from known values of f2,c1,c2,and c3. See Chapter 3, section 3.1, page 116.

A good way to collect arguments

Stack Flowrate Calculation

```
>   # subs(conds,[op(1,%)]); # an alternate way
>   subs(conds,[vars]);
```

$$[5280, 157.0447042, .5889176408, 2.355670563]$$

```
>   op(%);
```

$$5280, 157.0447042, .5889176408, 2.355670563$$

```
>   f1(%);
```

$$60.30456853$$

This is f1 as a function or command that uses a sequence as an argument; in this case the value of immediate interest is a simple number but Maple allows that we can easily adapt or adjust the inputs and redesign the chamber and the experiment.

```
>   f1(5280, 157.0447042, .5889176408, 2.355670563);
```

$$60.30456853$$

The Time Dependence

Initially, at time t=0, the concentration of CO_2 in the greenhouse air is c(0); at time t, the concentration is c(t). During the time interval from t to t+dt, the mass of CO_2 changes by the amount

```
>   V*diff(c(t),t)='rate of accumulation of CO2';
```

$$V\left(\tfrac{\partial}{\partial t}\,c(t)\right) = c1\,f1 + c2\,f2 - c(t)\,(f1 + f2)$$

```
>  subs(f1=f1(f2,c1,c2,c3),%);# substitute the full procedure
```

$$V\left(\tfrac{\partial}{\partial t}\, c(t)\right) = -\frac{c1\, f2\,(c2 - c3)}{c1 - c3} + c2\, f2 - c(t)\left(-\frac{f2\,(c2 - c3)}{c1 - c3} + f2\right)$$

```
>  ode:=%:
```

```
>  sol:=dsolve({ode,c(0)=c2},c(t));# initial concentration is c2
```

$$sol := c(t) = \frac{c3\, c2}{c2 - c1} - \frac{c3\, c1}{c2 - c1} + e^{\left(\frac{f2\,(c2 - c1)\,t}{(c1 - c3)\,V}\right)}(c2 - c3)$$

```
>  cr:=unapply(op(2,sol),V,f2,c1,c2,c3);# another procedure
```

$$cr := (V,\, f2,\, c1,\, c2,\, c3) \to \frac{c3\, c2}{c2 - c1} - \frac{c3\, c1}{c2 - c1} + e^{\left(\frac{f2\,(c2 - c1)\,t}{(c1 - c3)\,V}\right)}(c2 - c3)$$

```
>  op(subs(conds,[V,f2,c1,c2,c3])); # pertinent arguments
>  plot(cr(%,t),t=0..4, axes=boxed,
>  title='Greenhouse CO2 Concentrations',titlefont=[TIMES,BOLD,13],
>  labels=["time in hours","g/m3 "]);
```

2640, 5280, 157.0447042, .5889176408, 2.355670563

Greenhouse CO2 Concentrations

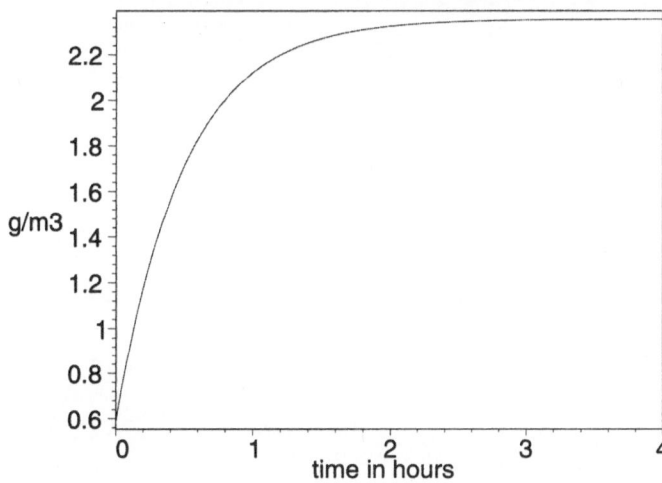

The figure shows changes in concentrations in the chamber with time from when the stack gas is introduced into the chamber. Clicking on the graph, and then specifically on the curve, allows that you can read off the graphical values; the lowest menu bar, the context menu, has a window on the left side. At 1 hour the concentration is 90% of the target concentration of 2.36 g/m^3.

```
>  0.9*2.36;
```

2.124

Exercise* 2.14 NO_2 and SO_2 Concentrations

Calculate the concentration of NO_2 and SO_2 expected in the greenhouse atmosphere by modifying the mass balance equation. Assume that no pollutants are present in the ambient air. Express each concentration as parts per billion by volume; this is equivalent to nanolitres per litre.

Exercise* 2.15 Synergistic Damage to Plants

Based on international literature and standards, a one-hour period of 200 ppb of SO_2 should not be exceeded to avoid acute leaf damage in plants. A similar level of 200 ppb for NO_2 is also recommended; however, the interaction between SO_2 and NO_2 is synergistic with combined concentrations of approximately 150 ppb causing leaf damage in soybean, oats, tomato and tobacco. Is the reclaimed oil burner suitable for CO_2 enrichment?

2.4 Worksheets and Documents

Maple is more than a mathematical tool; it is a logical formatting environment. see

```
>    ?worksheet,documenting,structure
```

2.4.1 Worksheets

You may start Maple by clicking on the Maple icon. Maple will presume you want a new worksheet. Otherwise, click on File/Open or your file name at the bottom of the File menu. You should start in the file at precisely the position you last stopped, except all the worksheet will be inactive. Depending on the .ini initialization file (see page 52) and the settings when you last used Maple, you may or may not have all the tools, e.g., the Expression Palette, available.

The Expression Palette is obtained from View/Palettes

Background Workings

Often you will want to do things that are separate from your present workings. Then, simply open up another worksheet using File/New, and carry on. This is particularly convenient when you are preparing a document and you don't want the reader to see all the detail. A good example is the Greenhouse Chamber figure in the last section. The plot may be presented more than one way: One is to simply use plot( ) and display the plot on the worksheet, with the plotting detail visible. Another assigns the plot a name so that it can be reused and saved, say, PP:=plot( ); this is awkward because Maple will spew forth all the detail or 'guts' of the plot, without plotting. Better to use PP:=plot( ): but this will show no output; followed by %; however, it names the plot and displays it in the worksheet. This is messy in a report and it is better to put the whole statement, with the ditto %; on a separate worksheet. The second worksheet generally communicates with the first but is saved separately. The plot is reproduced by the simple statement PP; and confusion is avoided if , after the graphic appears, the input, PP; is erased. Another possibility is to set Options/Plot Display/Window and collect the plot in another window, for separate use and separate printing. A cleanest possibility is to copy the entire plot from the second worksheet or window as an inactive form and paste it into the original worksheet.

The same holds for PLOT structures, plot( ) and display( )

Data

With real data there is always a subjective element in capture and interpretation. Often data have been misread, affected by electronic noise, need calibration or simple sorting. This is burdensome and can be done on an offside worksheet so that the story is not lost by the multitudinous operations. In any case, it is well to keep a copy of the original data as well as a copy of the operations done; a special worksheet can serve this purpose.

2.4.2   Execution Groups

Within a worksheet the active grouping is the execution group. This grouping may include many statements, Maple operations with correct syntax ending with a semicolon or colon. An execution group must be at least one line long and all statements within an execution group are done at the same time. When we take up procedures, in Chapter 3, we will see that a procedure may be a combination of execution groups that are executed at one time. You might want to place all such operations on a single line; a line could serve the same purpose as an execution group. However, such a line is difficult to read and organise, and difficult to comment. Also there is an automatic roll-around of lines which depends on the width of the present window. To get around this problem, line breaks are firmly positioned with Ctrl-Enter (or Ctrl-CR). The start and stop of a given execution group is shown by the vertical bars on the left side. These may be turned off with View/Show Group Ranges.

Break the line with Ctrl-Enter

An execution group is obtained whenever the [> button is pressed; or when the Insert/Execution Group menu item is activated. It appears after the execution group in which the cursor appears. The cursor can be positioned by clicking with the mouse. The cursor is repositioned at the start of the next execution group when the [> button is pressed. When the cursor is positioned in one execution group, Shift-Tab steps back to the last one. With two worksheets, Ctrl-Tab steps between them.

See ?hotwin

2.4.3   Sections and Subsections

Sections are formed in a directory manner using either the Insert Menu or the buttons to the right of the [> button. Sections and Subsections can be collapsed or expanded using the square, active button at the left of the line or View/Expand or View/Collapse all Sections. Used in conjunction with a Contents or Index, collapsed sections can make it easy to find something. Collapsed sections also don't require time for computer display; an effective way of dealing with complex algorithms or animated plot displays.

2.4.4   Headings and Listing

A component of text can have a different font size or style or be in bold or in italics with underlines, as shown in the lower, context bar. The right buttons along the context bar also allow left, center, and right justification. These items also appear under the Format menu; a large number of styles are available under Format/Styles/Character and in the centre window in the context bar. Also the manuscript styles presented in the left context menu box allow that sections and subsections have consistent formatting. Please note that style changes under Format/Styles may be saved and the original style lost (see Keeping a Stable Style, page 59). The original styles are:

Try the right mouse button in Windows or Unix

Author	Times New Roman	12	List Item	Times New Roman	12
Bullet Item	Times New Roman	12	Maple Output	Times New Roman	12
Dash Item	Times New Roman	12	Maple Plot	Times New Roman	12
Diagnostic	Courier New	10	Normal	Times New Roman	12
Error	Courier New	10	Text Output	Courier New	10
Fixed Width	Courier New	10	Title	Times New Roman	18
Heading 1	Times New Roman	18	Warning	Courier New	10
Heading 2	Times New Roman	14	2D Comment	Courier New	10
Heading 3	Times New Roman	12	2D Input	Courier New	10
Heading 4	Times New Roman	10	2D Output	Courier New	10

2.4.5 Cross Referencing, HyperLinks

Linkages are possible to another paragraph, another worksheet or even another internet connection, via the world wide web; these are *hyperlinks*. Within your worksheet this starts with a bookmark with View/Bookmarks/Edit Bookmark. Click the left mouse button in the paragraph where you want your bookmark; then go to View/Bookmarks/Edit Bookmark. Type in a convenient name for the bookmark, for linkage referencing; click OK. Go back to View/Bookmark; the bookmark should have been added to the menu list. The link to this paragraph is the Bookmark 'Referencing'.

> The Bookmark 'Referencing' is an anchour to this paragraph

Elsewhere, at the position of the desired linkage 'input', choose the word or words of consequence to make active, something with meaning. Highlight the active contact, the *reference*. Cut with Ctrl-x; in the menu chose Insert/Hyperlink. A Hyperlink Properties requester will appear; using Ctrl-v, paste the words into the top line (Link Text). This will form the active 'input' or Hyperlink.

> drag the mouse over

The link is not complete until the Link Target is established; this requires both a location and a Bookmark. Proceeding, select the location; URL, Worksheet or Help Topic. Chosing Worksheet and Browse brings forth a Browse requestor in which to select the current worksheet. Clicking on OK returns to Hyperlink Properties. The arrow in the bottom line chooses the convenient Bookmark (as allocated above). Click OK. The linkage should be complete; the Hyperlink should appear as active text, probably in blue. Two paragraphs back this text was 'hyperlinked' through the bookmark 'Referencing' with the word *reference*.

Corrections to hyperlinks can be made with the right mouse button using properties. Simply edit the material that comes up in the requester.

> ⌘ click on the Mac

2.4.6 Exporting: ASCII, LaTeX and HTML

Maple worksheets can be exported in many ways, including simple ASCII text, LaTeX and HTML. The ASCII text is a simplest mode of operating Maple. Raw text in lines with semicolons is transfered into a worksheet with the *read*

Use the .mpl
extension for
ASCII statements
in Maple

statement. As mentioned on page 6, an interactive method of using Maple is to use multiple windows and read the statements into Maple; the **command line version of Maple** works well in this mode; see section 2.4.9, on page 96. A raw editor like vi or Edit (in DOS) as well as Notepad or Wordpad can serve this purpose when the file is saved in ASCII, usually with an extension of .txt or .prn. It is awkward to set up Microsoft Word so that it will properly deal with an ASCII text without corrupting it.

LaTeX

For extensive document manufacture it is difficult to beat TEX, the typesetting language, and for mathematical quality one is drawn to LaTeX, the lay version of TEX. With insertions of postscript files (.ps or .eps) and .gif files, the most professional of documents can be manufactured. This book is written in LaTeX, with Maple .mws worksheets being exported into LaTeX. **LaTeX** is a compiled language that is **structured** so that the entire style is redone by changing a few lines. It is easiest to use through one of the LaTeX books, like

- *LaTeX: A Document Preparation System*, User's Guide and Reference Manual, by Leslie Lamport, 272 pages, Addison-Wesley (1994)

- *The LaTeX Companion*, by Michel Goossens, Frank Mittelbach and Alexander Samarin, 528 pages, Addison-Wesley (1994)

The author's web site and a follow-up volume to this book should contain a brief document, *LaTeX for the Complete Idiot* , by Andrew Corbyn.

The essentials are a command line system or a specialised editor, like OzTeX or Winedit or Scientific Workplace. With a command line in Windows, or better, Linux, one forms a directory, say 'latex'. Inside this directory are put all necessary style, class, include and package files. A complete directory is supplied with the CD version of this book; the style files are obtained either from a ctan site or from the Waterloo Maple web site, Share Library (www.maplesoft.com, in the Etc directory). The version of LaTeX in present use, latex2e, is also available from the ctan site; it is a part of most Linux Systems; the whole lot is essentially free software. An alternative under Windows is MikTeX; one just downloads the package (~21Mb) and runs a setup wizard that fills out the directory structure.

http://www
.miktex.de

When set up and in the appropriate directory, the worksheet is exported as a .tex file, say work.tex   Then, within the command line, or within ssystem(" ") one types

latex work

and, provided there are no errors or missing .sty files, a .dvi or device independent file is formed, work.dvi   This can be view with various viewing editors

xdvi in UNIX

or converted to postscript files with Thomas Rikiki's dvips program. This conversion allows for page selection. The command

dvips -pp 230-250 -o work.ps work

produces an output file work.ps from pages 230 to 250 of the original LaTeX document work.tex, stored as work.dvi

postscript

Linux allows for font generation and can print postscript files directly. Windows generally requires that the file work.ps be printed through one of the postscript viewers, like ghostscript or ghostview. The size of the file tends to grow, however. A worksheet file of a few kilobytes with figures can easily grow 100 fold in size if it has a number of figures. The raw worksheet .mws or the ASCII equivalent is most efficient in size, especially if the output is removed.

Inside your .tex files Maple will insert figures in several formats, including postscript .ps files and .gif This is automatically done in Mapleplot{ } format if the plot is inline within your worksheet. The normal default files appear as encapsulated postscript files work01.eps, work02,eps, work03.eps, etc.; one for each plot. See sections 1.6, page 53, and 3.6, page 149.

Note that these files can be altered because they are in ASCII format; care should be taken to follow the pattern of the postscript language. One alteration, for presentation, is to remove the border around the plot. Maple may do this for a single .eps plot, but not with multiple plots. The solution is to go into a simple editor like vi; the files are in ASCII format. You do need advice on the postscript language however, and you may corrupt the file. The heading of the .eps file may follow the example:

Some .eps files have the extension .ps

```
%!PS-Adobe-3.0 EPSF
%%Title: Maple plot
%%Creator: MapleV
%%Pages:  1
%%BoundingBox: 72 72 719 540
%%DocumentNeededResources: font Helvetica
%%EndComments
20 dict begin
gsave
/drawborder true def
/m {moveto} def
       .    .    .    .    .
```

To remove the border, simply change the true to false.

Note the presence of the BoundingBox instruction, commented in the header as part of the postscript setup; it specifies the lower left and upper right of the image size in pts, as x, y values; the BoundingBox gives the size and makes it encapsulated .eps Other commands rotate the image, translate it and scale it. These figures may be changed but be cautious–keep the old line intact; copy and paste the line to the mext line, comment one of the lines out with a %, and then change the remaining line. Alternately, it is possible that Maple can move or rotate the image, see section 3.9 on page 174. Then a new .eps file is generated through Maple. A useful reference is

- *PostScript Language, Tutorial and Cookbook*, by Adobe Systems Incorporated, 243 pages, Addison-Wesley (1986)

Early versions of Maple rotated images to present them in landscape mode. Removal of this rotation can bring them back into portrait mode.

HTML

File/Export As in the menu

Export from Maple to **HyperText Markup Language** HTML produces a .htm or .html file that can be view on a web browser, like Netscape or Internet Explorer. The format is ASCII and, like LaTeX, can be edited with a simple editor. Export As also produces a large number of .gif images of figures and icons that appear with the .htm output; these images can be separated and used, variously, in the preparation of other documents. It is also possible to organise interactive worksheets from Maple on the web using the **Tech Explorer**. This is essentially free software, an upgrade plug-in to your favorite browser.

2.4.7 Transferring: Powerpoint, Word, Scientific Workplace and Excel

Direct exchange is possible with Windows 2000

With ordinary Copy in the Edit menu, it is possible to copy Maple Output directly into Powerpoint and Scientific Word, in a OLE Bitmap form. Word requires that the Copy be followed by Insert/Object with Bitmap Image selected. Output from Maple spreadsheets can also be copied into Excel, see section 6.4, page 331; this is more transparent in Maple 6, with a simple copy and paste; Maple 6 commands can also be run within Excel.

2.4.8 Getting into the System

Your initialization files, the use of macro( ) and alias( ) should incorporate a convenience factor, particularly when you are operating beyond Maple, with another program like LaTeX, Photoshop or Excel. Defining a few procedures of common use to you with convenient operations can make your interactions smooth, inside and outside of Maple. The example, below, is the case of reading a Maple binary file from the default directory assigned by Maple in Windows. The system procedure os( ) makes this smoother. These are Windows commands though the same technique may be evoked on other systems. In command line Maple the explanation mark ! at the first position of a line

escapes to the operating system. Remember that what you do out there in the
operating system could be anything and Maple has put restrictions on these
activities. You may find that there is no return to Maple.

> Something
> wrong, look at
> the current
> directory

```
>  read 'caple.m';

Error, could not open 'caple.m' for reading

>  cd:=currentdir():  currentdir();
       "C:\\Program Files\ \Maple V Release 5.1\\BIN.WNT"

>  system("dir ca*"):  printf(op(2,%));


  Volume in drive C has no label

  Volume Serial Number is 3D48-10DE

  Directory of C:\Program Files\Maple V Release 5.1\BIN.WNT


CAPEL    M         332,365  08-06-99  4:47p capel.m

         1 file(s)         332,365 bytes

         0 dir(s)      143,097,856 bytes free
```

The 'jump out to the operating system' is direct and properly formatted
with os( ).

```
>  os := proc(ar); ssystem(ar):  printf(op(2,%)); end;

       os := proc(ar) ssystem(ar); printf(op(2, %)) end proc

>  os("dir ca*");


  Volume in drive C has no label

  Volume Serial Number is 3D48-10DE

  Directory of C:\Program Files\Maple V Release 5.1\BIN.WNT


CAPEL    M         332,365  08-06-99  4:47p capel.m

         1 file(s)         332,365 bytes

         0 dir(s)      143,097,856 bytes free
```

> The name is
> spelled right now

```
>  read 'capel.m';
```

2.4.9   Command Line Maple

Graphical User
Interface

Following from above, it is possible to run Maple only on the command line, without a GUI; the Maple supplied is 'command line maple' and is entered on the command line or DOS-prompt as, simply, maple or cmaple. This removes the need for Maple to produce a GUI and can improve speed; it is a way to run Maple transparently. Maple can also be run remotely with redirected files or piped information flow. Provided you are in the proper Maple directory or your path is correct, command line might be:

Directory
"Program Files
\Maple 6\BIN.WNT"
in DOS

```
cmaple <infile  >outfile
```

The ASCII file infile should contain the statements Maple normally receives. The convention has been that this file has the extension   .mpl   If so, the file name should be enclosed in full quotes as "infile.mpl" when referred to inside of Maple. Inside the file, statements must have proper syntax and the file may contain a   quit. The file outfile will contain the ASCII equivalent of the GUI output, in a 'prettyprint' format. If other files are read/written during the running of maple, the operation may be totally transparent and quick, a remote running without all those keystrokes or mouse jerks.

The placing of a path to a program in DOS is accomplished by insertion into the initialization file autoexec.bat, to add a path to the executable command line maple, cmaple.exe. That is, go to the DOS-prompt and step back to your main disk partition, c:

```
cd ..
edit autoexec.bat
```

Having opened edit, add a path statement into the the file autoexec.bat. If there is already a path statement, put this line just after the other path statement. The %path%; allows that your former path is preserved. Use the File menu to Save the altered file, and Exit edit. In the DOS-prompt, type exit and CR to return to Windows. Close down and restart your machine. Returning to the DOS-prompt and typing cmaple on the command line should start command line maple.

Be careful with
autoexec.bat

```
path %path%;"c:\Program Files\Maple 6\BIN.WNT"

C:\WINDOWS> cmaple
    |\^/|      Maple 6 (IBM INTEL NT)
._|\|   |/|_. Copyright (c) 2000 by Waterloo Maple Inc.
 \  MAPLE  /  All rights reserved. Maple is a registered trademark of
 <____ ____>  Waterloo Maple Inc.
      |       Type ? for help.
> quit;
```

2.5 AAArgh–>Self Help

Guidewords:

> ssystem( ), readme, macro( ), attributes( ), find, looksee( ), print( ),
> lprint( ), printlevel, tracelast( ), infolevel, trace( ), verboseproc

It all has gone adrift. You can't cope. Some people call Maple friendly, but it is mostly unfriendly for those unaware. In the quest for mathematical exactitude, the makers of Maple have put barriers in the way. Some of these may seem unnecessary and useless and leave us hopeless. A piece of paper would be better! Sometimes the most simple things can seem impossible with Maple. Often, however, it is a flaw in the way we tackle problems, come to light under the scrutiny of some of the world's finest mathematicians, through Maple. The structure of thought and the mind is various, but, with exact definition, tends to be flawed–at least a little. Still you feel Maple has done you wrong and you want no more of it. Be positive. Consider how you can best do the task you have to do. Maple, and computers, maybe, should have no part; most things don't require either. Still, we have hopes for Maple and we want to see the best side of it. Here we consider many levels of help that build your own commiserate, critical library and information you need to know. A library that can be accessed at a simple level to record your anguish and, perhaps, how you coped with it. A listing of attributes can help you can keep track of your variables. Maple supplies information as it is in operation; this can be viewed through infolevel or trace.

Use the index

There are also indices of commands and exercises

You can evolve your own help files and view them through Maple's Help. See the Appendix AAArgh, page 358. You can make your own library that is a part of the Maple library system. You can even build your own special package.

2.5.1 Help through a simple readme file

This is derived through use of your own editor. You can simply start a readme file and write whatever you want in it. In your particular configuration the file may be accessible from the Maple directory; but more likely it needs to be placed in your own, working directory. If it were to reside with the Maple executable, it would be there regardless of your own directory structure. However, you may not have full control of it; you may not be able to delete it. If you put it in your own working directory, you retain the file with your job and can preserve it whenever you copy or backup your system. The file would normally be a simple ASCII text file but could be arranged as a file in any simple editor; one you are comfortable with. You might simply record your anger and frustration but, hopefully, you eventually find a solution; then you will want to record your success in the editor, a record of how you coped and succeeded, to be used to prevent any further, similar problems. You can simply open an editor, quite separate from Maple, and write into it and save it. Here is an alternative, accessing the 'outside system' inside of Maple, to evoke the DOS editor.

```
>   ssystem("edit readme",3)[2];
```

The second parameter wasn't used in this instance

```
Warning, timeout parameter to ssystem ignored by this operating system
```

""

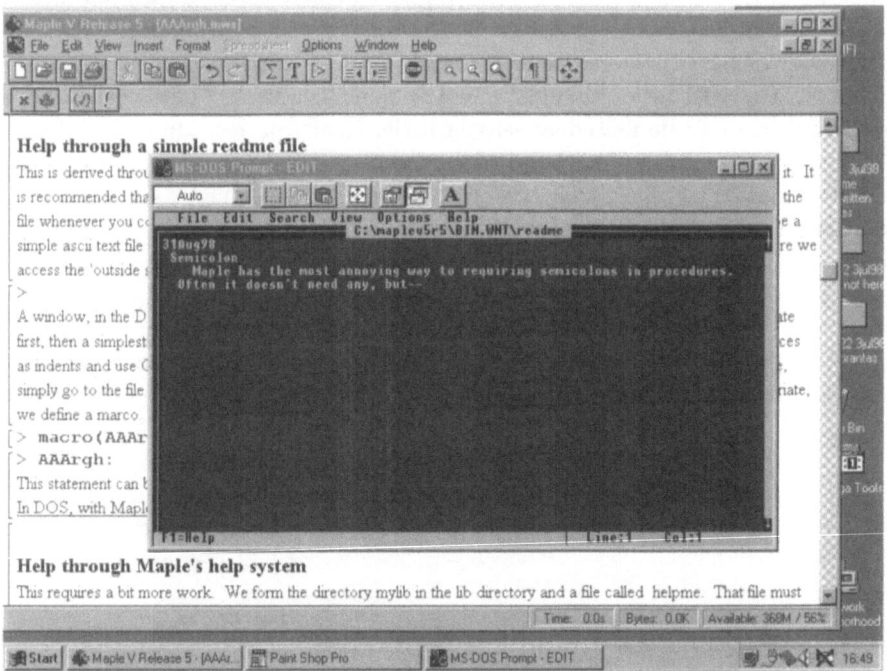

A window, in the DOS editor appears. Simply type in whatever you please. It is recommended that you use the date first, then a simplest title, for future reference. Remember that this is a simplest of editors so you need to put in spaces as indents and use CR at the ends of line, when you come to the limit of the 'seeable' line. Try to keep them less than 70 or so characters long. When you are done, go to the File menu and click on save. Then Exit the editor and/or click the close button, and carry on. The readme file, probably titled Readme, will appear in your current directory. Check this. If you started Maple from a Maple Icon, the readme in DOS will appear in a directory like c:/Program Files/Maple V Release 5/Bin.wnt/. Remember that this particular directory has controlled access.

The ssystem( ) command second argument (optional) allows 3 seconds for the edit command to be executed. Two elements are returned from the command; the first is the return code from executing the command; the second [2] is the result of executing the command; here, the edit window.

To make it easier to use, we define a macro.

```
>   #macro(AAArgh=ssystem("edit readme",3)[2]);
```

```
>   AAArgh;
```

Warning, timeout parameter to ssystem ignored by this operating system

The line is inactivated after use with a #

"()"

Once the macro is defined, simply typing AAArgh; on the command line is sufficient to evoke the editor, in the outside operating system. This statement can be inserted into your mapleini file so it is part of your startup. The file is maple.ini in Windows. It may be convenient to locate the file in your current directory. In DOS, in Windows, this means the location of the file on which you double clicked, to evoke Maple. This file is accessed easily, again, through Maple. Make sure you have the syntax perfect, following the macro statement above without the # and ending with a ;

```
>   ssystem("edit maple.ini")[2];
```

If the .ini file gets corrupted, you may not be able to use Maple. This can be corrected by removing the .ini file temporarily to the desktop.

"()"

2.5.2 Help by including attributes

These are specific qualities or comments you want to make about a variable, to keep order or simply soothe your mind.

setattribute(e, a) attaches the attributes a to the expression e. attributes( ) can be used to keep track of a variable or change the characteristics of a name.

```
>   restart:
```

```
>   attributes(chi);
```

```
>   setattribute(chi,Hard_Work,man!);
```

$$\chi$$

```
>   attributes(chi);
```

Hard_Work, man!

```
>  setattribute(chi);
```

$$\chi$$

It is best to use a
backward quote

```
>  setattribute(chi,'Well, do something else');
```

$$\chi$$

```
>  attributes(chi);
```

Well, do something else

Expression sequences can assign multiple attributes to an object. Attributes can be assigned to a name, list, set, procedure, unevaluated function call or a float. The attribute is attached to the object, though names are modified in place.

2.5.3   Edit/Find

Find

The use of Edit/Find can be a big help in locating important information within your worksheet, the help files or other files. In addition, the following set of statements gives a preview of Maple procedures, one that supplies a direct access to the reference list, section 1.5, page 40, the listing of commands that has been compiled into a file called "goodies" or "goodies.tex" that can be viewed on another worksheet and searched for examples. Goodies is an ASCII file with, approximately, a Maple text format so that it can be copied directly into the worksheet. It is a good idea for you to add items from time to time, items your find valuable. Try to keep it alphabetical.

For this to work, you must put the file "goodies" in an appropriate directory, as represented below, or change the Directory Structure shown so that the file can be found. The whole procedure can be placed in your Maple initialization file, see section 1.6, page 52. In any case, if mylib lies on your libname path, the binary file "looksee.m" will be found and read directly into Maple; in turn, if the file "goodies" is in the directory mylib, it will also be read.

```
>  restart;
```

```
>   fl :="c:\\mplbook\\mylib\\goodies";
```

$$fl := \text{``c:}\backslash\backslash\text{mplbook }\backslash\backslash\text{mylib}\backslash\backslash\text{goodies''}$$

```
>   looksee := proc( ) global fl;
>   interface(prettyprint=1); # 0 removes the spaces
>   while not feof(fl) do
>   print();
>   convert(readline(fl),name);
>   printf("%A",%); od; fclose(fl);
>   interface(prettyprint=3); # The default
>   end;
```

$looksee := \mathbf{proc}()$

   $\mathbf{global}\, fl;$

      $\text{interface}(prettyprint = 1)\,;$

      $\mathbf{while\ not}\ \text{feof}(fl)\,\mathbf{do}\,\text{print}()\,;\,\text{convert}(\text{readline}(fl),\ name)\,;\,\text{print}("\%A", \%)\,\mathbf{od}\,;$

      $\text{fclose}(fl)\,;$

      $\text{interface}(prettyprint = 3)$

   $\mathbf{end}$

```
>   #save looksee,"c:\\mplbook\\mylib\\looksee.m";
```
Notice that, again, the # has been used to inactivate the statement after it was used. This prevents unintential overwriting of the file. Also, if a directory structure is not used, the file will be in your current directory and, if you started Maple from the desktop, looksee.m would be on the desktop. With save, the command is available at call, provided we read it into a worksession.

To use the looksee( ) command we go to another worksheet using File/New; then we type

```
>   #looksee(); # inactivate to keep the screen clean
```
or, perhaps in another session– since we have saved the Maple binary–we can read in the procedure.

```
>   read "c:\\mplbook\\mylib\\looksee.m";
```

```
>   # looksee(); # runs the procedure
```
and, within the new worksheet we will see a short listing of commands, from "goodies". They stay there while you are working; you can minimise the size to keep them out of the way. This scratch pad concept is always valuable when you are trying out some ideas. Using Edit/Find allows you to 'browse' through the listing and Ctrl-c and Ctrl-v, or Edit/Copy and Edit/Paste copies the items into the first worksheet. Try the command out and add your own variations.

2.5.4　Active Worksheets

Much of the material in Chapter 3 and beyond is available on CD as Maple worksheets, accessible through your favorite browser, Netscape or Internet Explorer. Your browser is started and the file "mplbook.htm" is opened. The various active sections are linked to active Maple worksheets, which you can activate and alter as your curiosity allows. Included are some active bookmarks that link to sites with Maple information, see section 1.5, page 34. this rote technique of learning is limited, however; it is best to persist with tactile learning–using your fingers and mind as much as possible– to get into deep learning.

2.5.5　More Detail from Maple

no need to read
further

In the internal structure of Maple, there are a number of ways to get information on how Maple is working. These include print, printlevel, infolevel, and trace.

Print

Debugging loops, conditional if statements and procedures are often easiest to debug by simply printing variables 'on-the-fly'. The following loop has a problem. The loop is supposed to pick off the position of the last / in the directory structure string. substring checks each letter, beginning from the right side.

```
>  s := "help/example"; n := length(s);
```

$$s := \text{"help/example"}$$

$$n := 12$$

```
>  for j from 1 to n do;
>  ss := substring(s,-j);
>  if ss="/" then BREAK fi;
>  od:

>  j;
```

$$13$$

The routine goes through the whole string, without detecting the /. We add a print statement, in the if statement, to find out what happened.

```
>   j:='j';
```

$$j := j$$

```
>   for j from 1 to n do;
>   ss := substring(s,-j);
>   if ss="/" then print(j); BREAK fi;
>   od:
```

8

It is detecting the / but the BREAK is ineffective. Try a lower case break.

```
>   for j from 1 to n do;
>   ss := substring(s,-j);
>   if ss="/" then print(j); break fi;
>   od:
```

8

```
>   j;
```
Now just remove the print statement, reactivate; it is a working routine.   Fixed!

8

lprint

A variant of print, lprint, gives a 'raw' ASCII output that can be cut and pasted more easily. This is convenient in utilising structures within Maple.

```
>   readlib('student/simpson'):lprint(%);
```

No need to read
further

```
proc (F, dx) local a, b, f, loci, h, n, rg, x; global i, j; option
'Copyright (c) 1990 by the University of Waterloo. All rights
reserved.'; if type(dx,equation) and type(op(1,dx),name) and
type(op(2,dx),range) then x := op(1,dx); rg := op(2,dx) else
ERROR('usage: simpson( f(x) , x=a...b , iterations)') fi; if 2 < nargs
and not type(args[3],{name, integer}) then ERROR('usage: simpson( f(x)
, x=a...b , iterations)') elif nargs = 2 then n := 4 else n :=
args[3]; if type(n,numeric) and not 'mod'(n,2) = 0 then ERROR('must
have an even number of intervals') fi fi; if not type(n,{name,
numeric}) then RETURN('procname(args)') fi; a := op(1,rg); b :=
op(2,rg); h := (b-a)/n; f := student[makeproc](F,x); if not
assigned(i) then loci := i elif not assigned(j) then loci := j else
loci := readlib('tools/genglob')('i') fi;
1/3*h*(f(a)+f(b)+4*Sum(f(a+(2*loci-1)*h),loci = 1 ..
1/2*n)+2*Sum(f(a+2*loci*h),loci = 1 .. 1/2*n-1)) end
```

This is 1D or simply printed output. For 2D output, the following works
well

```
> interface(verboseproc=2); readlib('student/simpson'):
> print(%);
```

Printlevel and Tracelast

Displays all information produced by conditional or repetition statements, depending on the level of the operation. The values are integers; the ordinary, interactive level is 0; each conditional or repetition block adds one level; a procedure adds 5 levels. Hence, a repetitive do loop in a procedure would have information displayed if the printlevel were set at 6. An if block with an internal do loop would reveal its secrets at printlevel=2. With execution errors and printlevel > 2, a summary of calling routines is provided. The debug facility tracelast will also show your last error.

The following procedure searches several Directory Structures for the topic, the name following the last /

```
> restart:
```

```
> directories := 'help/text/mtex',
> 'help/text/tex', 'help/tex/text/usage', 'help/TeX/text/example';
```

$$directories := help/text/mtex, \ help/text/tex, \ help/tex/text/usage, \ help/TeX/text/example$$

```
> topics:= proc(sequence) local i,j,n,m,s,ss;
> n := nargs; # arguments or number of directories
> ss := NULL; # the NULL sequence
> for i from 1 to n do
> s:=sequence[i];
> for j from 1 to length(s[i])
> while (substring(s,-j)<>"/")
> do; m:=j; od;
> ss := ss,substring(s,-m..-1);
> od;
> ss;
> end;
```

$topics := \mathbf{proc}(sequence)$

    $\mathbf{local}\,i,\,j,\,n,\,m,\,s,\,ss;$

        $n := \mathrm{nargs};$

        $ss := NULL;$

        $\mathbf{for}\,i\,\mathbf{to}\,n\ \mathbf{do}$

            $s := sequence_i;$

            $\mathbf{for}\,j\,\mathbf{to}\,\mathrm{length}(s_i)\,\mathbf{while}\,\mathrm{substring}(s,\,-j) \neq \text{``/''}\,\mathbf{do}\,m := j\,\mathbf{od};$

            $ss := ss,\,\mathrm{substring}(s,\,-m..-1)$

        $\mathbf{od};$

        ss

    $\mathbf{end}$

```
>   topics(directories);

Error, (in topics) string or symbol expected for substring

>   tracelast;

 topics called with arguments: help/text/mtex, help/text/tex,
help/tex/text/usage, help/TeX/text/example
 #(topics,5): for j to length(s[i]) while substring(s,-j) <> "/" do
... od;

Error, (in topics) string or symbol expected for substring

 locals defined as: i = 1, j = 1, n = 4, m = m, s =
'help/text/mtex'[1], ss =
```

Something has gone wrong with the innerworkings of the procedure. Further detail may be revealed with a printlevel greater than 5 (for the procedure)+1 (for the outer do) + 1 (for the inner do) + 1 (for the conditional if).

```
>   printlevel:=8;
```

$$printlevel := 8$$

```
>   topics(directories);

{--> enter topics, args = help/text/mtex, help/text/tex,
help/tex/text/usage, help/TeX/text/example
```

$$n := 4$$

$$ss :=$$

$$s := help/text/mtex_1$$

```
<-- ERROR in topics (now at top level) = string or symbol expected for
substring}

 topics called with arguments: help/text/mtex, help/text/tex,
help/tex/text/usage, help/TeX/text/example
 #(topics,5): for j to length(s[i]) while substring(s,-j) <> "/" do
... od;

Error, (in topics) string or symbol expected for substring

 locals defined as: i = 1, j = 1, n = 4, m = m, s =
'help/text/mtex'[1], ss =
```

It seems that the sequence of directories is being confused with the argument list to the procedure. Substitute the reserved variable args for sequence; make it a list. Also, the if condition is not properly being defined since directories is a sequence of symbols (with ' '), not strings (with " "). It works so the printlevel is lowered to the default. Note also the effect of the colon : at the end of the statement that calls the procedure; with the colon and the high printlevel, only the extraordinary information, the entries and exits to procedures are shown. The information relates to all entries and exits to procedures, both inside and outside of the procedure.

```
> printlevel:=0;
```

$$printlevel := 0$$

```
> topics:= proc(sequence) local i,j,n,m,s,sl,ss;
> n := nargs; # arguments or number of directories
> ss := NULL; # the NULL sequence
> for i from 1 to n do
> sl:=[args];
> s:=sl[i];
> for j from 1 to length(s[i])
> while (substring(s,-j)<>'/')
> do; m:=j; od;
> ss := ss,substring(s,-m..-1);
> od;
> ss;
> end:

> topics(directories);
```

mtex, tex, usage, example

See what happens at a higher print level.

```
>   printlevel:=8;
```

$$printlevel := 8$$

```
>   topics(directories):
```

```
{--> enter topics, args = help/text/mtex, help/text/tex,
help/tex/text/usage, help/TeX/text/example
```

```
<-- exit topics (now at top level) = mtex, tex, usage, example}
```

```
>   printlevel:=0; # Return printlevel to normal
```

$$printlevel := 0$$

infolevel

If the particular package has built-in statements of the sort userinfo(3,'Finished first Convolution') at various points in its progress, it is possible to simply view what Maple is doing. Here int is the indefinite integration routine and we want to see how Maple copes with a difficult integral.

```
>   infolevel[int]:=4;
```

$$infolevel_{int} := 4$$

```
>   F:=int(1/(x^5+2*x+2),x);
```

```
int/indef1:   first-stage indefinite integration
```

```
int/ratpoly:   rational function integration
```

```
int/rischnorm:   enter Risch-Norman integrator
```

```
int/rischnorm:   exit Risch-Norman integrator
```

```
int/risch:   enter Risch integration
```

```
int/risch:   the field extensions are
```

$$[x]$$

```
unknown:   integrand is
```

$$\frac{1}{x^5 + 2\,x + 2}$$

int/ratpoly/horowitz: integrating

$$\frac{1}{x^5 + 2\,x + 2}$$

int/ratpoly/horowitz: Horowitz' method yields

$$\int \frac{1}{x^5 + 2\,x + 2}\, dx$$

int/risch/ratpoly: starting computing subresultants at time   6.457

int/risch/ratpoly: end of subresultants computation at time   6.527

int/risch/ratpoly: Rothstein's method – factored resultant is

$$\left[\left[z^5 - \frac{80}{3637}\,z^3 - \frac{20}{3637}\,z^2 - \frac{15}{29096}\,z - \frac{1}{58192}, 1\right]\right]$$

int/risch/ratpoly: result is

$$\sum_{\_R=\%1} \_R \ln(x - \frac{3724288}{625}\,\_R^4 + \frac{465536}{625}\,\_R^3 + \frac{23728}{625}\,\_R^2 + \frac{17514}{625}\,\_R + \frac{512}{625})$$
$$\%1 := \mathrm{RootOf}(58192\,\_Z^5 - 1280\,\_Z^3 - 320\,\_Z^2 - 30\,\_Z - 1)$$

int/risch: exit Risch integration

$$F := \sum_{\_R=\%1} \_R \ln(x - \frac{3724288}{625}\,\_R^4 + \frac{465536}{625}\,\_R^3 + \frac{23728}{625}\,\_R^2 + \frac{17514}{625}\,\_R + \frac{512}{625})$$
$$\%1 := \mathrm{RootOf}(58192\,\_Z^5 - 1280\,\_Z^3 - 320\,\_Z^2 - 30\,\_Z - 1)$$

```
>   infolevel[int]:=0; # Return infolevel to
>   normal
```

$$infolevel_{int} := 0$$

trace and untrace

These are, perhaps, the most powerful tools to see what is going on. They need to be turned on and off, however, because they produce a lot of output. We consider expand.

```
>   (sin(2*x)+cos(3*x))^3;
```

$$(\sin(2\,x) + \cos(3\,x))^3$$

```
>   trace(expand);
```

expand

```
>   expand(%%);
execute expand, args = (sin(2*x)+cos(3*x))^3

execute expand, args = 2*x

execute expand, args = 2*sin((n-1)*y)*cos(y)-sin((n-2)*y)

execute expand, args = x

execute expand, args = x

execute expand, args = 3*x

execute expand, args = 2*cos((n-1)*y)*cos(y)-cos((n-2)*y)

execute expand, args = 2*x

execute expand, args = 2*cos((n-1)*y)*cos(y)-cos((n-2)*y)
```

$$8\sin(x)^3\cos(x)^3 + 48\sin(x)^2\cos(x)^5 - 36\sin(x)^2\cos(x)^3 + 96\sin(x)\cos(x)^7$$
$$- 144\sin(x)\cos(x)^5 + 54\sin(x)\cos(x)^3 + 64\cos(x)^9 - 144\cos(x)^7 + 108\cos(x)^5$$
$$- 27\cos(x)^3$$

```
>   untrace(expand);
```

expand

Maple's Source Code

The very 'blood' which flows through Maple's arteries is revealed with verboseproc and interface( ). The normal value for verboseproc is 1. More information is revealed at higher levels.

```
>   interface(verboseproc=2);
```

```
>   readlib(value);
```

proc(f)
    local *funs, procs*;
    option '*Copyright (c) 1992 by the University of Waterloo. All rights reserved.*';
        if nargs $\neq 1$ **then** ERROR('*incorrect number of arguments*') **fi**;
        funs := [op(map2(*op*, 0, indets(*f*, '*function*')))] ;
        if *funs* = [] **then** RETURN(*f*) **fi**;
        procs := map('*value/define*', *funs*) ;
        ASSERT(nops(*funs*) = nops(*procs*)) ;
        eval(*f*, zip(**proc** (x, y) **if** $x = y$ **then** *NULL* **else** $x = y$ **fi end**, *funs*, *procs*))
    end

A command to Maple is a part of an input statement; a procedure follows through and generates the output

The command value( ) only arrived in 1992; it activates Maple commands starting with an upper case letter, commands which are 'frozen' or unevaluated and appear in Maple output as symbolic, mathematical expressions.

```
>   Int(sin(exp(x)^2),x=0..xi);
```

$$\int_0^\xi \sin((e^x)^2)\, dx$$

```
>   Xi:=value(%);
```

$$\Xi := \frac{1}{2}\,\mathrm{Si}((e^\xi)^2) - \frac{1}{2}\,\mathrm{Si}(1)$$

```
>   subs(xi=Pi,%):  evalf(%);
```

$$.3122154718$$

```
>   readlib(int);
```

proc()
    local *answer*;
    option '*Copyright (c) 1997 Waterloo Maple Inc. All rights reserved.*';
        answer := readlib('*int/int*')([args], *Digits*, _*EnvCauchyPrincipalValue*) ;
        if *answer* = *FAIL* **then** ERROR('*wrong number (or type) of arguments*') **fi** ;
        answer
    end

```
> readlib(copy);
```

proc(*A*)
    option '*Copyright (c) 1990 by the University of Waterloo. All rights reserved.*';
        if type(*A*, {*array*, *table*}) **then**
            if type(*A*, *name*) **then** map(**proc**() args **end**, eval(*A*))
            else map(**proc**() args **end**, *A*)
            fi
        else *A*
        fi
    end

If we lower verboseproc, a security veil covers the Copywrite.

```
> interface(verboseproc=1);
> readlib(copy);
```

$$\text{\textbf{proc}}(A) \dots \textbf{end}$$

```
> readlib('student/simpson');
```

$$\text{\textbf{proc}}(F, dx) \dots \textbf{end}$$

```
> interface(verboseproc=3); # largest value for verboseproc
```

Not much
information here

```
> print(eval);
```

Even with the largest value of verboseproc, the builtin routines are not revealing.

$$\text{\textbf{proc}}() \text{ \textbf{option} } builtin;\ 151 \text{ \textbf{end proc}}$$

Back to the default level, verboseproc=0.

```
>   readlib(value);
```

$$\mathbf{proc}(f) \ldots \mathbf{end}$$

Appendix AAArgh

At the end of this volume, on page 358, an appendix extends help. On loading a library, help information can become available through Maple's help system. Alternately, you can attach your own library information.

> Clearly, Maple has an ability to do things other than mathematics. Catenation puts things together; op( ) pulls them apart. Functions and commands do things. Types have properties suitable for a given purpose; commands and types must be compatible; conversion between types is useful. A sequence as an argument can be ambiguous; a list made from a set has no duplicates. Assignment limits the flexability of a name or symbol. Symbols can 'talk to Maple' and retrieve information but Illogic disrupts the pattern. Units put life into environmental numbers which Maple can make into documents. With examples and help, a working knowledge of Maple is emerging.

Acknowledgements for Chapter 2

Kelvin Maybury of Environmental Science at Murdoch contributed much of the section on Environmental Numbers. Greg Fee of CECM, Simon Fraser University, exposed the details of help. Scott Rabuka of Maple Waterloo Inc. suggested the use of 2D printout.

Chapter 3

Manipulations, Procedures and Plots

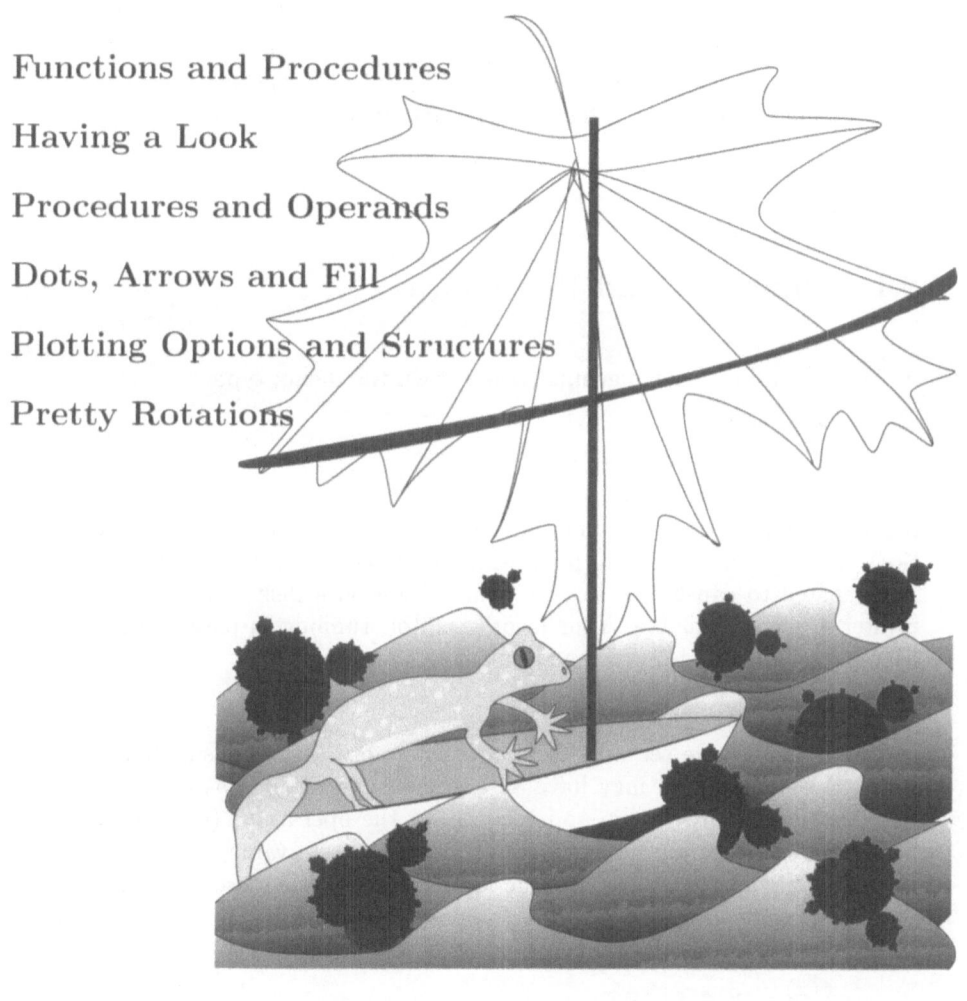

Manipulations, Procedures and Plots

Übersicht

The generation of functions and procedures, what they do and how
they can be viewed; making them do the hard work and making
manipulations smooth; the twists and turns of plotting, plotting
options, PLOT Structures and some pretty plots.

Guidewords:

functions, -> , proc( )...end; if..fi; plot( ), options, discont,
sample; limit, map( ), lhs( ), rhs( ), plot3d, fractals

```
>  restart;
```

3.1 Functions and Procedures

As before, turn on your machine. Bring down the menu; type

```
>  ?function
```
or

```
>  help( procedure);
```
preview the definition of a function, the structure of procedures and plot options, and view the beautiful plots. Copy them across and make them active.
Be prepared to capture or to save your session on a disk or even to use the
printer if one is available. Your favorite editor, running in parallel with your
Maple session, can also prove convenient.

Functions

Note that Maple sees the form f(x) as simply a symbolism that tells it
(or you) that the form may have an argument, x. Maple has no connection
with that argument and no appreciation for the operations (multiplications,
additions, etc) that are necessary to realise a concrete value, symbolic or numerical. The functional form or function call f(x) is little more than a complex
name that you can use to purpose. Suppose you assign it to be an expression
in x.

```
>  f(x):= 3*x+2;
```
$$f(x) := 3\,x + 2$$

```
>  f(1); # has no associated activity
```
$$f(1)$$

```
>  subs(x=2,f(x));
```

8

But, if you substitute an integer value for x, the expression is automatically evaluated and there is an integer value as a result. The realisation of the value 8, above, involved a recognition that f(x) was an expression, along with a substitution of the value 2 for x.

In Maple, these operations are performed by function calls to Maple, as placed on the command line; a statement is executed and a function is called. If Maple recognises the function call as a defined procedure, with assigned symbolism, it will process the argument and produce a result. It is probably easiest for you to simply define (or draw-up) a procedure which Maple will recognise through the function call.

A Maple statement is a line ending in a ;

Function calls, like sin( ) and exp( ), for known or defined functions are followed through; the substitution and evaluation are completed, with the given arguments. The Maple Language allows you to make your own definitions for function calls; these are assignments of function calls; the idea is presented here as 'writing procedures'.

Alternately, you can simply combine function calls that are known to Maple and generate new defined (or assigned) function calls.

```
>   sc := sin + cos;
```

$$sc := \sin + \cos$$

```
>   sc(x+y);
```
$$\sin(x + y) + \cos(x + y)$$

```
>   expand(%);
```
$$\sin(x)\cos(y) + \cos(x)\sin(y) + \cos(x)\cos(y) - \sin(x)\sin(y)$$

As a Follow up, we can convert to another form. Consider

```
>   combine(ex,trig)
```
or

```
>   trigsubs
```
which maintains a library of trig substitutions.

A diversion into trig identities- playing with Maple

```
>   combine(%,trig);
```
$$\sin(x + y) + \cos(x + y)$$

```
>   trigsubs(sin(x+y)); # a list of trig identities
```

simplify(ex,trig) works with a suggested substitution

$$\left[\sin(x+y),\ -\sin(-x-y),\ 2\sin(\frac{1}{2}x+\frac{1}{2}y)\cos(\frac{1}{2}x+\frac{1}{2}y),\ \frac{1}{\csc(x+y)},\ -\frac{1}{\csc(-x-y)}, \right.$$

$$\left. 2\,\frac{\tan(\frac{1}{2}x+\frac{1}{2}y)}{1+\tan(\frac{1}{2}x+\frac{1}{2}y)^2},\ \frac{-1}{2}\,I\left(e^{(I\,(x+y))}-e^{(-I\,(x+y))}\right) \right]$$

Procedures

A procedure performs a specific task, like a subroutine. A procedure can be produced using the

```
->
```

Maple 6 uses
end proc; to end
a procedure

operator or, more formally, using

```
> proc(  ). . end;
```

```
> f := x -> (x^3-8)/(x-2);
```

$$f := x \to \frac{x^3 - 8}{x - 2}$$

The construction literally means *take the value of x, cube it, subtract 8, divide by the original value less 2; return the result.* In other words, convert the value of x to the value of the algebraic expression. x is the argument of the procedure; you put in x and get out the 'contorted' value given.

2 semicolons can
be omitted

```
> g := proc(x);(x^3-8)/(x-2); end;
```

$$g := \mathbf{proc}(x)\,(x^3 - 8)/(x - 2)\,\mathbf{end}$$

```
> g := proc(x) option operator,arrow;(x^2-9)/(sqrt(x)-2); end;
```

$$g := x \to \frac{x^2 - 9}{\sqrt{x} - 2}$$

```
> type(f,procedure),type(g,procedure);
```

$$true,\ true$$

Procedures f and g are the same but not identical. They perform the same operation on the variable x, however.

```
> evalb(f(x)=g(x));
```

$$false$$

The unapply( ) command makes procedures from expressions in a direct way.

```
> h := unapply((x^2-9)/(sqrt(x)-2),x);
```

$$h := x \to \frac{x^2 - 9}{\sqrt{x} - 2}$$

```
> type(h,procedure), evalb(f(x)=h(x));
```

$$true,\ false$$

Functions and procedures can be combined

```
> (sc+f)(x-3); simplify(%); expand(%);
```

$$\sin(x - 3) + \cos(x - 3) + \frac{(x - 3)^3 - 8}{x - 5}$$

$$\sin(x - 3) + \cos(x - 3) + x^2 - 4x + 7$$

$$\sin(x)\cos(3) - \cos(x)\sin(3) + \cos(x)\cos(3) + \sin(x)\sin(3) + x^2 - 4x + 7$$

```
> f(xx);
```

$$\frac{xx^3 - 8}{xx - 2}$$

```
> f(a+b); simplify(%);
```

$$\frac{(a + b)^3 - 8}{a + b - 2}$$

$$a^2 + 2ab + 2a + b^2 + 2b + 4$$

```
> f(3-x); expand(%);
```

Contains more
simple terms

$$\frac{(3 - x)^3 - 8}{1 - x}$$

$$9\frac{x^2}{1 - x} - \frac{27x}{1 - x} + \frac{19}{1 - x} - \frac{x^3}{1 - x}$$

Here we explore for discontinuities and singularities. It is clear that the procedure f(x) has a problem when x=2.

```
> subs(x=2,f(x));
```

```
Error, division by zero
```

There is a limit

```
> limit(f(x),x=2);
```

$$12$$

The function has a pole in it at x=2; it appears with careful careful plotting.

A pole is a
missing point, or
hole

3.2   Plotting

Plotting enables a visual depiction of mathematical quantities. It can show shortcomings in the analysis, and graphing itself can corrupt the data, depending on how it is 'massaged'. Nonetheless, it is a quick and visual way to attack a problem. In Maple, as in life, it is best to 'have a go' and simply try it. When all else fails, use underlinesection 1.2, page 17

```
>   ?help
>   # or
>   help(plot);
```

Each version of Maple may have different ways of handling discontinuities.

Exploring for the 'hole'

Here we use Maple plot( ) to investigate the curve, above. In this, we make sure the discontinuity is sampled.

```
>   plot(f(x),x=0..4):
```

Replace the : with a ; use CR, and you see the plot

This shows nothing; we can look at the graph and view what was done. First, click on the graph. Second, click at roughly the point x=2 on the graph. The x,y values should appear in the little window on the lower left of the menu banner. Then click on the various sinuous icons; one should show the actual points plotted by Maple. Another way to do this is to use the option point in the plot command. Combining this with an appropriate value of numpoints should reveal the problem point.

```
>   numpoints=20
```

Maple has smoothed over the point

is a minimum.   Maple 'massages' the plot to make it continuous on a line graph.

```
>   plot(f(x),x=0..4,style=point, numpoints=51);
```

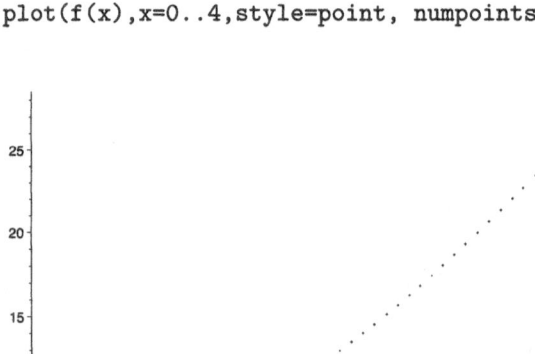
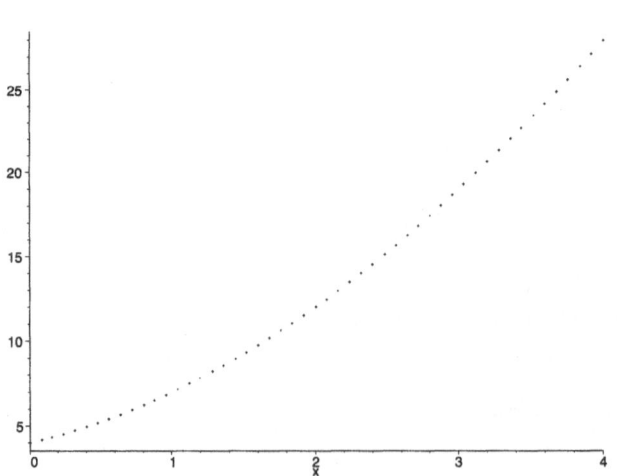

The plot is forced to be faithful by using the sample option in plot (see section Options, page 143) or

This includes the suspect point, x=2

```
>  ?plot,options
```

```
>  sa := [seq(i/10,i=0..40)];
```

$$sa := [0, \frac{1}{10}, \frac{1}{5}, \frac{3}{10}, \frac{2}{5}, \frac{1}{2}, \frac{3}{5}, \frac{7}{10}, \frac{4}{5}, \frac{9}{10}, 1, \frac{11}{10}, \frac{6}{5}, \frac{13}{10}, \frac{7}{5}, \frac{3}{2}, \frac{8}{5}, \frac{17}{10}, \frac{9}{5}, \frac{19}{10}, 2, \frac{21}{10}, \frac{11}{5}, \frac{23}{10}, \frac{12}{5}, \frac{5}{2}, \frac{13}{5},$$
$$\frac{27}{10}, \frac{14}{5}, \frac{29}{10}, 3, \frac{31}{10}, \frac{16}{5}, \frac{33}{10}, \frac{17}{5}, \frac{7}{2}, \frac{18}{5}, \frac{37}{10}, \frac{19}{5}, \frac{39}{10}, 4]$$

```
>  plot(f(x),x=0..4,sample=sa);
```

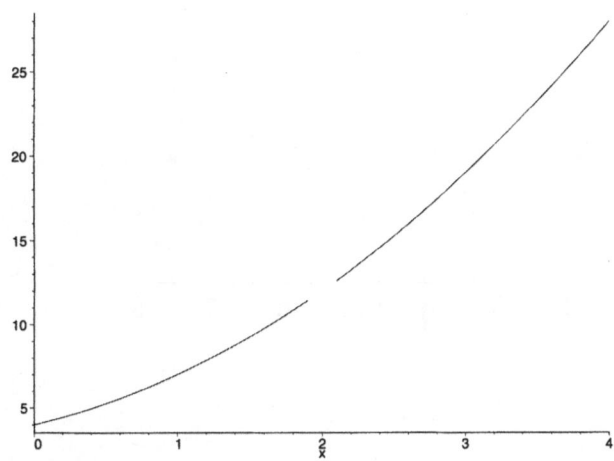

The discontinuity is easily found with discont.

```
>  readlib(discont):
```

```
>  discont(f(x),x);
```

$$\{2\}$$

Consider another function.

```
>  f(x) := (x - 4)/(sqrt(x) - 2);
```

$$f(x) := \frac{x - 4}{\sqrt{x} - 2}$$

```
>  discont(f(x),x);
```

$$\{0, 4\}$$

Options page 143
or ?plot,options

The last graphs had rather small fonts. Add a few more plot options.

```
> TIMES,BOLD:
> fts:=titlefont=[%,20], axesfont=[%,12], labelfont=[%,12];
```

fts := *titlefont* = [*TIMES, BOLD,* 20], *axesfont* = [*TIMES, BOLD,* 12],
labelfont = [*TIMES, BOLD,* 12]

```
> plot(f(x),x=0..8,2..5,title='No Hole',fts);
```

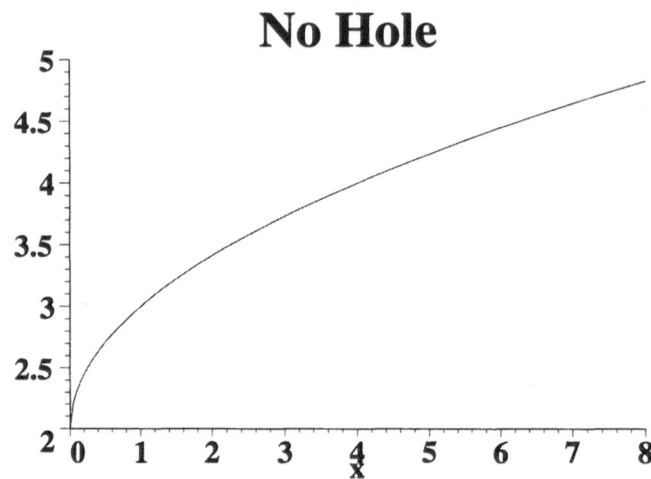

Breaking the curve into two separate curves reveals the flaw.

```
> thickness=2,discont=true,numpoints=101:

> opts:= thickness=2, title=Hole,
> discont=true, numpoints=101:

> p1:=plot(f(x),x=0..4,2..5,opts,fts):

> p2:=plot(f(x),x=4..8,2..5,opts,fts):

> with(plots):  display(p1,p2);
```

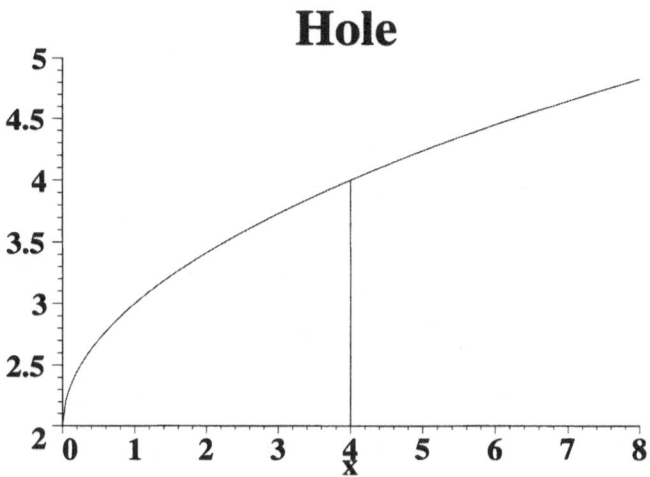

Exercise 3.1 Show the 'holes'

Use the sample option to show the holes in the above function.

Exercise 3.2 Discontinuities

Examine the function below; look for singularities. Use limit( ) and discont( ). Plot it in the range y = 1.5..2; change the numpoints, discont and sampling options to reveal the discontinuities.

```
> g(y) := (y+1)^2/(y^3-3*y);
```

$$g(y) := \frac{(y+1)^2}{y^3 - 3\,y}$$

Infinite Limit

The limit can even be infinity. Consider the function

```
> g:=x->sqrt(x^2-4*x)-x;
```

$$g := x \to \sqrt{x^2 - 4\,x} - x$$

```
> limit(g(x),x=infinity);
```

$$-2$$

```
> thickness=3, labels=["snirtle","feague"],
> labeldirections=[horizontal,vertical]; # Maple 6
```

The argument x is replaced by a label

```
> plot(g(x),x=0..50,-4..-1,fts,%);
```

thickness = 3, *labels* = ["snirtle", "feague"], *labeldirections* = [*horizontal*, *vertical*]

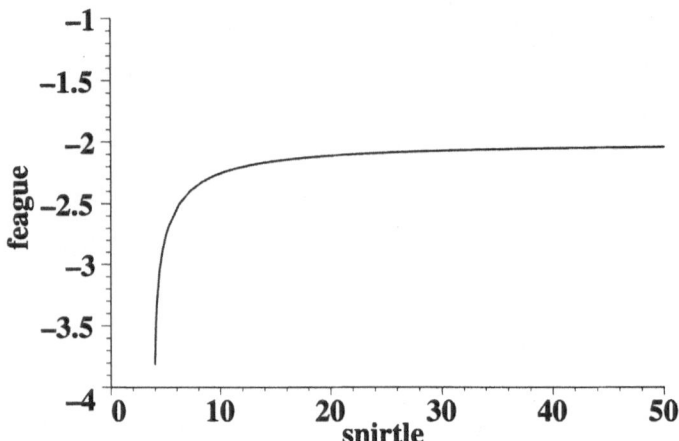

When x is less than 4, the function becomes imaginary. At large values of x there is a clear limit of -2.

3.3   Procedures

These have a more formal, block form, subroutine or process structure that has a large application. They are particularly useful for piecewise functions or conditional situations; they allow for breaks, loops and recursion. Consider that we simply want to reorganise the following equations.

```
>   e1 := x + y + z = 10 + c^2;
```
$$e1 := x + y + z = 10 + c^2$$
```
>   e2 := x^2 + y^2 = 1 + c + d;
```
$$e2 := x^2 + y^2 = 1 + c + d$$
```
>   e3 := 10*x + 20*z = 1 + d^2;
```
$$e3 := 10\,x + 20\,z = 1 + d^2$$
```
>   eqs := [e1,e2,e3];
```
$$eqs := [x + y + z = 10 + c^2,\ x^2 + y^2 = 1 + c + d,\ 10\,x + 20\,z = 1 + d^2]$$

Return or Enter   Type in the following lines as given. Do not use the CR because Maple will try to execute your statement. During Maple input, use Shift-CR if you want a CR at the ends of your lines.

Generally a procedure fits statements into a structure with **proc( ) end**:

```
#---- Define a procedure to do a swap ----#
swap := proc(x) local a,b;
   a := rhs(x); b := lhs(x); a=b; end;
#-------------------------------------#
```

```
> swap := proc (x)
> local a, b;
> a := rhs(x);
> b := lhs(x);
> a=b;
> end;
```
$$swap := \mathbf{proc}(x)\, \mathbf{local}\, a,\, b;\; a := \mathrm{rhs}(x)\,;\; b := \mathrm{lhs}(x)\,;\; a = b\, \mathbf{end}$$

Be careful when you type this in. Remember that Maple uses the CR to execute a command (or a procedure). It is easiest to form a large number of execution groups and keep them inactive (no CR's) until the whole procedure is typed in. You can then join the group of > using the edit menu so that they act as one. This means clicking to make new execution lines. Another way is to simply use CR and let Maple object! Maple will continue to make new executable > until you correct your mistake; again, it is easiest if you simply make many extra lines and fill in the holes. Using the CR along with the Shift key, as suggested, produces the equivalent of a broken line, wrapped to the size you want.

Maple is good at finding errors here

Within the procedure, Maple doesn't respond to ;   properly, it waits for end; Hence any errors you make can only be cleared by typing end; and Maple may even refuse to let you go further. If you are frustrated, don't hesitate to skip to another execution group, play around with structures and help or whatever, and come back when you have sorted out the difficulty. The routine normally returns the last line unless RETURN( ) is included; local is used for defining dummy variables only accessible inside the procedure. In this case the semicolons are only necessary in the assignments of a and b; the right hand side rhs and left hand side lhs of the dummy argument, the equality x.

```
> swap(e1);
```
$$10 + c^2 = x + y + z$$

```
> swap(e2);
```
$$1 + c + d = x^2 + y^2$$

```
> seqs := map(swap,eqs);# map works on the whole list
```
$$seqs := [10 + c^2 = x + y + z, \, 1 + c + d = x^2 + y^2, \, 1 + d^2 = 10\,x + 20\,z]$$

Try another procedure, with juncture from if statements.

The syntax is
more clear with
semicolons at the
ends·of lines

```
> f := proc(x)

> if   x < 0 then 0

> elif x < 1 then x

> elif x < 2 then 2-x

> else 0

> fi

> end;
```
$$f := \mathbf{proc}(x)\,\mathbf{if}\,x < 0\,\mathbf{then}\,0\,\mathbf{elif}\,x < 1\,\mathbf{then}\,x\,\mathbf{elif}\,x < 2\,\mathbf{then}\,2 - x\,\mathbf{else}\,0\,\mathbf{fi}\ \ \mathbf{end}$$

This function is a piecewise one, with discontinuities.

```
> plot(f(x),x=0..2);# Plot the discontinuous procedure
Error, (in f) cannot evaluate boolean
```

The error occurs before the plot function started executing

```
> f(x);
Error, (in f) cannot evaluate boolean
```

The logic here is flawed because Maple cannot evaluate $x < 0$ as x has
no numerical value. This problem is solved by ensuring that only the function name is passed to plot; evaluation occurs only when plot has assigned
numerical values. Two forms of the command do this:

```
>  #plot('f(x)',x=0..2);
```

```
>  plot('f'(something),something=0..2,fts);
```

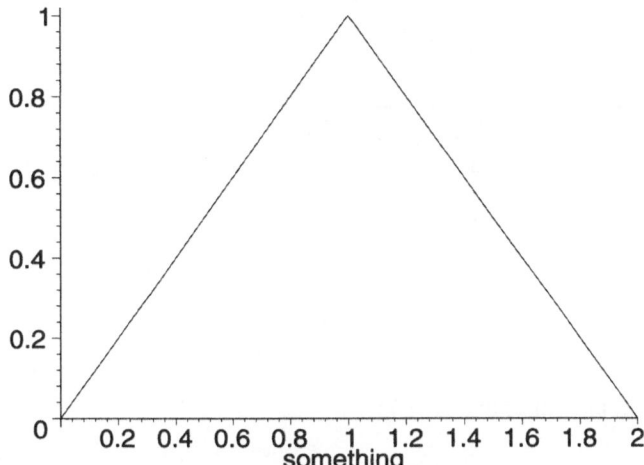

It is clear that Maple only needs to know that it has argument 'something' and that it should not worry about the functional value until it really needs to know. Also, the argument takes the name you want to be shown on the plot. The name can be anything in plot.

Try another discontinuous function.

This syntax works without semicolons

```
>  g := proc(x)
>  if x < 0 then x-1
>  else x^2
>  fi
>  end;
```

$$g := \mathbf{proc}(x) \, \mathbf{if} \, x < 0 \, \mathbf{then} \, x - 1 \, \mathbf{else} \, x^2 \, \mathbf{fi} \, \mathbf{end}$$

```
>  plot(g,-2..2,thickness=2,fts);
```

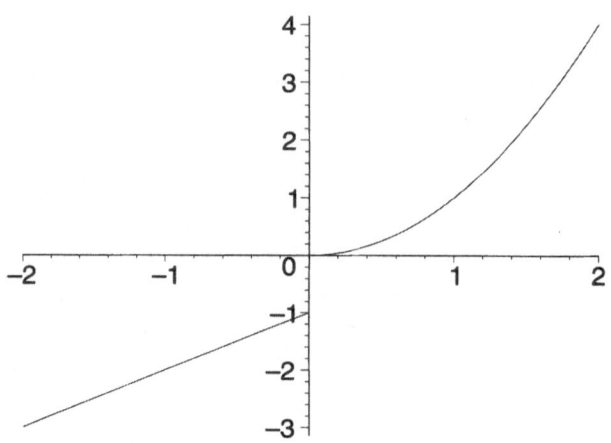

```
>  plot(f,0..2,thickness=3,fts);
```

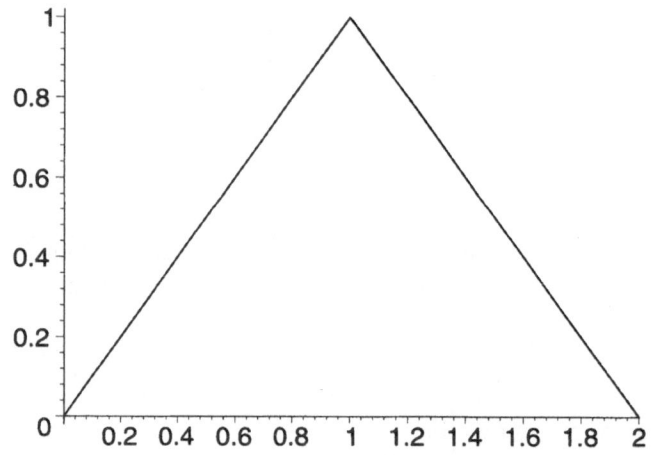

Here we are completely flummoxed. How can Maple know what the argument is? What about the discontinuity? The answer is clear when we realise that it doesn't need to know and, since it doesn't, it doesn't worry and doesn't evaluate until it needs to. The g plot clearly shows the discontinuity at x=0 with one-sided limits. Also,

```
>  g(-1.01), g(0), g(.01);
```

$$-1.01, 0, .0001$$

```
>  sa:=[-.01,seq(i/5,i=-10..10)];
>  opts:=style=point,adaptive=false,
>  symbol=diamond,font=[SYMBOL,14],
>  ytickmarks=[-2,1,2,3,4]:
>  plot('g'(z),z=-2..2,sample=sa,opts);
```

$$sa := [-.01, -2, \frac{-9}{5}, \frac{-8}{5}, \frac{-7}{5}, \frac{-6}{5}, -1, \frac{-4}{5}, \frac{-3}{5}, \frac{-2}{5}, \frac{-1}{5}, 0, \frac{1}{5}, \frac{2}{5}, \frac{3}{5}, \frac{4}{5}, 1, \frac{6}{5}, \frac{7}{5}, \frac{8}{5}, \frac{9}{5}, 2]$$

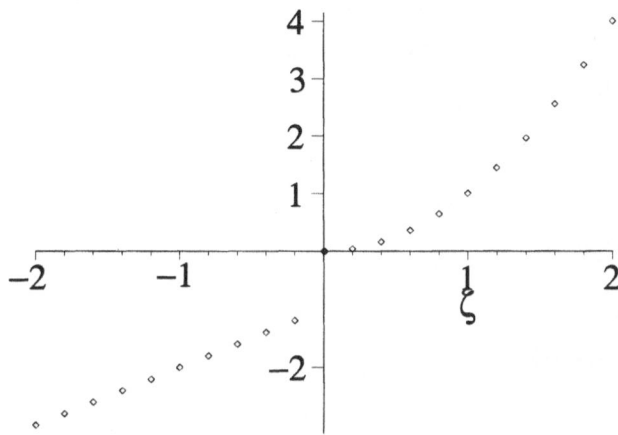

3.4 Operands

The construction of a procedure has six possible operands

1. a sequence of Arguments or Parameters
2. a sequence of local variables
3. a sequence of options
4. a remember table
5. a description string
6. a sequence of global variables

These operands can generate much flexibility in the procedure, but tight control must be exercised. For the more subtle features of procedures; the reader is referred to one of the many programming books for help. The following is summarized from Monagan *et al*, MapleV Release 5, Programming Guide [23]. Importantly, procedures have special syntax that may not use the semicolon,

Maple 6 uses
end proc;

they must end in end; and the operations inside are subject only to one level
of evaluation.

The procedure g, evaluated at one level, echos the detail that was typed in

> `eval(g);`

$$\mathbf{proc}(x)\,\mathbf{if}\,x < 0\,\mathbf{then}\,x - 1\,\mathbf{else}\,x^2\,\mathbf{fi}\,\mathbf{end}$$

Exploring the zero operand tells us that it is a procedure

> `op(0,eval(g));`

$$\textit{procedure}$$

The first operand is the parameter or argument x

> `op(1,eval(g));`

$$x$$

> `swap`

also uses the parameter or argument x

> `op(1,eval(swap));`

$$x$$

and local variables

> `op(2,eval(swap));`

$$a,\,b$$

Options in Procedures

Any symbol may be an option in a procedure but the following options
have a special meaning.

- 'Copyright 1998 by Monagan' Will not print the body of the procedure
  unless the interface variable verboseproc is 2 or greater.

- operator — Allows Maple to make simplifications to the procedure.

- arrow — Displays the procedure as an ->.

- remember — Maple remembers previous invocations of the procedure
  in a remember table and retrieves the result rather than executing the
  program again.

- system — Maple is allowed to remove entries from the remember table
  during garbage collection, an important part of Maple's management of
  its memory resource.

- decription — Appears as a header before any other part of the procedure. It is a way of attaching a one line comment to a procedure.

Redundant Entries in a Data Set

To demonstrate the use of a procedure and its associated remember table, consider Exercise 2.11. The procedure stalk( ) looks at a given argument and, if it is the same as the last previously used argument, it is rejected and a NULL sequence is output. The effect is that the element of the data set disappears.

The technique relates to forming and erasing remember tables. Note that, even without the remember option, a procedure can have a memory table. This is useful if you wish to override a calculation or 'force' a given value. It is even possible to make up for a singular point (discontinuity) in an expression. Specifically, taking 10 as the value of x and (44+a) as the corresponding value of $g(x)$, an assignment into the memory table of a procedure $g(x)$ is of the form

```
>   g:=proc(x) x+1 end;
```
$$g := \mathbf{proc}(x)\, x + 1 \,\mathbf{end\ proc}$$
```
>   g(10) := (44+a);
```
$$g(10) := 44 + a$$
```
>   op(4,eval(g)); # reveals the remember table
```
$$\text{table}([10 = 44 + a])$$

The entire memory table can be erased by using

```
>   forget(g);op(4,eval(g));
```

Alternately, the specific entry in the table is erased with the second argument to the command forget( ).

```
>   g(burserk) := pitlabo;op(4,eval(g));
```
$$g(burserk) := pitlabo$$
$$\text{table}([burserk = pitlabo])$$
```
>   forget(g,burserk);op(4,eval(g));
```
$$\text{table}([])$$

This takes us to the redundant data set. The extra entries need to be erased, without destroying the order. Again there are many ways of doing this, some with procedures, some without; some adequate, some not. This one is quite general and it may be possible to streamline it. Again, care should be taken with such procedures because they are recursive to some degree.

```
>   restart;
>   [179,204], [179,204], [176,204], [176,204], [174,204], [174,204],
>   [174,204], [173,204], [175,207], [175,207], [175,207], [175,207],
>   [179,212], [184,216], [188,218], [188,218], [192,219], [192,219],
>   [192,219], [194,218], [194,216], [192,214], [192,214], [189,212],
>   [186,209], [186,209], [182,207], [179,204]:

>   data:=[%];
```

$data :=$
[[179, 204], [179, 204], [176, 204], [176, 204], [174, 204], [174, 204], [174, 204],
[173, 204], [175, 207], [175, 207], [175, 207], [175, 207], [179, 212], [184, 216],
[188, 218], [188, 218], [192, 219], [192, 219], [192, 219], [194, 218], [194, 216],
[192, 214], [192, 214], [189, 212], [186, 209], [186, 209], [182, 207], [179, 204]]

```
>   forget(stalk); # removes any possible memory table
>   stalk:=proc(x) local rt,last;
>   if type(op(4,eval(stalk)),table) then
>   # check to see if there is a remember table
>   rt:=op(4,eval(stalk)); # the remember table
>   entries(rt);
>   last:=op(%); # the last entry
>   forget(stalk);
>   stalk([x]):=x; # the last entry is listed and the only entry
>   if last=x then NULL else x fi;
>   else
>   # put the first entry into rt
>   stalk([x]):=x; # entry into the remember table
>   x;
>   fi
>   end;
```

$$stalk := \mathbf{proc}(x)$$
$$\mathbf{local}\,rt,\ last;$$
$$\quad \mathbf{if}\,\mathrm{type}(\mathrm{op}(4,\ \mathrm{eval}(stalk)),\ table)\,\mathbf{then}$$
$$\quad\quad rt := \mathrm{op}(4,\ \mathrm{eval}(stalk))\,;$$
$$\quad\quad \mathrm{entries}(rt)\,;$$
$$\quad\quad last := \mathrm{op}(\%)\,;$$
$$\quad\quad \mathrm{forget}(stalk)\,;$$
$$\quad\quad \mathrm{stalk}([x]) := x\,;$$
$$\quad\quad \mathbf{if}\,last = x\,\mathbf{then}\,NULL\,\mathbf{else}\,x\,\mathbf{end\ if}$$
$$\quad \mathbf{else}\,\mathrm{stalk}([x]) := x\,;\ x$$
$$\quad \mathbf{end\ if}$$
$$\mathbf{end\ proc}$$

```
>   stalk(3);
```

3

```
>  op(4,eval(stalk));# the remember table
```
$$\mathrm{table}([[3] = 3])$$
```
>  stalk(11);
```
$$11$$

No Maple output
the second time

```
>  stalk(11);# same input as before
>  map(stalk,[1,2,2,2,3,4,5,1,1,1,6]);
```
$$[1, 2, 3, 4, 5, 1, 6]$$

The repetitions
disappear

```
>  op(4,eval(stalk));
```
$$\mathrm{table}([[6] = 6])$$
```
>  map(stalk,data);
```

$$[[179, 204], [176, 204], [174, 204], [173, 204], [175, 207], [179, 212], [184, 216],$$
$$[188, 218], [192, 219], [194, 218], [194, 216], [192, 214], [189, 212], [186, 209],$$
$$[182, 207], [179, 204]]$$
```
>  op(4,eval(stalk));
```
$$\mathrm{table}([[[179, 204]] = [179, 204]])$$

The x y pairs start and end with the same location, and there are no repetitions in the listlist.

A Recursive Procedure

Recursion, or a procedure that calls itself, it difficult to comprehend. But Maple procedures do it. Recursion is an effective way of 'filling-in' or 'digging-down', completing repetitive operations, as required. Take care, and it will work.

```
>  restart;

>  fat:=2;
```
$$fat := 2$$
```
>  fib:= proc(n::nonnegint)
>  description 'Recursive memory disaster';
>  global fat;
>  option remember, operator, arrow;
>  fib(n-1)+fib(n-2)+fat;
>  end;
```
$$\textit{fib} := n{::}nonnegint \rightarrow \mathrm{fib}(n-1) + \mathrm{fib}(n-2) + \textit{fat}$$

```
>  fib(0):= 3;
```

$$\mathrm{fib}(0) := 3$$

```
>  fib(1):= 5;
```

$$\mathrm{fib}(1) := 5$$

```
>  op(3,eval(fib)); # the options
```

remember, operator, arrow

```
>  op(4,eval(fib)); # the remember table
```

$$\mathrm{table}([$$
$$0 = 3$$
$$1 = 5$$
$$])$$

```
>  op(5,eval(fib)); # the description
```

Recursive memory disaster

```
>  op(6,eval(fib)); # the global variable
```

fat

```
>  fib(9);
```

$$341$$

```
>  op(4,eval(fib));
```

$$\mathrm{table}([$$
$$3 = 17$$
$$4 = 29$$
$$5 = 48$$
$$6 = 79$$
$$7 = 129$$
$$0 = 3$$
$$8 = 210$$
$$1 = 5$$
$$9 = 341$$
$$2 = 10$$
$$])$$

Exercise 3.3 Fib Table

Try to figure out what the above procedure is doing. Note that the remember table values were assigned during the assignments of values to fib (or, really, its remember table) but the calculation inside fib is recursive; fib completes the recursive addition, adding in the outside variable fat. Change the value of fat and have some more fun.

procname, args, nargs

Internal to the procedure, you can involve the special parameters procname, args and nargs to give details about the procedure; they are, respectively, the name of the procedure, the arguments of the procedure and the number of arguments.

```
>  restart:

>  glug := proc()
>  local x,n,na,newlist;
>  if nargs>1 then
>  x := [args]; na:=nargs;
>  n:=rand() mod na + 1;
>  newlist := subsop(n=NULL,x);
>  op(map(y->(', '.y),[op(2..na,x)]));
>  cat('Of the ',nargs,' fruits; ',op(1,x),%,'; ',
>  'procname',' has eaten the ',x[n],' leaving only the ');
>  print(%);
>  op(newlist);
>  else 'nothing left' fi; end;
```

The extra coding makes Glug( ) responsive, pretty printout

$$glug := \mathbf{proc}()$$
$$\mathbf{local}\, x,\, n,\, na,\, a,\, newlist;$$
$$\quad \mathbf{if}\, 1 < \text{nargs}\, \mathbf{then}$$
$$\quad\quad na := \text{nargs};$$
$$\quad\quad n := (\text{rand}() \bmod na) + 1;$$
$$\quad\quad x := [\text{args}];$$
$$\quad\quad a := x_n;$$
$$\quad\quad newlist := \text{subsop}(n = NULL, [\text{args}]);$$
$$\quad\quad \text{op}(\text{map}(y \to ', '.y, [\text{op}(2..na, x)]));$$
$$\quad\quad \text{cat}('Of\ the\ ', \text{nargs}, '\ fruits;\ ', \text{op}(1, x), \%, ';\ ', \text{procname}, '\ has\ eaten\ the\ ', a,$$
$$\quad\quad\quad '\ leaving\ only\ the\ ');$$
$$\quad\quad \text{print}(\%);$$
$$\quad\quad \text{op}(newlist)$$
$$\quad \mathbf{else}\, 'nothing\ left'$$
$$\quad \mathbf{fi}$$
$$\mathbf{end}$$

Maple 6 uses || for cancatenation

```
>  glug(apple,peach,orange);
```
Of the 3 fruits; apple, peach, orange; glug has eaten the apple leaving only the peach, orange

```
>  glug(%);
```
Of the 2 fruits; peach, orange; glug has eaten the peach leaving only the orange

```
>  glug(%);
```
nothing left

3.5   Dots, Arrows and Fill

As seen above, the view you have of a function or procedure is quite different if presented in points as a sample of the full curve or as a line. Some judgement is required as to what sort of line to draw. The presentation of the information contained in the function colours how you see it or the value you place in it. Some massage is generally necessary.

Points and Lines

Consider a square of single points, given as a list of lists.

```
>  pts := [[2,2],[2,5],[5,5],[5,2]];
```
$$pts := [[2, 2], [2, 5], [5, 5], [5, 2]]$$

```
>  plot(pts,0..7,0..7,thickness=4);  # only a range specified
```

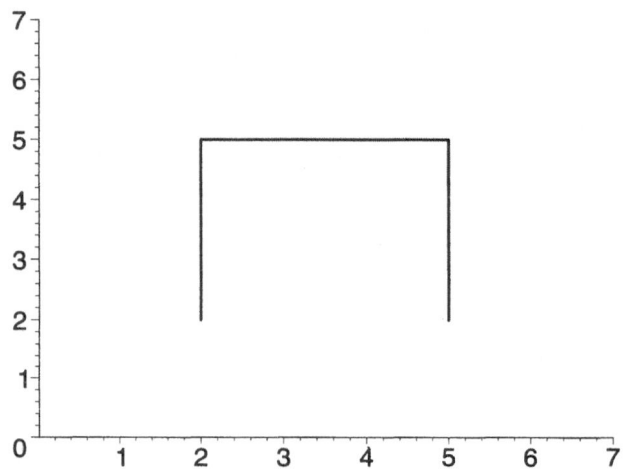

```
>  pp:=plot(pts,0..7,0..7,style=POINT):
```
With this simple plot of points, the symbols are too small to see.
```
>  op(1,pp);
```
CURVES([[2., 2.], [2., 5.], [5., 5.], [5., 2.]], COLOUR(RGB, 1.0, 0, 0))
```
>  with(plottools);
```

[*arc, arrow, circle, cone, cuboid, curve, cutin, cutout, cylinder, disk, dodecahedron, ellipse, ellipticArc, hemisphere, hexahedron, homothety, hyperbola, icosahedron, line, octahedron, pieslice, point, polygon, project, rectangle, reflect, rotate, scale, semitorus, sphere, stellate, tetrahedron, torus, transform, translate, vrml*]

```
> PTS := proc(pts::listlist,r,cc,pp)
> local ss,p,opts;
> ss := NULL;
> for p in pts do;
> opts:=colour=cc;
> ss := ss,plottools[disk](p,r,opts)
> od;
> subsop(1=ss,pp);# Substitute for the first operand
> end;
```

see page 153
about subsop( )

$PTS := \mathbf{proc}(pts::listlist, r, cc, pp)$
$\mathbf{local}\ ss, p, opts;$
$\quad ss := NULL;$
$\quad \mathbf{for}\ p\ \mathbf{in}\ pts\ \mathbf{do}\ opts := colour = cc;\ ss := ss, plottools_{disk}(p, r, opts)\ \mathbf{od};$
$\quad subsop(1 = ss, pp)$
$\mathbf{end}$

```
> macro(teal=COLOR(RGB,56/255,142/255,142/255));
> PTS(pts,0.1,teal,pp);
```

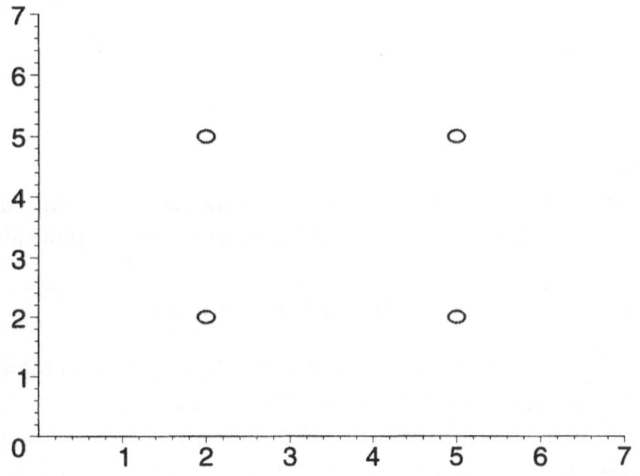

Fill

The PLOT Structures are many; see section 3.7, page 150 ; some are FONT driven; some are POINT driven; some are CURVES driven; some are POLYGON driven. The FONT and POINT arguments have no real dimensions whereas the CURVES are one dimensional and the POLYGONS are two dimensional. A filled structure must have a two dimensional quality and often a listlist of points will do; then substitution of CURVES for POLYGONS will fill in the figure.

```
>  lprint(plot(pts,0..7,0..7));
```

```
PLOT(CURVES([[2., 2.], [2., 5.], [5., 5.], [5.,
2.]],COLOUR(RGB,1.0,0,0)),AXESLABELS(``,``),VIEW(0 .. 7.,0 .. 7.))
```

```
>  lprint(subs(CURVES=POLYGONS,%));
```

```
PLOT(POLYGONS([[2., 2.], [2., 5.], [5., 5.], [5.,
2.]],COLOUR(RGB,1.0,0,0)),THICKNESS(4),AXESLABELS(``,``),VIEW(0 ..
7.,0 .. 7.))
```

```
>  %;
```

Overlaying Plots

As used above, display allows one to mix plots. One needs to bring in the package plots to use it. Logically, the shape and scales on the two plots should be the same.

```
>  with(plots);    # brings in the plots library
```

[animate, animate3d, animatecurve, changecoords, complexplot, complexplot3d, conformal, contourplot, contourplot3d, coordplot, coordplot3d, cylinderplot, densityplot, display, display3d, fieldplot, fieldplot3d, gradplot, gradplot3d, implicitplot, implicitplot3d, inequal, listcontplot, listcontplot3d, listdensityplot, listplot, listplot3d, loglogplot, logplot, matrixplot, odeplot, pareto, pointplot, pointplot3d, polarplot, polygonplot, polygonplot3d, polyhedra_supported, polyhedraplot, replot, rootlocus, semilogplot, setoptions, setoptions3d, spacecurve, sparsematrixplot, sphereplot, surfdata, textplot, textplot3d, tubeplot]

```
>  G := plot(tan,-Pi..Pi,-Pi..Pi):
```

```
>  pts := [[-3,3],[-2,2],[-1,1],[0,0],[1,-1],[2,-2],[3,-3]];
```
$$pts := [[-3, 3], [-2, 2], [-1, 1], [0, 0], [1, -1], [2, -2], [3, -3]]$$

```
>  H := plot(pts,-Pi..Pi,-Pi..Pi,style=POINT,symbol=box):

>  plots[display]({G,H},view=[-5..5,-4..4],title='Two Plots',
>  titlefont=[TIMES,BOLD,16]);
```

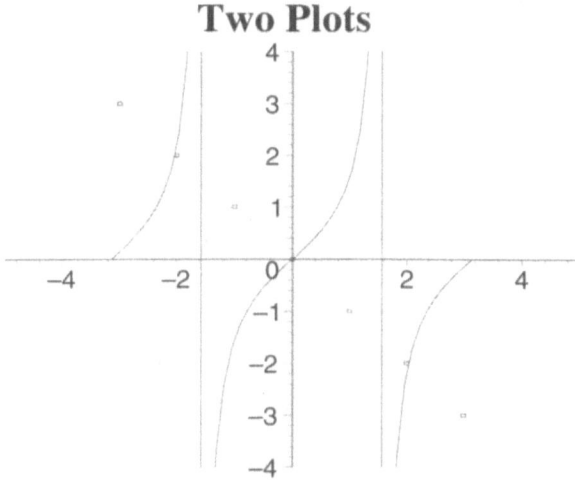

Polar Plots

A parametric plot is best represented as two functions of a parameter driving the x y coordinates. The parameter is usually considered time and the x y positions could specify the location on a planar map. Such a graphing technique is good for curves that double back on themselves. Note that the plot of a semicircle produces a part of an ellipse without scaling=CONSTRAINED.

```
>  plot([sin(t),cos(t),t=0..Pi], colour=khaki):  subs(CURVES=POLYGONS,%);
>  plot([x,x,x=0..3*Pi],-8..8,-10..10,coords=polar);
```

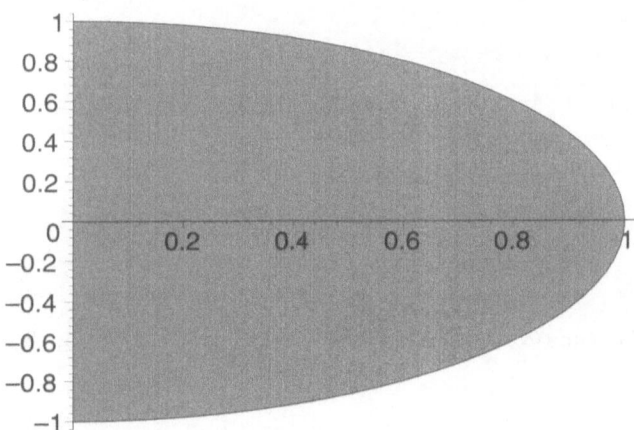

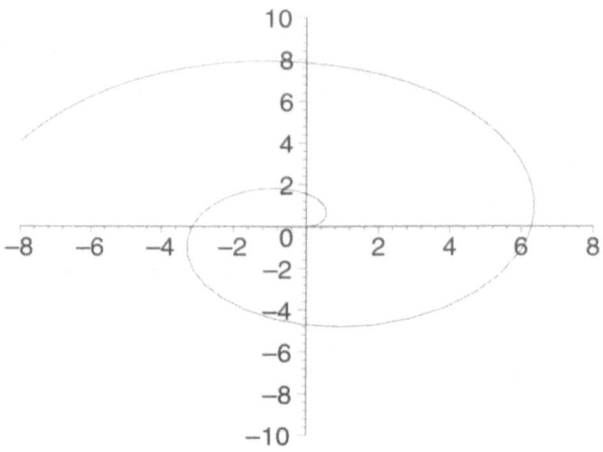

```
> plot([exp(sin(t)),exp(cos(t)),t=-Pi..Pi],colour=plum):
> subs(CURVES=POLYGONS,%);
```

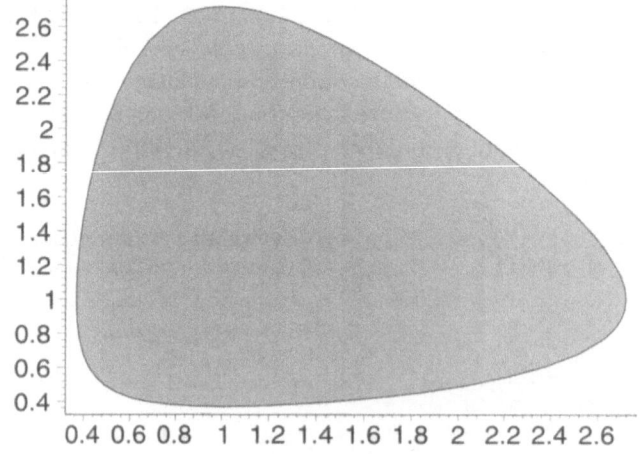

```
> plot({seq([i*cos(x),i*sin(x),x=0..2*Pi],i=1..3)},-5..5,-5..5);
```

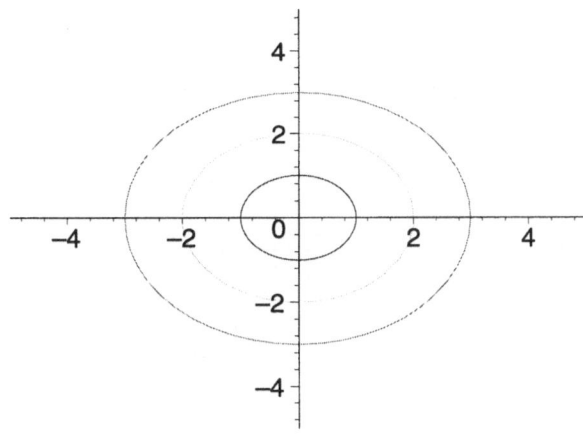

```
>   a := 1/2 :   b := 1/2 :   plot({seq([a*ln(r),t,t=0..2*Pi],r=1..5)},
>   -1..1,-1..1,coords=polar);
```

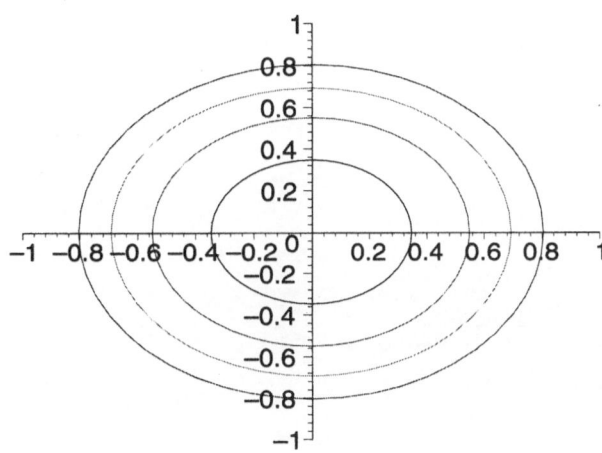

Change to x,y coordinates:

```
>   a := 1/4 :   b := 1/4 :
>   plot({seq([a*ln(r)*cos(t)+b*ln(r)*(cos(t)+.5),
>   a*ln(r)*sin(t)+b*ln(r)*sin(t),t=0..2*Pi],r=1..5)},
>   -1..1,-1..1);
```

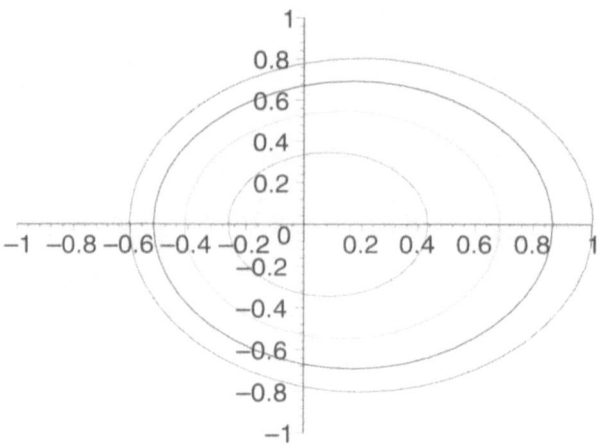

Straight Lines in Polar Coordinates

```
>   plot({[2,x,x=0..3*Pi],[4,x,x=0..3*Pi],[x,Pi/4,x=3..5],
>   [3*cos(Pi/4)/cos(x),x,x=Pi/4..Pi/3.5],
>   [3*sin(Pi/4)/sin(x),x,x=Pi/4..Pi/4.4]},
>   -4..4,-5..5,coords=polar,axes=NONE,thickness=2);
```

This reversal of presentation illustrates that straight lines can be drawn in a polar coordinate system.

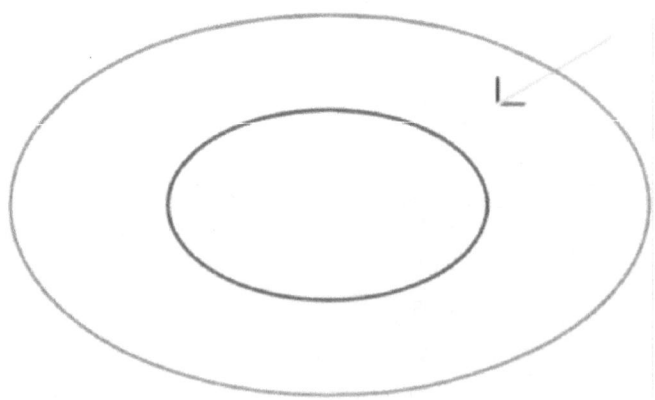

An Arrow Procedure

```
>  restart: with(linalg):
```

Make an arrow using linalg

```
Warning, new definition for norm
```

```
Warning, new definition for trace
```

```
>  arrow := proc(x,y,theta,l)
>  local dd,d1,d2,d3,p1,p2,p3,p4,p5;
>  vector([x,y]);
>  # presumes tip angle +- 20degs
>  d1:= vector([l*cos(theta)/2,
>  l*sin(theta)/2]);
>  p1:= matadd(%%,d1);
>  p2:= matadd(%%%,d1,1,-1);
>  # tail and head
>  dd := 1/5;
>  # barb measurement
>  theta-Pi/9;
>  d2:=vector([dd*cos(%),
>  dd*sin(%)]);
>  p3:= matadd(p2,d2);
>  # barb one
>  p4:= matadd(p2,d1,1,1/4);
>  # centre cross
>  theta+Pi/9;
>  d3:=vector([dd*cos(%),dd*sin(%)]);
>  # barb two
>  p5:= matadd(p2,d3);
>  map(convert,[p1,p2,p3,p4,p5,p2],list);
>  evalf(%);
>  end:
```

```
>  pp := arrow(1,1,Pi/4,1/2);
```

$$pp := [[1.176776695, 1.176776695], [.8232233047, .8232233047],$$
$$[.9138540834, .8654851309], [.8674174786, .8674174786],$$
$$[.8654851309, .9138540834], [.8232233047, .8232233047]]$$

```
>  plot(pp,thickness=3);
```

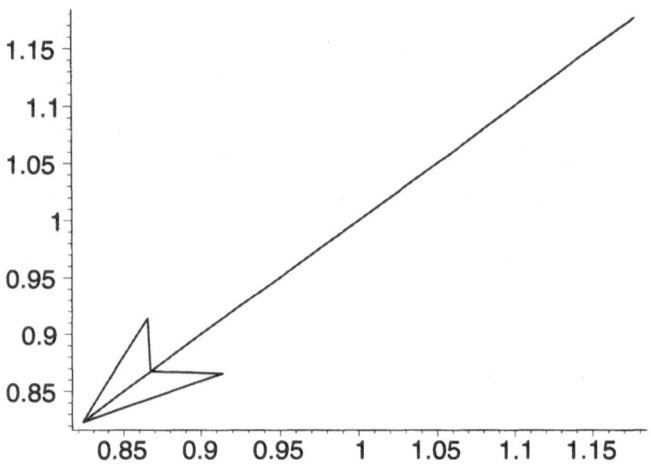

Circles and Arrows

```
>  i := 'i':
```

This is a problem   You use i as an index in loops; later, it is defined when reused.

```
>  [l*cos(t),l*sin(t),t,1/2]:op(subs({t=2*Pi*i/8,l=.5},%));
```

$$.5\cos(\frac{1}{4}\,\pi\,i),\ .5\sin(\frac{1}{4}\,\pi\,i),\ \frac{1}{4}\,\pi\,i,\ \frac{1}{2}$$

```
>  arrows1 := seq(arrow(%),i=1..8):
```

```
>  [l*cos(t),l*sin(t),t,1/4]:
>  op(subs({t=2*Pi*(i+1/2)/8,l=1.0},%));
```

$$1.0\cos(\frac{1}{4}\,\pi\,(i+\frac{1}{2})),\ 1.0\sin(\frac{1}{4}\,\pi\,(i+\frac{1}{2})),\ \frac{1}{4}\,\pi\,(i+\frac{1}{2}),\ \frac{1}{4}$$

```
>  arrows2 := seq(arrow(%),i=1..8):
```

```
>  circles := seq([i*cos(x)/2,i*sin(x)/2,x=0..2*Pi],i=1..2):
```

```
>  plot({arrows1,arrows2,circles},-1..1,-1..1,
>  axes=NONE,color=blue,thickness=1);
```

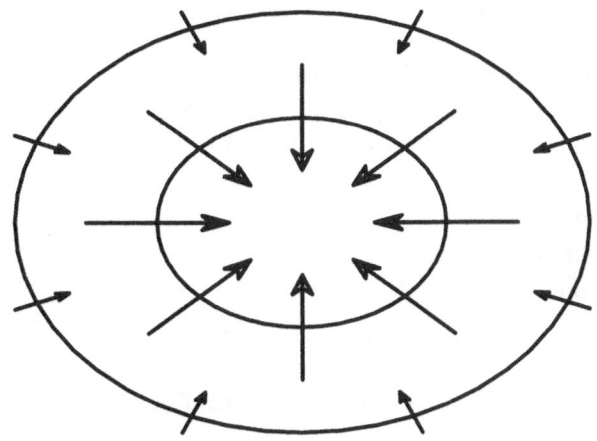

Exercise 3.4 Infinity Plot

This is an interesting concept. Is it really possible to plot something with an infinite size? Maple does it. The procedure uses an arctan conversion of the axis. Certainly useful if you want to look at something in a gross sense! Consider the following list of expressions:

```
>   [(x+x^2)/(x-1),x/(sqrt(x)-1),x^x-1,1/(x^x-1)];;
```

Plot them with the given command. Examine the singular points, the points of discontinuity.

```
>   plot(%,x=0..infinity,y=-infinity..infinity);
```

Exercise 3.5 Sun Icon

Define a procedure to make a quarter circle of radius 4 joined with two rays from 4 to 6 at 45 and 135 degrees. Plot the function. Note the use of the package plottools.

```
>   with(plottools):
>   pieslice([0,0],4,Pi/4, colour=brown):
>   plots[display](%, scaling=CONSTRAINED);
```

3.6   plot Options

Usually plot( ) or plot3d( ) generate either a PLOT( ) Structure or a PLOT3D( ) Structure; these Structures are displayed on evaluation or jointly with a call to display( ). In the following listing of options, the earlier listings are particular to plot( ) and plot3d( ) calls; the options are in lower case. Note that options may be globally set with setoptions( ) or setoptions3d( ) from the plots package.

The later listings relate to PLOT Structures. Importantly, the component calls in PLOT Structures are in upper case. PLOT Structures are ready to plot and may be easy to manipulate. See section 3.7, page 150.

```
>   ?plot,options
>   ?plot,structure
>   ?plots[setoptions]
>   ?plots[setoptions3d]

>   plot(real_function,abscissa=1..10,ordinate=0..5,Options)
```

All but the real function (or procedure) are optional. The Options are separated by commas; these are of the form option = value.

- adaptive=false disables the use of adaptive plotting.

  This means that Maple will not adjust the number of plotting points as the curves become more difficult.

BOXED is an old version of BOX

- axes=FRAME, BOX, NORMAL, or NONE

AXES in an old version of AXESSTYLE

AXESSTYLE(BOX)

- axesfont=[TIMES,ROMAN,12], [TIMES,BOLD,14], [TIMES,ITALIC,16], [TIMES,BOLDITALIC,14], [COURIER,BOLD,10], [COURIER,OBLIQUE,12], [COURIER,BOLDOBLIQUE,12], [HELVETICA,BOLD,10], [HELVETICA, OBLIQUE,12], [HELVETICA,BOLDOBLIQUE,12], or [SYMBOL,20]

- AXESLABELS("x","y")
  AXESLABELS("along","up",FONT(TIMES,BOLD,10))

Maple 6 syntax

AXESLABELS(x,y,FONT(TIMES,ROMAN,10),VERTICAL,HORIZONTAL)

- The tick marks on the coordinate axes

  AXESTICKS([1,2,5,10],[1='Jan', 5='May'])
  AXESTICKS(DEFAULT,DEFAULT,FONT(COURIER,BOLD,10))

- color=tan or colour=pink or color=yellow   ?plot,colour or ?plot[color]

aquamarine	black	blue	navy	coral	cyan	
brown	gold	green	gray	grey	khaki	
magenta	maroon	orange	pink	plum	red	
sienna	tan	turquoise	violet	wheat	white	yellow

```
> macro(palegreen=COLOR(RGB, .5607, .7372, .5607));
> macro(skyblue = COLOR(RGB, 0.1960, 0.6000, 0.8000));
> macro(randcolor = COLOUR(HUE, rand()/10^12));
> macro(gray = COLOUR(RGB, .8$3));
> macro(randgray = COLOUR(RGB, rand()/10^12$3));
> color=COLOR(RGB, rand()/10^12, rand()/10^12, rand()/10^12)
> color=COLOUR(RGB, .3, 0, 0)
> color=COLOUR(HUE, .5)
> color=COLOUR(HSV, .5, .9, 1)
```

COLOR(RGB,0,0.7,0.7) COLOR(HUE,.6)
COLOR(HSV,.77,.7,1) is a purple

- coords=polar (default is cartesian) bipolar, cardioid, cassinian, elliptic, hyperbolic, invcassinian, invelliptic, logarithmic, logcosh, maxwell, parabolic, polar, rose, and tangent.

  See ?coords for a full description of each of these coordinate systems.

- discont=true or discont=false

  true forces plot to call the function discont or fdiscont to locate discontinuities in the input. Plot then breaks the horizontal axis into appropriate intervals where there is continuity.

- font=[TIMES,ROMAN,12], [TIMES,BOLD,14], [TIMES,ITALIC,16], [TIMES, BOLDITALIC,14], [COURIER,BOLD,10], [COURIER,OBLIQUE,12], [COURIER, BOLDOBLIQUE,12], [HELVETICA,BOLD,10], [HELVETICA,OBLIQUE,12], [HELVETICA,BOLDOBLIQUE,12], or [SYMBOL,20]

  Fonts used for TEXT or titles of graphs

  FONT(TIMES,BOLD,10)

- labels = ["Metres into the Game","Speed"]

Defaults to the name (names) of the argument (arguments) in the function to be plotted.

LABELS is an old version of AXESLABELS

AXESLABELS("along","up",FONT(TIMES,BOLD,10))

- labeldirections = [horizontal,vertical]

Dictates how the name lies along the axes, defaults to horizontal. The letters are not 'stacked'.

Maple 6 syntax

AXESLABELS(x,y,FONT(TIMES,ROMAN,10),VERTICAL,HORIZONTAL)

- legend = [curve1,curve2,curve3,curve4]
  legend = "A Effect from the Gecko"

Stacks the legend for the curves below the graph under the word Legend. The listed curves correspond to the legend list.
plot([sin,cos],0..10,legend=[sin,cos]); # Maple 6
PLOT(CURVES([[.5,-3],[6,10],[-1,3]],COLOUR(RGB,0,0,1),
LEGEND("Standard")),SYMBOL(CROSS),STYLE(POINT)); Maple 6
LEGEND(sin) is one of the calls inside CURVES( )

- labelfont=[TIMES,ROMAN,12], [TIMES,BOLD,14],[TIMES,ITALIC,16],
  [TIMES,BOLDITALIC,14], [COURIER,BOLD,10], [COURIER,OBLIQUE,12],
  [COURIER,BOLDOBLIQUE,12], [HELVETICA,BOLD,10], [HELVETICA,
  OBLIQUE,12], [HELVETICA,BOLDOBLIQUE,12], or [SYMBOL,20]

These are the fonts used for the names of the axes.

- linestyle=1   for solid line
  linestyle=2   for dot
  linestyle=3   for dash
  linestyle=4   for dash-dot

LINESTYLE(2)

- numpoints=100

Minimum points to be generated. Adaptive plotting generates more points anyway. Default is 50.

PLOT Structures PLOT( ) and PLOT3D( ) See section 3.7

```
>   ?plot,structure
```
These structures include
POINTS([1,1],SYMBOL(CIRCLE))
POINTS([[0,0][0.5,-3][6,10][-1,3]],SYMBOL(CROSS))–not Maple 6
CURVES([[1,1],[0,1],[0,0]],THICKNESS(3),LINESTYLE(4))
POLYGONS([0,0][1,0][1,1][0,2][-1,1][-1,0],STYLE(LINE),THICKNESS(2))
TEXT([10,3],"soapiness",ALIGNLEFT,ALIGNBELOW)

plot(sin, 0..10, 0..1);
plot(sin(Russia), Russia=0..2*Pi, -1..1); # named abscissa
plot(sin, -10..10, Boabs=0..1); # named ordinate
plot([[0,0][0.5,-3][6,10][-1,3]], symbol=DIAMOND, style=POINT)

- Range–see view

- resolution=400

  The display resolution in pixels (the default is n = 200). Used in the adaptive plotting scheme.

- sample=[1,2,5,10,20,50,100]

  Initial parameter values of the function(s). When adaptive=false, this controls the arguments used for the function.

- scaling=CONSTRAINED or scaling=UNCONSTRAINED (the default)

  When unconstrained, Maple adjusts the length of the x and y axes.

  SCALING(CONSTRAINED)

- style=LINE   interpolates between points
  style=POINT   plots points only
  style=PATCH   uses patch style for polygons
  style=PATCHNOGRID   patch style without grid lines

  STYLE(POINT)

- symbol=BOX, CROSS, CIRCLE, POINT, and DIAMOND.
  symbolsize=20

  SYMBOL(DIAMOND)
  SYMBOL(BOX,20)

  TEXT([2,1],"Magnetic Dinosaurs",ALIGNLEFT,
  FONT(TIMES,ROMAN,12),COLOUR(RGB,.7,.8,0))

- thickness= 0, 1, 2, or 3   0 is the default thickness.

  THICKNESS(2)

- tickmarks=[5,10] or tickmarks=['default','default']

  AXESTICKS(5,10)
  AXESTICKS([1,2,3,4,5],[0=none.2=couple,3=few])

- title='The Murder of Worm'

  TITLE("Earwigs at Rest",FONT(TIMES,ROMAN,12))
  TEXT([0,0.5],'Dotted',ALIGNBELOW)
  Allows for plotting text at a given location

- titlefont=[TIMES,BOLD,14], [TIMES,ITALIC,16], [TIMES,BOLDITALIC,14],
  [COURIER,BOLD,10], [COURIER,OBLIQUE,12], [COURIER,BOLDOBLIQUE,12],
  [HELVETICA,BOLD,10], [HELVETICA,OBLIQUE,12], [HELVETICA,
  BOLDOBLIQUE,12], or [SYMBOL,20]

  Font for the title of the plot.

- view=[0..10, 1..5]

  Minimum and maximum coordinates of the curve. The default is the
  entire curve.

  VIEW(0..10,20..3)

- xtickmarks=5, ytickmarks=10 or
  xtickmarks=[1,3,5,10], ytickmarks=[1=small,5=medium,10=big]
  xtickmarks=[1/2=Jan,3/2=Feb,5/2=Mar,7/2=Apr,9/2=May,11/2=Jun]
  xtickmarks=[13/2=Jul,15/2=Aug,17/2=Sep,19/2=Oct,21/2=Nov,23/2=Dec]

AXESTICKS(5,10,FONT(COURIER,HELVETICA,10)
AXESTICKS([3='1 Mar',6='1 Jun',9='1 Sep'],DEFAULT),TEXT([5,.2],"1999")

Plotting to a file

The plotoutput must be set as well as the orientation, colour, size and border characteristics of the plot.

Visit pages 53 and 93

```
>  ?plot,options
>  ?plot,device,ps
>  ?setoptions
>  ?setoptions3d
```

setoptions is part
of the plots
package

```
>  plotsetup(help);
```

Usage plotsetup(device,options).

The following device types are known:

PostScript, X11, char, cps, default, dumb, gif, hpgl, hplj, inline,
jpeg, laserjet, pcx, postscript, ps, tek, tektronix, window, x11,
xwindow

Please see ?plotsetup for more information

```
>  plotsetup(jpeg,plotoutput="c:\plot.jpg",plotoptions="portrait,noborder");
```

```
>  plotsetup(ps,plotoptions="shrinkby=.7,colour=rgb,portrait,noborder");
```

This shrinkby
makes the plot
30% of the
original

Note that options for plotting may be changed globally with plots[setoptions]. The above setup is for one file which takes the default name postscript.eps Further plotting overwrites the file. The default orientation of the file may be landscape; it is best to include portrait as one of the plotoptions. The file should appear in the current directory on your main drive c: Importantly, the plot will not appear on the screen. Normally, one would follow this command with a return to the default.

```
>  plotsetup(default);
```

3.7   PLOT Structures

Underlying every Maple plot is a Plot Structure. As with most elements of
Maple, they are best shown by example. Go to the help file. One thing you
find is plottools. Make a POLYGON PLOT Structure.

```
>  with(plottools);
```

Warning, new definition for arrow

[*arc, arrow, circle, cone, cuboid, curve, cutin, cutout, cylinder, disk, dodecahedron,
ellipse, ellipticArc, hemisphere, hexahedron, homothety, hyperbola, icosahedron,
line, octahedron, pieslice, point, polygon, project, rectangle, reflect, rotate, scale,
semitorus, sphere, stellate, tetrahedron, torus, transform, translate, vrml*]

```
>  polygon([[0,0],[3,4],[3,2],[2.2,1.5],[3,1.4]], color=green,
>  thickness=3);
```

$$POLYGONS([[0, 0], [3., 4.], [3., 2.], [2.2, 1.5], [3., 1.4]],$$
$$COLOUR(RGB, 0, 1.00000000, 0), THICKNESS(3))$$

```
>  plots[display](%);
```

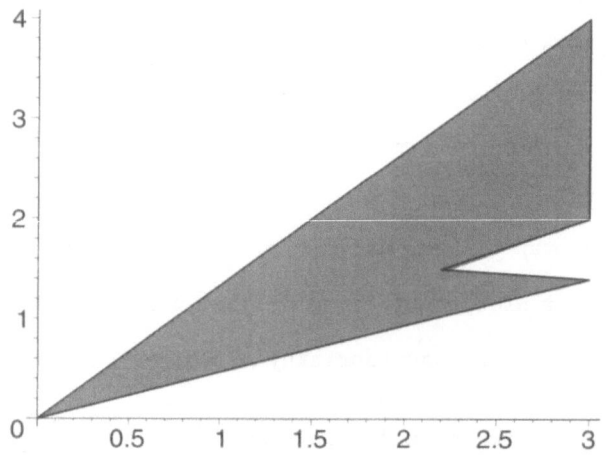

The workings of the plot are contained in the POLYGONS PLOT Structure. The structures are best always spelled with capitol letters though they
also work with only small letters. They can be typed in directly, as below, but
this is not recommended; it is easy to make errors.

```
>   TITLE('Faramir\'s Revenge',FONT(TIMES,ROMAN,20)),
>   AXESLABELS('Crikle','Gog',
>   FONT(COURIER,OBLIQUE,20)),
>   AXESTICKS([1,2,3,4,5],[1='a little',3='3',
>   5='something',8='whatever']),
>   AXESSTYLE(NONE),

>   CURVES([[-3,-3],[13,-3],[13,13],[-3,13],[-3,-3]],
>   COLOUR(RGB,1,0,0)),
>   POINTS([[0,1],[1,2],[3,4],[4,5],[6,7],[8,9]],
>   VIEW(-2..5,-3..10)),
>   POLYGONS([[0,0],[6,6],[6,12]],
>   STYLE(LINE),THICKNESS(3),COLOUR(RGB,0,1,0)),
>   TEXT([1.2,1],'Help',ALIGNRIGHT,ALIGNABOVE,
>   FONT(COURIER,OBLIQUE,20),THICKNESS(2)):
>   [%]:# This sequence of things is converted to a table.
```

CURVELIST is an old version of CURVES

In tackling these structures, it is best to remember that they have evolved through the efforts of many individuals to suit a number of purposes; these purposes are varied and interwoven. Some structures fit one into the other; none necessarily have a need for all the options, but they all have some requirements for their arguments. Some, by their very nature, are incompatible with others. Remember that the word DEFAULT can take the place of many elements.

```
>   %[1];
```
TITLE(*Faramir's Revenge*, FONT(*TIMES*, *ROMAN*, 20))
```
>   p := %%;
```

$p :=$ [TITLE(*Faramir's Revenge*, FONT(*TIMES*, *ROMAN*, 20)),
AXESLABELS(*Crikle*, *Gog*, FONT(*COURIER*, *OBLIQUE*, 20)),
AXESTICKS([1, 2, 3, 4, 5], [1 = *a little*, 3 = 3, 5 = *something*, 8 = *whatever*]),
AXESSTYLE(*NONE*),
CURVES([[−3, −3], [13, −3], [13, 13], [−3, 13], [−3, −3]], COLOUR(*RGB*, 1, 0, 0)),
POINTS([[0, 1], [1, 2], [3, 4], [4, 5], [6, 7], [8, 9]], VIEW(−2..5, −3..10)),
POLYGONS([[0, 0], [6, 6], [6, 12]], STYLE(*LINE*), THICKNESS(3),
COLOUR(*RGB*, 0, 1, 0)), TEXT([1.2, 1], *Help*, *ALIGNRIGHT*, *ALIGNABOVE*,
FONT(*COURIER*, *OBLIQUE*, 20), THICKNESS(2))]

```
>  with(plots);
```
The plots package is needed for display and more, now

```
>  display(p[7],PLOT(p[1],p[2],p[3],p[5],p[8]));
```

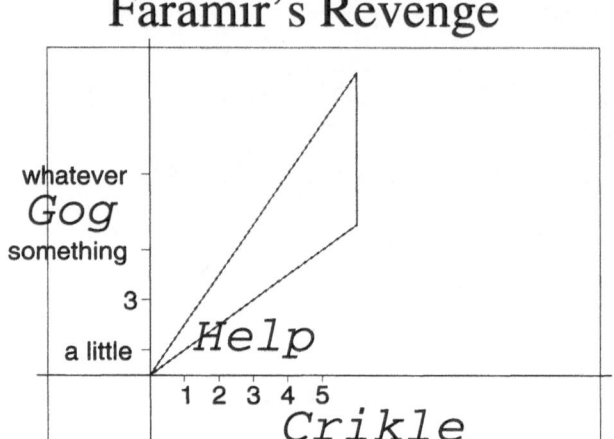

3.7.1   Direct Editing

Suppose you have a PLOT Structure and you don't like the look of it. Direct editing and the use of subsop can either remove the offending part , alter it or allow you to add to it.

```
>  plot([[1,1],[2,2],[2,1],[1,1]],style=line);
```

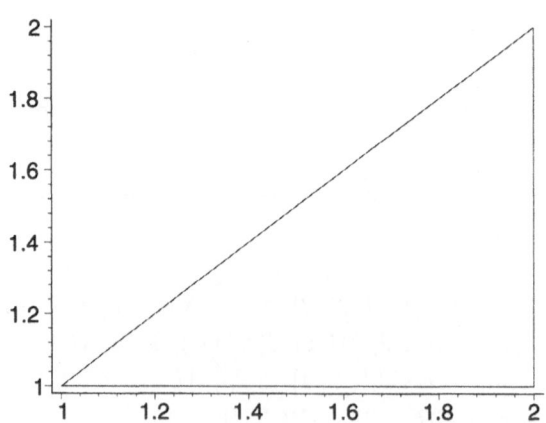

```
>  P :=%;
```

$P := \text{PLOT}(\text{CURVES}([[1., 1.], [2., 2.], [2., 1.], [1., 1.]], \text{COLOUR}(RGB, 1.0, 0, 0)),$
$\text{AXESLABELS}(,), \text{STYLE}(LINE), \text{VIEW}(DEFAULT, DEFAULT))$

The principle of 'Last Name Evaluation' leaves the right hand side not fully evaluated and you see the 'guts' of the PLOT Structure. Highlight it and copy it to the next execution group. The # has been added so it is inactive.

See page 227 of Monagan's book [24]

```
>  #INTERFACE_PLOT(CURVES([[1., 1.], [2., 2.], [2., 1.], [1., 1.]],
   COLOUR(RGB, 1.0, 0, 0)), STYLE(LINE), AXESLABELS('','''), VIEW(DEFAULT,
   DEFAULT))
```

Now, edit it as you see fit.

Remember to use UPPER CASE

```
>  INTERFACE_PLOT(CURVES([[1, 1], [2, 2], [2, 1.5], [1.8, 1.4],
   [2, 1.3], [2, 1], [1, 1]], COLOUR(RGB, 1.0, 0, 1)), STYLE(LINE),
   AXESLABELS('Hi', 'Chow'), VIEW(DEFAULT, DEFAULT));
```

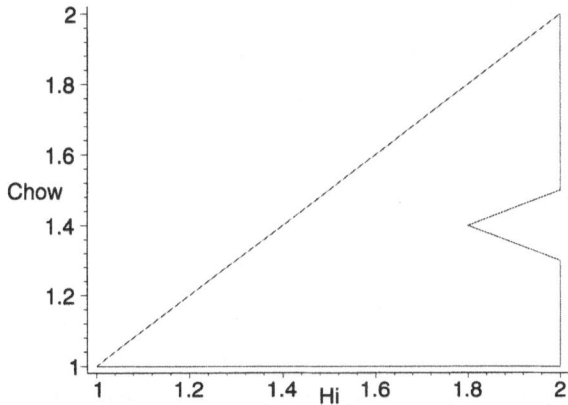

3.7.2 Using subsop

Alternatively, subsop can 'slip in ' the structure you want.

```
>  op(0,%);
```

$$\text{PLOT}$$

```
>  op(1,%%);
```

$\text{CURVES}([[1, 1], [2, 2], [2, 1.5], [1.8, 1.4], [2, 1.3], [2, 1], [1, 1]],$
$\text{COLOUR}(RGB, 1.0, 0, 1))$

```
>  op(2,%%%);
```

$$\text{STYLE}(LINE)$$

The line print version can also be used as a reference. It will not reveal the parsing Maple has done.

```
> lprint(P);
```

```
PLOT(CURVES([[1., 1.], [2., 2.], [2., 1.], [1.,
1.]],COLOUR(RGB,1.0,0,0)),AXESLABELS('','),STYLE(LINE),VIEW(DEFAULT,D
EFAULT))
```

The word
COLOUR must
be used

```
> subsop(1=CURVES([
> seq([2*i/30,i/30],i=1..30),
> seq([2,i/20],i=1..20)],
> COLOUR(RGB,0,0,1)),2=STYLE(POINT),P);
```

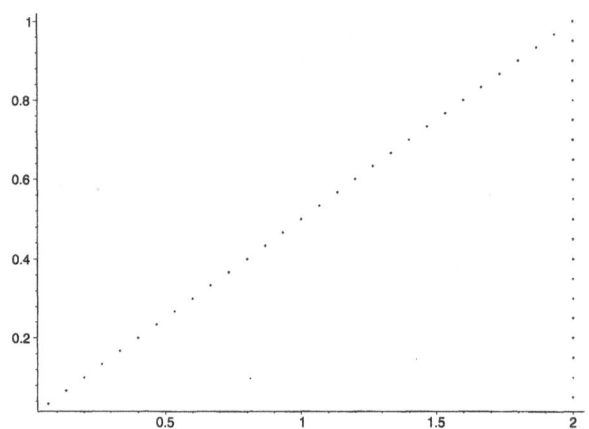

```
> lprint(%);
```

```
PLOT(CURVES([[1/15, 1/30], [2/15, 1/15], [1/5, 1/10], [4/15, 2/15],
[1/3, 1/6], [2/5, 1/5], [7/15, 7/30], [8/15, 4/15], [3/5, 3/10], [2/3,
1/3], [11/15, 11/30], [4/5, 2/5], [13/15, 13/30], [14/15, 7/15], [1,
1/2], [16/15, 8/15], [17/15, 17/30], [6/5, 3/5], [19/15, 19/30], [4/3,
2/3], [7/5, 7/10], [22/15, 11/15], [23/15, 23/30], [8/5, 4/5], [5/3,
5/6], [26/15, 13/15], [9/5, 9/10], [28/15, 14/15], [29/15, 29/30], [2,
1], [2, 1/20], [2, 1/10], [2, 3/20], [2, 1/5], [2, 1/4], [2, 3/10],
[2, 7/20], [2, 2/5], [2, 9/20], [2, 1/2], [2, 11/20], [2, 3/5], [2,
13/20], [2, 7/10], [2, 3/4], [2, 4/5], [2, 17/20], [2, 9/10], [2,
19/20], [2,
1]],COLOUR(RGB,0,0,1)),STYLE(POINT),STYLE(LINE),VIEW(DEFAULT,DEFAULT))
```

Note that this has not altered P. A new PLOT Structure has evolved.

Suppose, instead, you want to simply remove the x and y values from around the plot below, and change the design.

```
>   fx := sin(x^2 + abs(y)); r:= 3*Pi:
```

```
> opts :=
> axes=boxed,grid=[50,50],scaling=constrained,style=patchnogrid:
> densityplot(fx, x=-r..r,y=-r..r, opts); G:=%:
```

$$fx := \sin(x^2 + |y|)$$

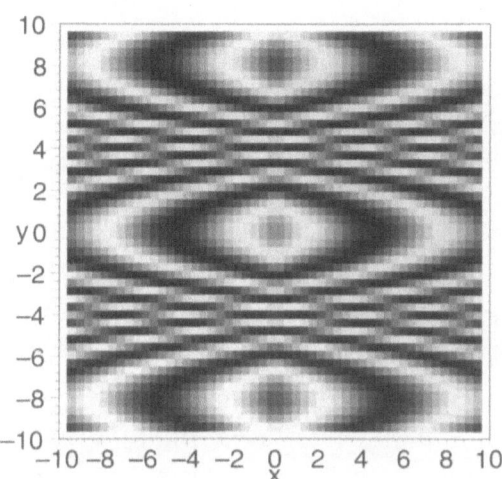

This is a complex PLOT Structure. We organise a search procedure to minimise the 'filling of the screen with rubbish'.

```
> Search := proc(x,Y)
> local i,bit,y;
> for i from 1 to nops(x) do;
> bit := op(i,x);
> if has(%,Y) then;
> RETURN(bit,
> ' is in op('.i.')');
> fi;
> od;
> 'nothing';
> end:
> Search(P,AXESLABELS);
```

The lower evaluation level in the procedure can deal with plot

$AXESLABELS(\ ,\ ),\ is\ in\ op(3)$

```
> Search(G,AXESLABELS);
```

$AXESLABELS(x, y),\ is\ in\ op(2502)$

```
> Search(G,AXESSTYLE);
```

$AXESSTYLE(BOX),\ is\ in\ op(2503)$

```
> fx := sin(abs(x) + abs(y));
```

```
> G := densityplot(fx, x=-r..r,y=-r..r,
> opts):
> subsop(2502=NULL,2503=AXESSTYLE(NONE),G);
```

$$fx := \sin(|x| + |y|)$$

3.7.3 More Structures, Colour

Coloured Squares

Search for the
following
examples in the
help file!

From the Maple help files.

```
> N:=50; i:=k/N: j:=1-i:
> PLOT(CURVES(
> seq([[i,i],[i,j],[j,j],[j,i],[i,i]],
> k=1..N)),
> TITLE(SQUARES),COLOR(HUE,0.5),
> AXESSTYLE(BOX),AXESTICKS(10,
> [0='0',0.5='1/2',1='1']));
> PLOT3D(POLYGONS([[0,0,0],[1,0,0],[1,1,0],[0,1,0]],
> [[0,0,0],[0,1,0],[0,1,1],[0,0,1]],
> [[1,0,0],[1,1,0],[1,1,1],[1,0,1]],
> [[0,0,0],[1,0,0],[1,0,1],[0,0,1]],
> [[0,1,0],[1,1,0],[1,1,1],[0,1,1]],
> [[0,0,1],[1,0,1],[1,1,1],[0,1,1]]),
> LIGHT(0,0,0.0,0.7,0.0), LIGHT(100,45,0.7,0.0,0.0),
> LIGHT(100,-45,0.0,0.0,0.7), AMBIENTLIGHT(0.4,0.4,0.4),
> TITLE(CUBE),STYLE(PATCH),COLOR(ZHUE));
```

$$N := 50$$

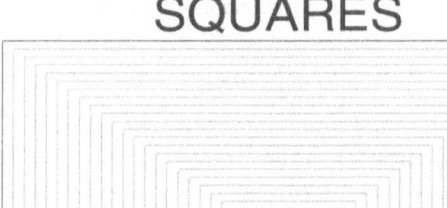

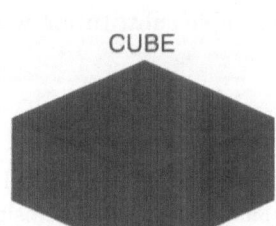

Aspects of Colour

There are many alternative ways of colouring objects in Maple. Here we consider some, though a list is presented in section 3.6, page 145. Also, refer to

?plot, color

?plot, structure

The colours of PLOT Structures are presented in three ways.

COLOUR(RGB,0,.7,.7)

COLOUR(HUE,0.6)

COLOUR(HSV,.7,.9,1), similar to HUE as above

The RGB argument is followed by three fractions for the relative amounts of the three colours; Red, Green and Blue. The HUE argument is a fraction that represents the light reflected from or transmitted through an object.

HSV is followed by three fractional numbers representing the Hue, Saturation and lightness Value. The saturation is the strength or purity of the colour or chroma. The lightness is the relative lightness or darkness of the colour.

Generally the spelling may be either COLOUR or COLOR, colour of color but an ocasional package will only acknowledge one or the other spelling. Equivalent presentations for the command plot( ), for the same effects are

```
colour=COLOUR(RGB,0,.7,.7)
colour=COLOUR(HUE,0.6)
colour=COLOUR(HSV,.7,.9,1), which is the same as above
colour=cyan and any number of other named colours
```

With 2D plotting, if it is desired to fill an object, convert the CURVES structure to a POLYGON. A simple substitution will do.

```
>  subs(CURVES=POLYGON,%);
```

Otherwise, some colour guides are presented below, with coloured circles and coloured bars. The size, shape and relative positions of objects affect the perception of colour so it is worthwhile to alter these as well as the background.

Selecting Colours

A difficulty in dealing with colours is selection. Colour patterns allow a discriminate gauging of colours and Colour Charts may pin down a particular colour. It is unclear how one should relate the colours; the perception of colour depends on light intensity, background and one's general ability to see and discern colour. One way is to stack colours together and selecting the colour of interest; then check to see what parameters were used to create it. Below, some circles are stacked. Note converting CURVES to POLYGONS fills in the circles with colour.

```
>  PLOT(seq(plottools[circle]([i mod
>  5,ceil(i/5)],.5,colour=COLOUR(HUE,i/20)),i=1..20),
>  SCALING(CONSTRAINED),AXESSTYLE(NONE)):
```

```
>  subs(CURVES=POLYGONS,%);
```

RGB Colour Chart

Armed with circles of colour, we can make a proper RGB colour chart. The effect uses the active plotting of Maple, with a layered animation. The first colour is red, written in the upper left. The other two colors are found in the little window at the left of the Context Menu bar, the lowest level of menus. Clicking on the centre of a circle gives a rough value of the decimal fractions needed to produce that particular colour. clicking on the ->| cycles through the decimal fractional values for red.

```
>    with(plottools):with(plots):
>    for i from 0 to 10 do; sss:=NULL:
>    for j from 0 to 10 do;
>    for k from 0 to 10 do;
>    sss:=sss,circle([j/10,k/10],1/20,
>    color=COLOUR(RGB,i/10,j/10,k/10));
>    od; od;
>    convert(evalf(i/10,1),string);
>    PLOT(sss,TEXT([0,1.1],
>    "(RGB, ".%,FONT(TIMES,ROMAN,10)));
>    colorS[i]:=subs(CURVES=POLYGONS,%)
>    od:

>    display(seq(colorS[i],i=0..10),insequence=true,
>    scaling=CONSTRAINED,axes=NONE,
>    title="RGB Color Chart",
>    titlefont=[TIMES,BOLD,16]);
```

RGB Color Chart

(RGB, 0

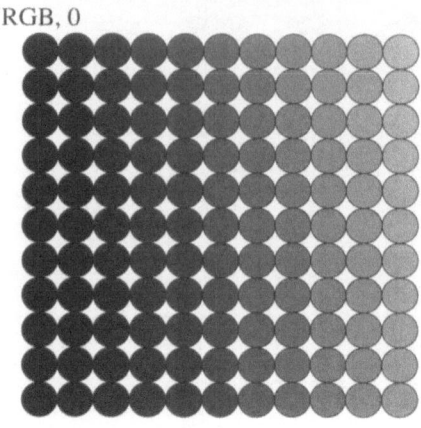

HUE Colour Chart

A simpler pattern is produced for the Hue argument. In the figure one can read the value directly off the abscissa or click on the figure, and collect the first number in the little window.

```
>   PLOT(seq(CURVES([[i/100,0],[i/100,1]],
>   COLOR(HUE,i/100),THICKNESS(3)),i=0..100),
>   TITLE("HUE Color Chart",FONT(TIMES,BOLD,16)),
>   AXESTICKS(DEFAULT,[]));
```

HUE Color Chart

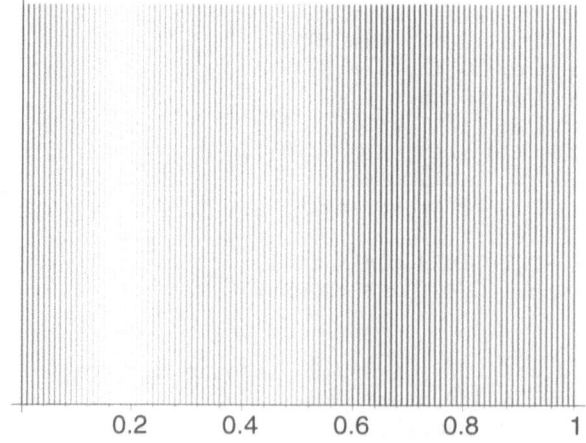

HSV Colour Chart

Last, the HSV chart reveals the rather non-linear features of saturation and lightness. The subtitle on the graph shows the Lightness and the scales give a direct measure of Hue and Saturation. Clicking on a colour gives the Hue and Saturation in the little window. Notice that, if the brightness is less than about .5, there isn't much colour and the picture is dreary. If the saturation is low, the colours are all washed out; if more than about .6, there is little difference in the colours. Of course, you might see this quite differently; it depends on your particular view of the world and the background that is used.

```
>   sssP:=seq(PLOT(seq(seq(CURVES(
>   [[i/100,(2*j-1)/50],[i/100,(2*j+1)/50]],
>   COLOR(HSV,i/100,j/25,k/10),
>   THICKNESS(3)),i=0..100),
>   j=0..25),TEXT([.15,1.1],
>   cat("Lightness ",convert(evalf(k/10,1),string)),
>   FONT(TIMES,ROMAN,10))),k=0..10):
```

```
>   plots[display](sssP,insequence=true,
>   labels=[H,S], title="HSV Color Chart",
>   titlefont=[TIMES,BOLD,16]);
```

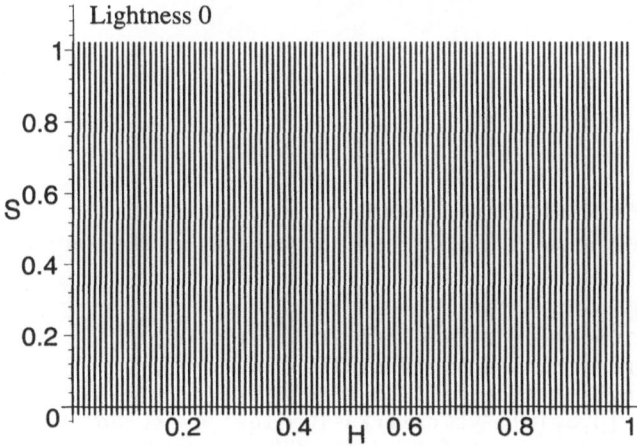

PLOT Structures Revisited

Various examples of PLOT Structures appear in the following montage, partly from Maple's help.

```
> Succotash;
> PLOT(POINTS([0,1],SYMBOL(DIAMOND)),
> TEXT([0,0.1],''Origin'',ALIGNABOVE,ALIGNRIGHT,
> FONT(COURIER,OBLIQUE,10)),
> CURVES([[-3,1.5],[3,1.5]],THICKNESS(2),LINESTYLE(4)),
> TEXT([0,1.4],''Dotted'',ALIGNBELOW),
> TEXT([3.14,1],'p',FONT(SYMBOL,12)),
> TEXT([-3.14,1],'P',FONT(SYMBOL,12)),
> POLYGONS([[-2,.7],[-2,0.5],[2,0.5],[2,0.8]],COLOR(HUE,0.5)),
> TEXT([0,.6],''Red'',COLOR(RGB,1,0,0)),
> AXESSTYLE(FRAME),VIEW(-4..4,0..2),
> TITLE(Succotash,FONT(SYMBOL,20)));
```

Succotash

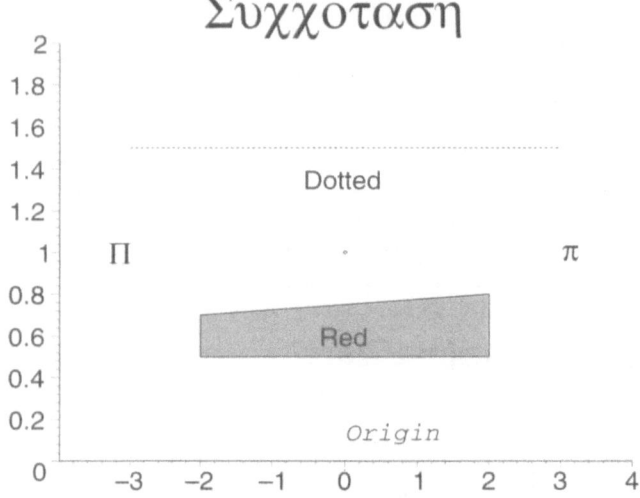

Exercise 3.6 Curves Galore in Colour

Redefine the plot below as PLOT Structures so that all the components work properly.

```
> with(plots):  i := 'i':
> plot({seq(10*cos((x^2+i)/4)/(x^2+i),i=1..10),
> title='Figgle',titlefont=[HELVETICA,BOLD,30],
> axes = BOXED, axesfont=[TIMES,ROMAN,15],
> xtickmarks=[0,5,10], ytickmarks=[0='low',4='so-so',8='high'],
> labels=['X','W'], font=[HELVETICA,OBLIQUE,20]},x=-5..10);
```

3.8   Fractals

The idea of a broken line having different properties than a straight one is intriguing. One can imagine a situation where a line is scribbled across a page in such a way as to 'cover' the whole page. Then one might say 'the line has two dimensions'. Here we look at a special construction, called a 'Koch Curve'; it breaks a line in a special way. The amazing thing about it is that the line is completely determined by the construction technique, that is, it is 'deterministic'; yet it shows some randomness and has statistical properties. In fact, following from the idea of dimensions, we might say the broken line has more than one dimension.

Self-Similarity

This type of construct has the property of independence of scale, called 'self-similarity'. That is, viewed at a large scale, it exhibits quantities and qualities (or statistics) that are the same at ever-smaller scales. Importantly, natural phenomena exhibit the same property; coastlines are a good example.

A Koch Curve

This is constructed by breaking a straight line into 3 sections, and making an equilateral triangle having one side that is the centre section. The centre section of the original line is removed, producing four equal line segments that are joined and broken, making an 'omega' like shape. The operation is then repeated with each line segment, making 16 broken and connected line segments. Repeating the operation again produces 64 connected line segments, and so forth. The 5th or 6th repetition or 'order' produces a 'random-looking' curve. Of course, depending on one's computing and 'printing power' one can take the calculation to any order.

Here we take a line and divide it into 3 parts with the procedure sel( ). The line is represented by the two points p and q which are lists of the x and y values; it is broken into four points to make the three line segments.

```
> sel := proc(p,q)
> local dx, dy;
> dx := (q[1]-p[1])/3;
> dy := (q[2]-p[2])/3;
> p,[p[1]+dx,p[2]+dy],
> [p[1]+2*dx,p[2]+2*dy],q;
> end;
```

$$sel := \mathbf{proc}(p, q)$$
$$\quad \mathbf{local}\, dx,\, dy;$$
$$\quad\quad dx := 1/3 \times q_1 - 1/3 \times p_1;$$
$$\quad\quad dy := 1/3 \times q_2 - 1/3 \times p_2;$$
$$\quad\quad p,\, [p_1 + dx, p_2 + dy],\, [p_1 + 2 \times dx, p_2 + 2 \times dy],\, q$$
$$\quad \mathbf{end}$$

```
> sel([0,0],[1,0]);
```

$$[0, 0], [\frac{1}{3}, 0], [\frac{2}{3}, 0], [1, 0]$$

```
> sel([-1,1],[4,3]);
```

$$[-1, 1], [\frac{2}{3}, \frac{5}{3}], [\frac{7}{3}, \frac{7}{3}], [4, 3]$$

The equilateral triangle is found by considering the intersections of two circles with centres at the middle two points. The radius of both circles is the length of the middle line segment or the distance between the two middle points. This produces two real solutions, one for a triangle above the line and one for a triangle below the line. This forms a list of the two possible points, a listlist. Here the effect is illustrated by forming equilateral triangles on a line of length one on the x axis, with the left end at the origin. The general solution is the intersection of two identical circles at two different locations, [x1,y1] and [x2,y2]; we substitute in specific values for the two points and solve for the intersections.

```
> x^2+y^2=1;
> # circle of radius 1 at the origin
```

$$x^2 + y^2 = 1$$

```
> c1 :=subs(x=x-x1,y=y-y1,%);
> # same circle at [x1,y1]
```

$$c1 := (x - x1)^2 + (y - y1)^2 = 1$$

```
> c2 :=subs(x=x-x2,y=y-y2,%%);
> # same circle at [x2,y2]
```

$$c2 := (x - x2)^2 + (y - y2)^2 = 1$$

```
> conds := {x1=0;y1=0;x2=1.0;y2=0};
```

The decimal is to keep solve from producing RootOf( ) output

$$conds := \{x1 = 0, y1 = 0, y2 = 0, x2 = 1.0\}$$

```
> opts:=thickness=2,colour=green,scaling=CONSTRAINED;
```

$$opts := thickness = 2, colour = green, scaling = CONSTRAINED$$

```
> p1:=subs(conds,c1);
```

$$p1 := x^2 + y^2 = 1$$

```
> p2:=subs(conds,c2);
```

$$p2 := (x - 1.0)^2 + y^2 = 1$$

```
> with(plots: implicitplot({p1,p2},x=-2..2,y=-2..2,opts);
```

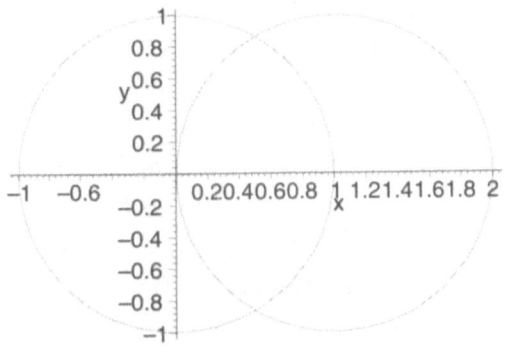

Proceeding, solve for the vertices of the triangles.

```
> ss :=solve({p1,p2},{x,y});
```

$ss := \{y = -.8660254038, x = .5000000000\}, \{x = .5000000000, y = 8660254038\}$

> ss[1],ss[2];

$\{y = -.8660254038, x = .5000000000\}, \{x = .5000000000, y = .8660254038\}$

The algebraic form is too lengthy to be useful

```
> solve({c1,c2},{x,y}):  allvalues(%):
> length(%);
```

6815

Efficiency is best if Digits is some multiple of 4

```
> Digits:=8;
```

$Digits := 8$

```
> evalf(sqrt(3));
```

1.7320508

Process fixpt( ) automatically gives the two solutions.

```
> fixpt := proc(p,q)
> local c1,c2,d2,ss,i,ra,rb,x,y;
> (p[1]-q[1])^2+(p[2]-q[2])^2;
> d2:= evalf(%);
> c1 := (x-p[1])^2 + (y-p[2])^2=d2;
> c2 := (x-q[1])^2 + (y-q[2])^2=d2;
> [solve({c1,c2},{x,y})];
> ss := remove(has,%,I);
```

Some imaginary solutions can creep through

```
> map(subs,ss,[x,y]);# listlist of
> solutions

> end;
```

$$\textit{fixpt} := \textbf{proc}(p,\, q)$$
$$\textbf{local}\, \textit{c1},\, \textit{c2},\, \textit{d2},\, \textit{ss},\, i,\, \textit{ra},\, \textit{rb},\, x,\, y;$$
$$(p_1 - q_1)^2 + (p_2 - q_2)^2 \,;$$
$$\textit{d2} := \mathrm{evalf}(\%)\,;$$
$$\textit{c1} := (x - p_1)^2 + (y - p_2)^2 = \textit{d2}\,;$$
$$\textit{c2} := (x - q_1)^2 + (y - q_2)^2 = \textit{d2}\,;$$
$$[\mathrm{solve}(\{\textit{c2},\, \textit{c1}\},\, \{y,\, x\})]\,;$$
$$\textit{ss} := \mathrm{remove}(\textit{has},\, \%,\, I)\,;$$
$$\mathrm{map}(\textit{subs},\, \textit{ss},\, [x,\, y])$$
$$\textbf{end}$$

```
>  fixpt([0,0],[1,0]);
```
$$[[.50000000,\, -.86602541],\, [.50000000,\, .86602541]]$$

```
>  fixpt([-1/54,48/27],[-2/27*sqrt(3),42/27]);
```
$$[[.11904080,\, 1.5715931],\, [-.26585938,\, 1.7617403]]$$

```
>  fixpt([0,0],[0,1]);
```
$$[[-.86602541,\, .50000000],\, [.86602541,\, .50000000]]$$

The procedures can be combined in various ways. It is useful to augment a sequence. Here, in pik( ) we arbitrarily pick the term n of the solution.

```
>  pik := proc(ss::listlist,n)
>  local i,s,sss;
>  global sel,fixpt;
>  sss:=NULL,ss[1];
>  for i from 2 to nops(ss) do;
>  s:=[sel(ss[i-1],ss[i])];
>  sss:=sss,s[2],op(n,fixpt(s[2],s[3])),s[3],s[4];
>  od;
>  RETURN(sss);
>  end;
```

$pik := \mathbf{proc}(ss::listlist,\ n)$
 $\mathbf{local}\ i,\ s,\ sss;$
 $\mathbf{global}\ sel,\ fixpt;$
  $sss := NULL,\ ss_1\,;$
  $\mathbf{for}\ i\ \mathbf{from}\ 2\ \mathbf{to}\ \mathrm{nops}(ss)\ \mathbf{do}$
   $s := [\mathrm{sel}(ss_{i-1},\ ss_i)]\,;\ sss := sss,\ s_2,\ \mathrm{op}(n,\ \mathrm{fixpt}(s_2,\ s_3)),\ s_3,\ s_4$
  $\mathbf{od};$
  $\mathrm{RETURN}(sss)$
 $\mathbf{end}$

```
>  rs :=pik([[-1,1],[3,4]],1);
```

$rs := [-1,\ 1],\ [\dfrac{1}{3},\ 2],\ [1.8660254,\ 1.3452995],\ [\dfrac{5}{3},\ 3],\ [3,\ 4]$

```
>  plot([rs],-1..4,0..4,style=line,scaling=constrained);
```

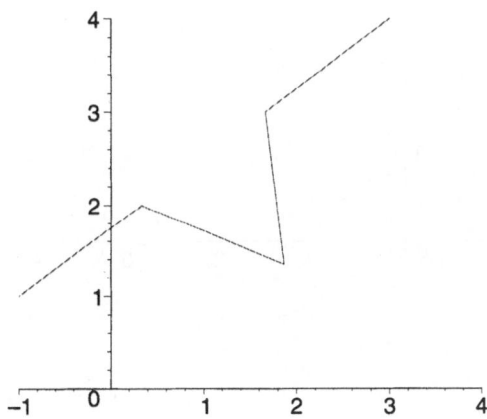

```
>  rs;
```

$[-1,\ 1],\ [\dfrac{1}{3},\ 2],\ [1.8660254,\ 1.3452995],\ [\dfrac{5}{3},\ 3],\ [3,\ 4]$

```
>  pik([%],2);
```

$[-1, 1]$, $[\frac{-5}{9}, \frac{4}{3}]$, $[-.62200847, 1.8849002]$, $[\frac{-1}{9}, \frac{5}{3}]$, $[\frac{1}{3}, 2]$, $[.84423069, 1.7817665]$, $[1.2886751, 2.1150998]$, $[1.3551280, 1.5635330]$, $[1.8660254, 1.3452995]$, $[1.7995725, 1.8968663]$, $[2.2440170, 2.2301997]$, $[1.7331196, 2.4484332]$, $[\frac{5}{3}, 3]$, $[\frac{19}{9}, \frac{10}{3}]$, $[2.0446582, 3.8849002]$, $[\frac{23}{9}, \frac{11}{3}]$, $[3, 4]$

```
>   plot([%],-1..4,0..4,style=line,scaling=constrained);
```

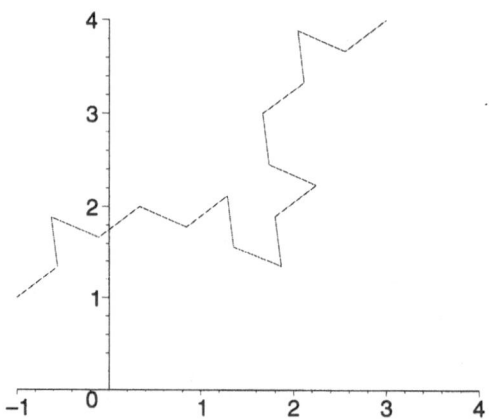

The entire curve generator koch( ) organises the solutions using the change in the gradient of the lines in the first two line segments (see the operator form of the conditional, 'if'). This value is consistent with the 'normal sense' of the curves, provided no lines become vertical or have negative slopes. Then, there is a tendency for the 'peaks' to 'break' or roll off to the left. This tendency is enhanced if the overall slope of the curve is large.

```
>   koch := proc(ss::listlist,n,mm)
>   local cond,i,m,p1,p2,ra,rb,s,s1,sss;
>   global sel,fixpt;
>   s1:=ss;
>   for m from 1 to mm do;
>   sss := s1[1];
>   for i from 2 to nops(s1) do;
>   s:=sel(s1[i-1],s1[i]);
>   fixpt(s[2],s[3]);
>   ra:=op(1,%); rb:=op(2,%%);
>   p1:=s[1]; p2:=s[2];
>   cond:=2*(ra[2]-p2[2])>(p2[2]-p1[2]);
>   'if'(cond,[ra, rb],[rb, ra]);
>   p2,op(n,%),s[3],s[4];
>   sss :=sss,%;
>   od;
>   s1:=[sss];
>   od;
>   end;
```

$$koch := \mathbf{proc}(ss::listlist, n, mm)$$
$$\mathbf{local}\ cond,\ i,\ m,\ p1,\ p2,\ ra,\ rb,\ s,\ s1,\ sss;$$
$$\mathbf{global}\ sel,\ fixpt;$$
$$s1 := ss;$$
$$\mathbf{for}\ m\ \mathbf{to}\ mm\ \mathbf{do}$$
$$\qquad sss := s1_1;$$
$$\qquad \mathbf{for}\ i\ \mathbf{from}\ 2\ \mathbf{to}\ \mathrm{nops}(s1)\ \mathbf{do}$$
$$\qquad\qquad s := \mathrm{sel}(s1_{i-1},\ s1_i);$$
$$\qquad\qquad \mathrm{fixpt}(s_2,\ s_3);$$
$$\qquad\qquad ra := \mathrm{op}(1,\ \%);$$
$$\qquad\qquad rb := \mathrm{op}(2,\ \%\%);$$
$$\qquad\qquad p1 := s_1;$$
$$\qquad\qquad p2 := s_2;$$
$$\qquad\qquad cond := p2_2 - p1_2 < 2 \times ra_2 - 2 \times p2_2;$$
$$\qquad\qquad \text{`if`}(cond,\ [ra,\ rb],\ [rb,\ ra]);$$
$$\qquad\qquad p2,\ \mathrm{op}(n,\ \%),\ s_3,\ s_4;$$
$$\qquad\qquad sss := sss,\ \%$$
$$\qquad \mathbf{od};$$
$$\qquad s1 := [sss]$$
$$\mathbf{od}$$
$$\mathbf{end}$$

```
>   k2 := koch([[-1,1],[3,4]],1,2):
>   plot(k2,-1..4,0..4,style=line,scaling=constrained);
```

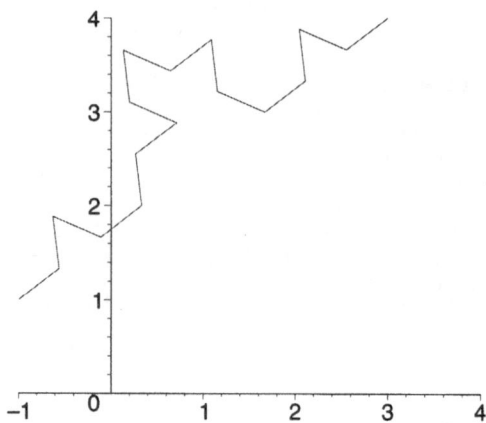

```
>  k3 := koch([[-1,1],[3,4]],1,3):
```

```
>  plot(k3,-1..4,0..4,style=line,scaling=constrained);
```

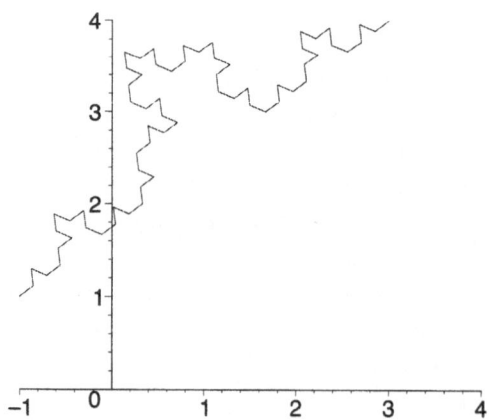

```
>  k4 := koch([[-1,1],[3,4]],1,4):
```

```
>  plot(k4,-1..4,0..4,style=line,scaling=constrained);
```

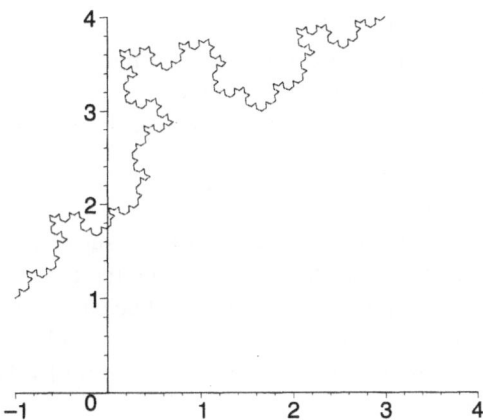

```
>   k5 := koch([[-1,1],[3,4]],1,5):nops(k5);
```

1025

This calculation
takes about 5
minutes

```
>   plot(k5,-1..4,0..4,style=line,scaling=constrained);
```

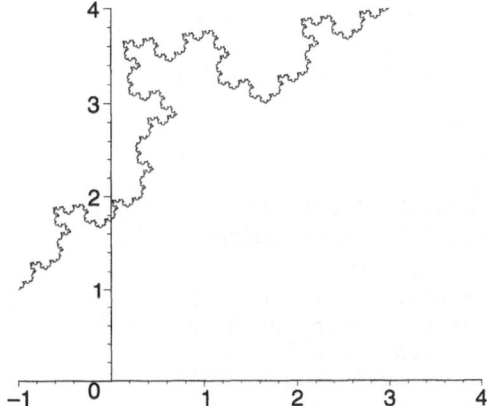

Exercise 3.7 One-Sided Fractal

Redo the procedure koch and select triangles 'only on one side'. The above use of the variation of slope leads to a 'flipping over' of the sense when the line goes vertical. Additional 'if' statements or the use of the vector cross product will work. The cross product produces another vector whose magnitude is proportional to the sine; the sine has a different sign for 120 and 240 degrees.

Exercise 3.8 Random Fractal

Use the random number generator rand( ) to randomise the selection of the sense of the triangle. This produces a 'random fractal' with a quite different look and different statistics. Reference to Chapter 9 may be helpful.

Exercise 3.9 Koch Snowflake

Analyse the two routines below. They produce a snowflake-like form that is a variation of the Koch curves. Routines are from Greg Fee, with a variation to produce different primary colours in each curve. The use of a hardware floating point calculation, through hfarray and evalhf gives a substantial increase in the efficiency, with time requirements decreased by about a factor of 10.

```
> Koch := proc(n)
> local e,x,y,d,ox,oy,i,j,L,cp;
> e := 4;
> x := hfarray(1..e); y := hfarray(1..e);
> x[1]:=-1; x[2]:=.5; x[3]:=x[2]; x[4]:=x[1];
> y[1]:=0; y[2]:=evalhf(sqrt(.75)); y[3]:=-y[2];y[4]:=0;
> for i from 2 to n do
> d := e;
> ox := hfarray(1..d); oy := hfarray(1..d);
> for j to d do ox[j] := x[j]; oy[j] := y[j]; od;
> e := 4*d-3; x := hfarray(1..e);
> y := hfarray(1..e);
> evalhf(next_Koch(d,ox,oy,e,x,y));
> od;
> L := [seq([x[i],y[i]],i=1..e)];
> cp :=[0,0,0]:  # for varied colour;
> cp[1 + (n mod 3)]:=1;
> SCALING(CONSTRAINED),AXESSTYLE(NONE);
> PLOT(CURVES(L,COLOUR(RGB,op(cp))),%);
> end:
> next_Koch := proc(d,ox,oy,e,x,y)
> local j,tr,ux,uy,i,px,py,qx,qy,dx,dy;
> j := 0; tr := evalhf(1/3);
> ux := .5; uy := evalhf(sqrt(.75));
> qx := ox[1]; qy := oy[1];
> for i from 2 to d do
> px := qx; qx := ox[i]; py := qy; qy := oy[i];
> j := j+1; x[j] := px; y[j] := py;  j := j+1;
> dx := qx-px; dy := qy-py; dx := tr*dx; dy := tr*dy;
> x[j] := px+dx; y[j] := py+dy;
> x[j+1] := x[j]+ux*dx-uy*dy; y[j+1] := y[j]+ux*dy+uy*dx;
> x[j+2] := x[j]+dx; y[j+2] := y[j]+dy; j := j+2;
> od;
> j := j+1; x[j] := qx; y[j] := qy;
> end:

> assign(k.l=Koch(l),l=1..7):
```

```
>  with(plots):  display(k.(1..7));
```

3.9 Pretty Rotations

In a given situation, many views are possible. And a figure often will not look
right in the wrong plane. One likes to see, for instance, the vertical distance
replicated by a vertical axis and, often we want north at the top of the page.
A few simple rotations can fix this, hopefully to our satisfaction.

```
>  with(plottools):
>  #help(polygon);
>  PP := polygon([[0,0],[3,4],[3,1]],
>  color=green, thickness=3);
```

$PP :=$ POLYGONS($[[0, 0], [3., 4.], [3., 1.]]$, COLOUR($RGB$, 0, 1.00000000, 0),
  THICKNESS(3))

```
>  with(plots):
>  display(PP);
```

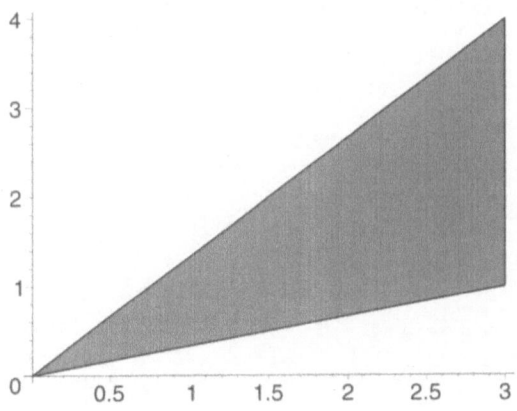

```
>  #help(rotate);
>  rotate(PP,Pi/3);
```

POLYGONS($[[0, 0], [-1.9641016, 4.5980762], [.63397460, 3.0980762]]$,
  COLOUR(RGB, 0, 1.00000000, 0), THICKNESS(3))

```
>   display(%);
```

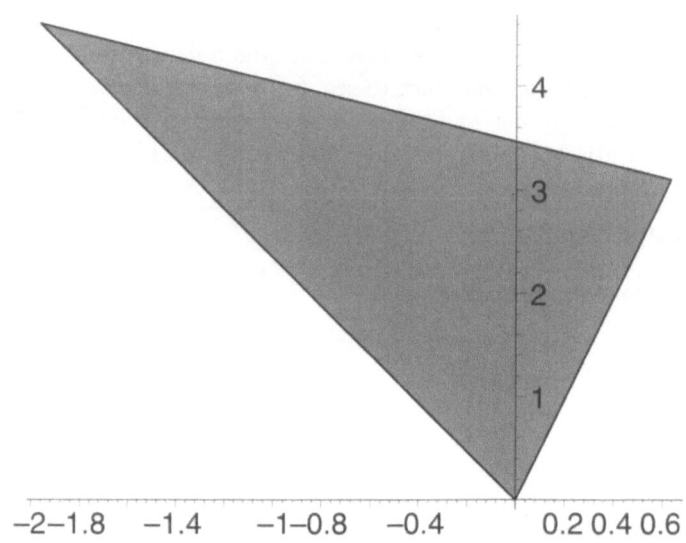

```
>   pp := plot([x*sin(2*x^2),x,x=0..2*Pi]):%;
```

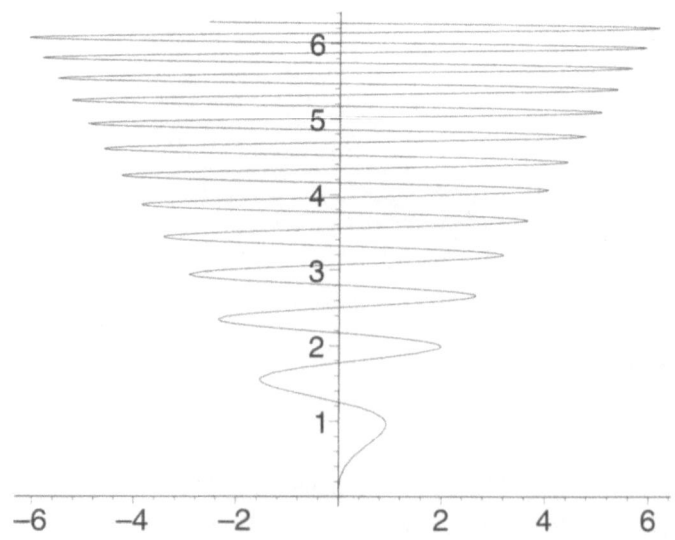

```
>   rotate(pp, -Pi/3);
```

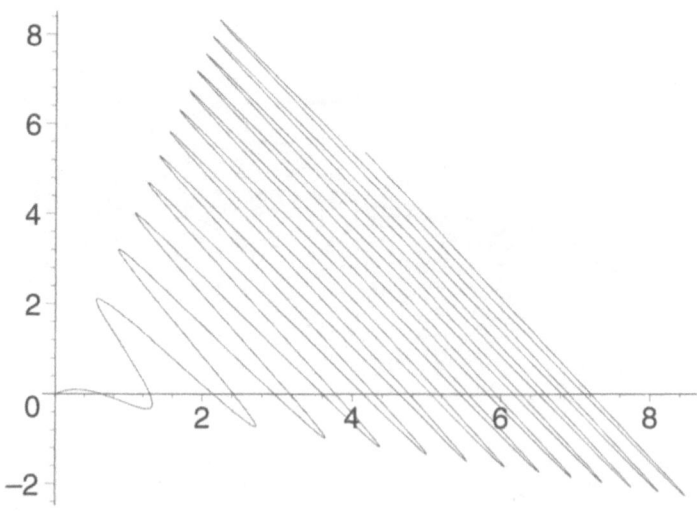

```
>   pp := plot3d([(1.3)^x *
>   sin(y),x,y],x=-1..2*Pi,y=0..Pi):%;
```

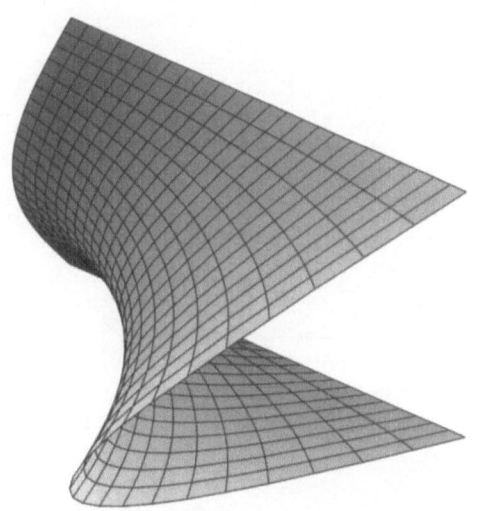

```
> rotate(pp,Pi/3,Pi/4,Pi/5);
```

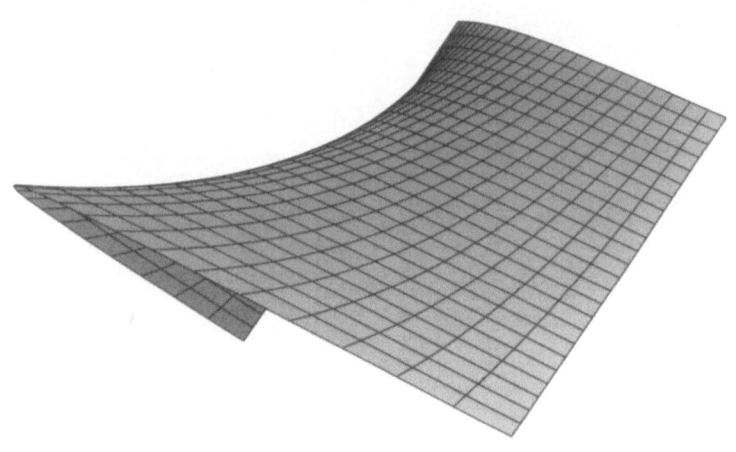

```
> display(hyperbola([0,0],1,1,-2..2));
```

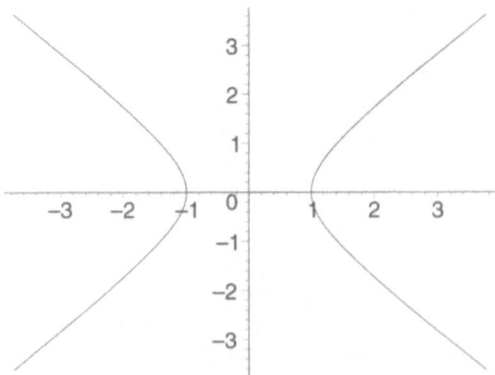

```
> display(rotate(%,Pi/4));
```

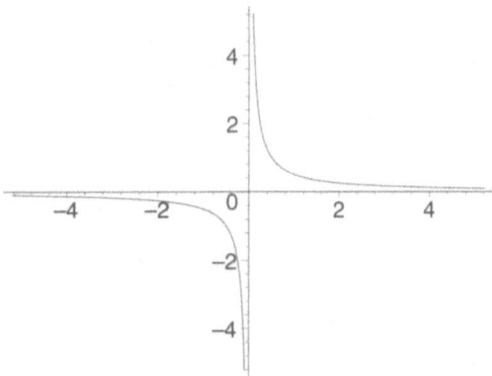

Stretched Globes

Rotations make some splendid picture patterns; here are two. 3D super-position of eight snapshots of a spheroid produces eight globes. The full 3D rotation of a petal makes a flower.

```
>   with(plottools):  n := 1:  fPi := evalf(Pi):
>       r[0] := sphere([3,0,0],1.):  a := fPi/4.:
>   while a-2*fPi < 0.  do; # floating boolian
>       r[n] := rotate(r[0],a,[[0,0,0],[0,0,1]]);
>       a+fPi/4.;  a:=%;  n+1; n:=%;
>   od:
>   plots[display3d]([seq(r[i],i=0..n-1)]);
```

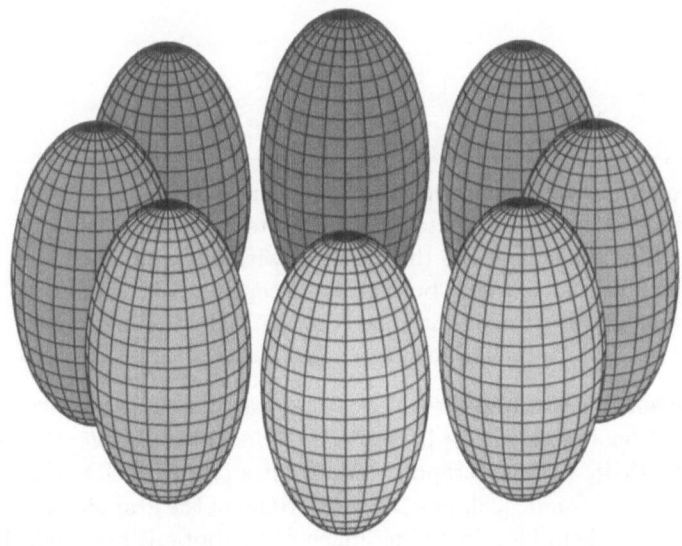

Colored Flower

```
> with(plots):        rnge := x=-Pi..Pi,y=-Pi/2..Pi/2:
> setoptions3d(scaling=constrained, style=patchnogrid,
>             projection=.5);  rnge := x=-Pi..Pi,y=-Pi/2..Pi/2:
>     osy := 1+sin(y):  Xc := sin(x)*cos(y)*osy:
>     Yc := cos(x)*cos(y)*osy:  Zc := .6*(sin(y)-1):
> centre:=plot3d([Xc,Yc,Zc],rnge,color=[.8,.9,.1]):  # the centre
> Yp:=Yc:  Xp:=2*sin(y)+1.85:  Zp:=.1*sin(x)*cos(y)*osy + y-y^2:
> petal:=plot3d([Xp,Yp,Zp],rnge):  # one petal
>     rotation := [Xp*cos(k*Pi/3)+Yp*sin(k*Pi/3),
>                         Yp*cos(k*Pi/3)-Xp*sin(k*Pi/3),Zp]:
> p:=k->plot3d(rotation,rnge,colour=[sin(2*y)/2,-cos(y)/2,.2],
>             ambientlight=[.1,.3,.1], light=[75,55,1,.9,.2]):
> petals := display([seq(p(k),k=1..6)]):  # petal rotated
> display3d({centre,petals},  orientation=[-20,60]);
```

We have looked at functions with strange or singular points; we will see that, indeed, real data contains such faults and that it is often piecemeal. Plotting emerges as the interface with our eyes, and gives a broad view of what is there, but can be deceptive. We can always draw a pretty picture; making it correct is the really difficult thing.

Acknowledgements for Chapter 3

Some material on functions has its roots in Wade Ellis' *Maple V Flight Manual* [12], pgs 45-48. The Operands section relies on Michael Monagan's *Maple V Programming Guide*, [23], pgs 181-206. Many of the graphics are variously from the Maple help files, including much of the options material. The Snowflake Fractal is from Greg Fee of the Centre for Experimental and Constructive Mathematics, Simon Fraser University. The flower is from Klimek and Klimek, *Discovering Curves and Surfaces with Maple* [17], pg 71; the densityplot square pattern derives from their page 24.

Chapter 4

Applications: Field Data

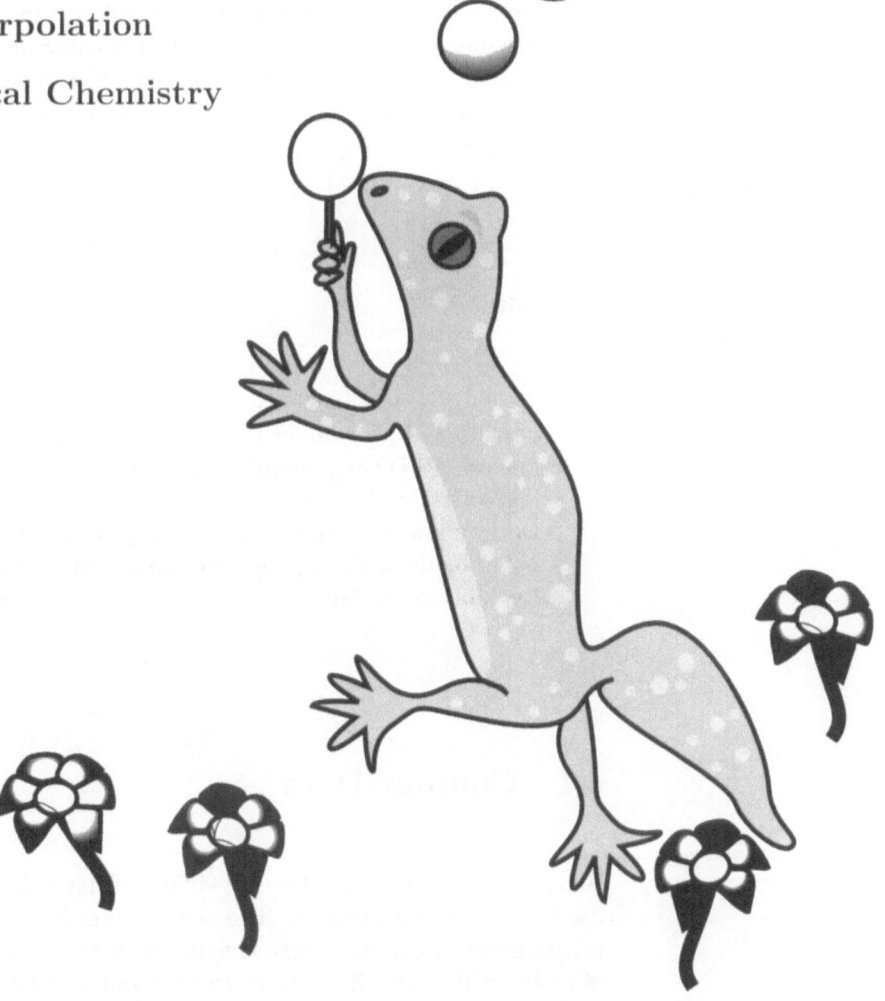

Applications: Field Data

Übersicht

Plotting the contours of lakes and the positions of boreholes. Then bubbles in three dimensions, moving up a column in animation. A look at Darcy's law for groundwater flow into lakes. Lastly, sulfur dioxide, carbon dioxide and ammonia contribute to the formation of acid rain. Data input to the computer and output to graphs, with procedures; this is the hub of analysis in the environmental sciences.

Guidewords:

data, fopen( ), readline( ), readdata( ), fclose( ), readlib( ); bubbles, lookup, PLOT3D, bubble sequences, map( ), colouring, streaking, resizing, insequence, animation; interpolation, formulation, Darcy's law, hydraulic conductivity, contourplot( ); acid rain, carbon dioxide, sulfur dioxide, ammonia, equilibrium, electroneutrality, kinetics, sulfate, pH

4.1   Contour Data

Here we dig deeply into the nitty-gritty of data. Indeed, it is a challenge to use Maple for this purpose. The simple spreadsheet is easier and a simplest programmer would know that such manipulations can be done with a few lines of code. Still, here *WE* are in control and we can make Maple simple and convenient. With experience, and a few procedures to do gross data handling, Maple can be rewarding; it can keep us from making errors.

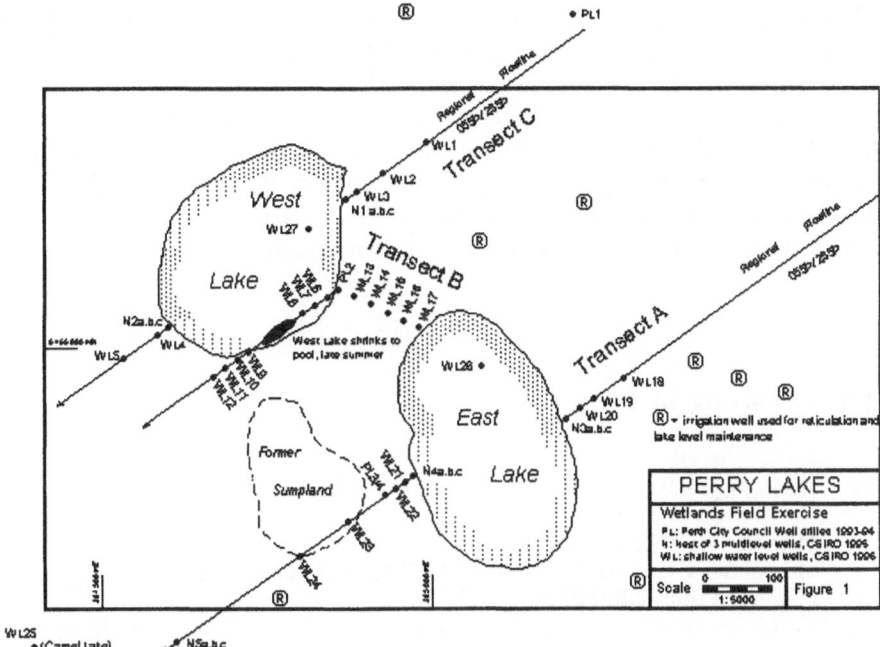

The example data set is from Perry Lakes, to the west of Perth in Western Australia. There have been a number of environmental difficulties with the parkland, including dredging the lakes for the 1960 Commonwealth games and gross pumping around the lakes, partly to maintain the grass and partly to maintain water in the lakes during summer. The area of West Lake containing open water is very small, particularly during summer. Irrigation wells are used to irrigate the lawns and maintain water levels in East Lake over summer. Water may also be added to West Lake during summer to maintain water for long necked tortoises, *Chelodina oblonga*.

During the summer with artificial level maintenance, East Lake becomes a local groundwater mound. The water going into the lake is pumped from adjacent bores, so the lake level rises while the surrounding water table is lowered. In summer, therefore, water tends to seep from the East Lake back into the adjacent unconfined aquifer; the lake recharges the aquifer and is a recharge lake. During this same period West Lake is reduced to a small 10x30m sump and partly becomes a discharge lake, receiving some water from the adjacent and mounded East Lake.

With winter storms, both lakes receive water from storm drains, become perched and function as recharge lakes. During late winter, levels in the unconfined aquifer rise, and both lakes become flow-through lakes. Groundwater flows into the lakes through the lake bed close to the upgradient shore; close to the opposite shore, water flows from the lakes back into the aquifer.

In this section we look at plotting the lake boundaries, posting the bore holes, and generally viewing the surrounds of the lakes. In section 4.3, water table heights around this lake are constructed. The boundary and posting data were extracted from a surveyed map of the area and 'digitised' into files by matching the image using a measuring cursor on the screen. On section 6.2, page 316 a method of doing this is outlined. Here we read in the digital data and plot them, to make maps of the lake boundaries. A second data set includes the sampling bores, from which the water table measurements were made.

Digitised using ghostview from Gnu

4.1.1   Reading in Data

```
>   restart:

>   filenme :='contours.prn';
>   #filenme :='c:/mplbook/scottVOL1/contours.prn';
>   fopen(filenme,READ);
```

The open gets the file ready for reading

With an appropriate directory structure, e.g., c:mplbook/perrylakes/ fopen will find the file and respond with zero or another number. That number can be used as a surrogate for the file name, 'a unit number'.

On all platforms use the Unix slash

filenme := contours.prn

$$0$$

```
>   readline(filenme);
```

Continue to read until the data appear. The routine counts the unread lines

"WestLake70m   long"

```
>   i := 1:  ex := 1:  while ex <> 0 do
>   ex := readline(filenme):i:=i+1:  od:  i-1;
```

$$110$$

```
>   fclose(filenme):
```

Closing is like closing the door. When the file is opened again, it reads from the start. readdata( ) reads all the data at once.

The data must be in a particular format

```
>   readlib(readdata):fopen(filenme,READ):
```

```
>   readdata(filenme,[string,float,float]);
```

[["WestLake70m"], ["2", 179., 204.], ["3", 176., 204.], ["4", 174., 204.], ["5", 173., 204.],
    ["6", 175., 207.], ["7", 179., 212.], ["8", 184., 216.], ["9", 188., 218.],
    ["10", 192., 219.], ["11", 194., 218.], ["12", 194., 216.], ["13", 192., 214.],
    ["14", 189., 212.], ["15", 186., 209.], ["16", 182., 207.], ["17", 179., 204.],
    ["bank"], ["19", 114., 216.], ["20", 107., 227.], ["21", 106., 252.],
    ["22", 112., 274.], ["23", 122., 295.], ["24", 137., 307.], ["25", 158., 318.],
    ["26", 176., 327.], ["27", 192., 332.], ["28", 208., 326.], ["29", 216., 320.],
    ["30", 227., 316.], ["31", 227., 315.], ["32", 224., 299.], ["33", 224., 282.],
    ["34", 223., 254.], ["35", 220., 240.], ["36", 217., 231.], ["37", 213., 225.],
    ["38", 207., 216.], ["39", 194., 211.], ["40", 180., 202.], ["41", 164., 199.],
    ["42", 153., 196.], ["43", 138., 197.], ["44", 126., 205.], ["45", 114., 216.],
    ["EastLake"], ["47", 272., 120.], ["48", 270., 124.], ["49", 268., 131.],
    ["50", 267., 139.], ["51", 270., 143.], ["52", 272., 146.], ["53", 270., 149.],
    ["54", 261., 161.], ["55", 261., 174.], ["56", 262., 189.], ["57", 270., 204.],
    ["58", 281., 216.], ["59", 289., 222.], ["60", 302., 224.], ["61", 314., 221.],
    ["62", 326., 215.], ["63", 339., 201.], ["64", 349., 190.], ["65", 358., 176.],
    ["66", 365., 154.], ["67", 371., 142.], ["68", 376., 129.], ["69", 377., 114.],
    ["70", 376., 101.], ["71", 372., 84.], ["72", 364., 73.], ["73", 345., 61.],
    ["74", 325., 62.], ["75", 308., 66.], ["76", 294., 76.], ["77", 283., 89.],
    ["78", 281., 99.], ["79", 277., 108.], ["80", 272., 120.], ["swamplanline"],
    ["82", 197., 67.], ["83", 181., 76.], ["84", 171., 88.], ["85", 168., 114.],
    ["86", 164., 140.], ["87", 164., 153.], ["88", 175., 168.], ["89", 187., 168.],
    ["90", 196., 165.], ["91", 201., 155.], ["92", 205., 146.], ["93", 210., 138.],
    ["94", 226., 129.], ["95", 238., 116.], ["96", 237., 99.], ["97", 232., 86.],
    ["98", 221., 77.], ["99", 209., 71.], ["100", 197., 67.], ["irrigwelR"],
    ["102", 266., 417.], ["103", 314., 269.], ["104", 382., 293.], ["105", 184., 40.],
    ["106", 416., 51.], ["107", 431., 156.], ["108", 454., 192.], ["109", 482., 181.],
    ["110", 511., 172.]]

```
>   AA := op(%):
>   op(AA[11]); op(1,AA[11]); whattype(%);
```
                    "11", 194., 218.
                        "11"
                        string

```
>   fclose(filenme):
```
The data are in the form of a listlist; sequential numbers appear at the
start of each list. After that is a position, the x, y coordinates on the edge
of the lake. The data have been acquired serially around the edge of the lake
so that they can be used to plot the contours of the lake perimeter; they are

in arbitrary units but 100 metres corresponds to 44; converting to kilometers requires multiplying by the ratio 0.1/44=1/440.

```
>  Fix := proc(x::string,y::float,z::float)
>  [y/440,z/440] end;
```

$$Fix := \mathbf{proc}(x\text{::}string,\ y\text{::}float,\ z\text{::}float)\ [1/440 \times y,\ 1/440 \times z]\ \mathbf{end}$$

```
>  WestLakewater :=
>  seq(Fix(op(AA[i])),i=2..17);
```

$WestLakewater := [.4068181818, .4636363636], [.4000000000, .4636363636],$

$[.3954545454, .4636363636], [.3931818182, .4636363636],$

$[.3977272727, .4704545454], [.4068181818, .4818181818],$

$[.4181818182, .4909090909], [.4272727273, .4954545454],$

$[.4363636364, .4977272727], [.4409090909, .4954545454],$

$[.4409090909, .4909090909], [.4363636364, .4863636364],$

$[.4295454545, .4818181818], [.4227272727, .4750000000],$

$[.4136363636, .4704545454], [.4068181818, .4636363636]$

Similarly, the other data are put into bins.

```
>  WestLakebank :=
>  seq(Fix(op(AA[i])),i=19..45):
>  EastLake := seq(Fix(op(AA[i])),i=47..80):
>  swampland :=
>  seq(Fix(op(AA[i])),i=82..100):

>  irrigwells := seq(Fix(op(AA[i])),i=102..110):
```

The WestLake water will be considered further and is shown by a full line. The WestLake banks, EastLake, and the swampland will be shown with dashed lines and the irrigation wells with symbols made from 'overprinting' R and O. In plot, we note that linestyle controls the dash pattern used to render the plot; n=1 makes a solid line; 2 makes a series of dots; 3 makes a dash, and 4 makes a dash-dot pattern. See Options, section 3.6, page 143. Also, the PLOT structures require UPPER CASE letters as shown below; though some symbols inside a PLOT structure may work with lower case letters, it is safer to stick with UPPER CASE.

LINESTYLE(3)
is a dashed line

```
>  TTT :=
>  op(map(TEXT,[irrigwells],'R',FONT(TIMES,ROMAN,10))):
>  CCC := op(map(TEXT,[irrigwells],'O',FONT(TIMES,ROMAN,15))):
>  PPP := PLOT(TTT,CCC,
>  CURVES([WestLakewater],THICKNESS(2),LINESTYLE(1)),
>  CURVES([EastLake],THICKNESS(1),LINESTYLE(3)),
>  CURVES([WestLakebank],THICKNESS(1),LINESTYLE(3)),
>  CURVES([swampland],THICKNESS(1),LINESTYLE(3)),
>  AXESTICKS(DEFAULT,DEFAULT,FONT(HELVETICA,BOLD,10)),
>  AXESSTYLE(FRAME),VIEW(0..1,0..1)):
```

```
>  %;
```

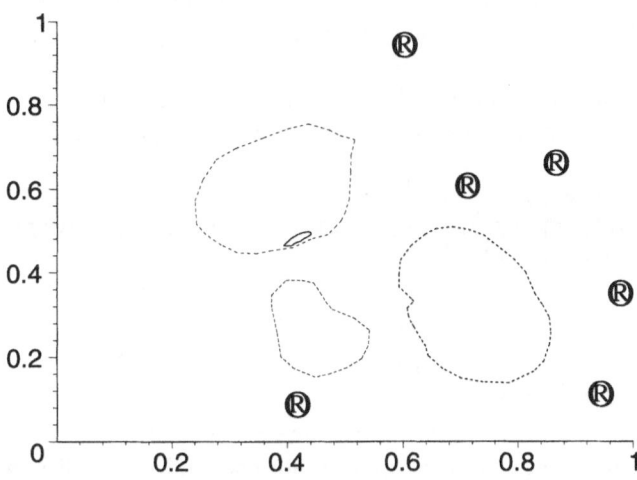

4.1.2   The Monitoring Boreholes

The well data are individually measured (with the mouse cursor) and entered below by hand. Some data always seem to require hand massage; section 6.2, page 316, reconsiders the problem of digitization.

The data could read from the graph directly

```
>  xy := table();
```

$$xy := \text{table}([$$
$$])$$

```
>  xy[WL1]:=279,333:   xy[WL2]:=251,313:   xy[WL3]:=235,300:
>  xy[WL4]:=105,208:   xy[WL5]:=83,194:   xy[WL6]:=215,233:
>  xy[WL7]:=207,227:   xy[WL8]:=199,222:   xy[WL9]:=165,198:
>  xy[WL10]:=158,193:  xy[WL11]:=149,188:  xy[WL12]:=141,182:
>  xy[WL13]:=232,233:  xy[N1a]:=227,295:  xy[N2a]:=112,214:
>  xy[PL2] :=252,106:  xy[lake]:=184,211:
```

All this is a trial and error, in stages, to check that the values are accepted by Maple.

```
>  xy[WL9];
```

$$165, 198$$

```
>  plot(xy);
```

Plotting error, empty plot

It seems Maple won't plot the bore data directly. The table xy has all the information; however, the individual entries are not numeric and they do not suit the needs of plot; lists of the x and y values are required. Shown here is a composing of lists using the operands of the xy table. Check your use of execution groups; the output of each group should be scrutinized before carrying on. Use separate execution groups until the progression is clear.

Expect the table items to have different orders each time you open Maple

```
>   nops(op(2,op(xy)));
>   op(1,op(1,op(2,op(xy)))); # could use op([2,1,1],op(xy))
>   op(2,op(1,op(2,op(xy))));
```

$$17$$
$$N2a$$
$$112, 214$$

Exercise 4.1 index and entry

Use the Maple commands index and entry to view the entries of the table xy. Proceed to use these data to make appropriate lists and lists of lists with the Maple command zip. Use plot with the same scale as the above figure to view the results.

Exercise 4.2 Enlarging WestLake

Anticipating a need for the detail of the water-containing region of West-Lake, blow up the area to a sensible size. Put an appropriate title on the graph and make the eastings (W-E position) and northings (S-N position) into metre measurements.

Procedures come to the rescue and all is 'small' and 'simple' again.

```
>   Anmr := proc(x)  nops(op(2,op(xy)))  end;
>   Anme := proc(i,x) op(1,op(i,op(2,op(xy)))) end;
>   Apos := proc(i,x) op(2,op(i,op(2,op(xy)))) end;

>   Anmr(xy), Anme(1,xy), Apos(1,xy);
```

$$17, N2a, 112, 214$$

An extra 440 acts to calibrate the points in kilometers

```
>   positions :=
>   [seq([op(1,[Apos(i,xy)])/440,op(2,[Apos(i,xy)])/440],
>   i=1..Anmr(xy))];
```

$$positions := [[\frac{14}{55}, \frac{107}{220}], [\frac{79}{220}, \frac{193}{440}], [\frac{63}{110}, \frac{53}{220}], [\frac{149}{440}, \frac{47}{110}], [\frac{141}{440}, \frac{91}{220}], [\frac{29}{55}, \frac{233}{440}],$$
$$[\frac{279}{440}, \frac{333}{440}], [\frac{23}{55}, \frac{211}{440}], [\frac{251}{440}, \frac{313}{440}], [\frac{47}{88}, \frac{15}{22}], [\frac{21}{88}, \frac{26}{55}], [\frac{83}{440}, \frac{97}{220}], [\frac{43}{88}, \frac{233}{440}],$$
$$[\frac{207}{440}, \frac{227}{440}], [\frac{199}{440}, \frac{111}{220}], [\frac{3}{8}, \frac{9}{20}], [\frac{227}{440}, \frac{59}{88}]]$$

```
> xynames := [seq([Anme(i,xy)],i=1..Anmr(xy))]:
```

```
> DP := plot(positions,0..1,0..1,style=point,symbol=box,
> axesfont=[HELVETICA,BOLD,10]):
> DP;
```

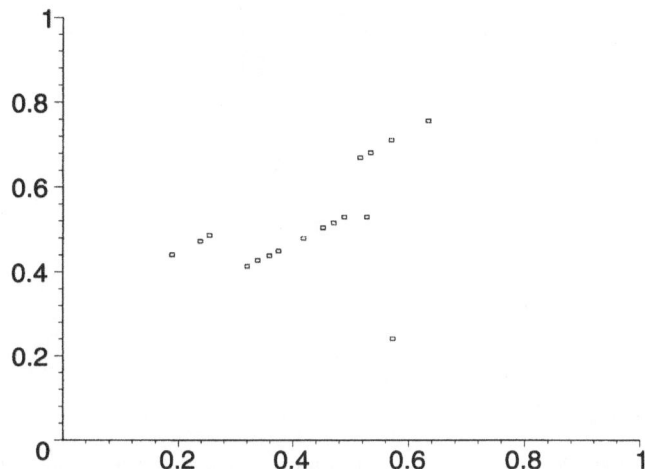

/bin/bash: point: command not found PLOT structures together. TP then becomes the full contour plot, with boreholes. Notice that when the data were digitised one of the positions is notional middle of the lake. The package plots supplys command display( ).

The 'constrained' keeps the x and y dimensions the same

```
> TP :=plots[display]({PPP,DP},axes=boxed,
> scaling=constrained,title='Perry Lakes with Boreholes',
> titlefont=[HELVETICA,BOLD,20]):  %;
```

Perry Lakes with Boreholes

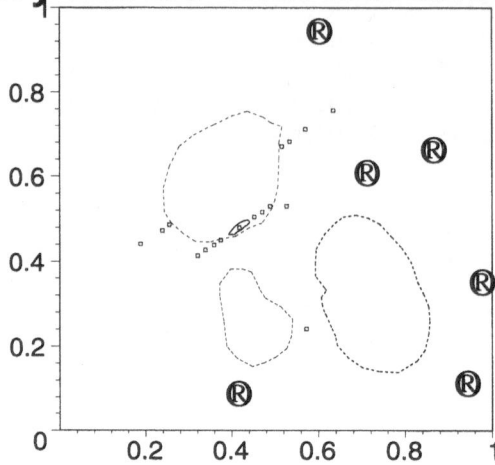

4.1.3 Saving Data

The lake contour and the bore data are needed later in section 4.3.1, page 204. Saving it can be simple or complex or, simply robust, like a pencil and paper. The simplest is to leave this worksheet active and carry on until there is no longer a need. This can be burdensome, awkward and inconvenient and doesn't allow for change.

On exiting Maple, the algebra on the worksheet becomes inactive. Replaying of the worksheet can retrieve the values, is a valuable technique of saving and a good learning experience. But it can be time consuming. Another, smooth alternative is to save the worksheet as a .m file in binary Maple internal format, or a text file, with a possible extension .mpl

Don't forget the space after save

```
>  save PPP,DP,"contour.m"; # save in internal Maple format

>  save PPP,DP,"contour.mpl"; # save in ASCII text format

>  save PPP,DP, contour; # simple name, in currentdir( )
```

Only material on the active worksheet is saved

These statements save the PLOT structures for the plots. Perhaps the positions or other variables will also be needed later. If everything in your Maple session is to be saved, give only a filename. If the file contains punctuation; periods, spaces, directories or device names; the name must be in double or backward quotes, a string or a symbol.

```
>  #save "c:/mplbook/scottVOL1/contours.m";
>  # save everything in binary

>  save "contours.m"; # save all workings in binary
```

These variables are saved as names so once the file is read, they are available for use as they were during the operation of the worksheet. The Maple internal format is preferred as it is the most efficient and fast; however, it is not very transportable between versions and platforms. It has been a custom to use the extension .mpl for the ASCII Maple language format, to present the input to Maple as one would type on the keyboard, without the prompt. It is easy to edit and read into Maple with a read statement.

Also move between windows with Ctrl-x, Ctrl-c and Ctrl-v

```
> #read "c:/mplbook/contour.mpl"; # read whatever
```

An alternative, presented here, is the simple use of the Maple output as a record. If the output is small and not activated, it remains and can be copied. This is particularly easy with lprint. In this case we will need PPP and DP; we organise an output that can be recopied; this is reasonably robust and is kept with the worksheet. Disadvantages include that it takes a little room and can be wiped out by careless use of the worksheet; a real advantage is that it records exactly the structure in use when the worksheet was abandoned.

The organisation of the save with lprint( ) insures that it can be copied from an inactive sheet, into section 4.3. PPP is the structure of the WestLake contours and DP, a PLOT structure of the postings of monitoring bores. The small area covered by water will be magnified when we look at the water table around the lake. All the operations are redone below to make the transfer totally clear.

```
> opts:=THICKNESS(2),LINESTYLE(1):
> west_water := evalf([WestLakewater],4):
> lprint(west_water);
```

```
[[.4068, .4636], [.4000, .4636], [.3955, .4636], [.3932, .4636],
[.3977, .4705], [.4068, .4818], [.4182, .4909], [.4273, .4955],
[.4364, .4977], [.4409, .4955], [.4409, .4909], [.4364, .4864],
[.4295, .4818], [.4227, .4750], [.4136, .4705], [.4068, .4636]]
```

Use the format

```
> PPP := PLOT(CURVES(west_water,opts)):
> #PPP;
```

```
> opts:= STYLE(POINT),SYMBOL(BOX):
> bore_points := evalf(positions,4):
> lprint(bore_points);
```

```
[[.2545, .4864], [.3591, .4386], [.5727, .2409], [.3386, .4273],
[.3205, .4136], [.5273, .5295], [.6341, .7568], [.4182, .4795],
[.5705, .7114], [.5341, .6818], [.2386, .4727], [.1886, .4409],
[.4886, .5295], [.4705, .5159], [.4523, .5045], [.3750, .4500],
[.5159, .6705]]
```

Use the format

```
> DP := PLOT(CURVES(bore_points,COLOUR(RGB,1,0,0),opts)):  #DP;
```

4.2 Bubbles: A Visualization

The movement of bubbles has broad application to environmental science, and is important in waste treatment for the transfer of oxygen and other gases in and out of water. Enclosed are some data on the rise of bubbles in a water column, acquired by observing the bubbles rise in water with a set of video cameras placed on two sides of the column, for a full 3D view. Unfortunately, bubbles are often poorly detected because the images are not perfect. Small disparities in the visualisation leads to mixed identification or loss of bubbles. The bubble movement usually follows a helical path but this effects a seemingly erratic behavior.

A reanimation shows the paths of the bubbles in a 3D sense and inconsistencies in the paths. The notional allocation of bubble names, corrupted in the data set, are sorted by colouring the bubbles, watching their movement, and streaking of the images.

4.2.1 Reading the Data File

First read in the data, have a look, and select the right bits.

```
> restart:
> filenme :='/mplbook/v1/450.prn';
> #filenme :='450.prn';
> fopen(filenme,READ);
```

$$filenme := /mplbook/v1/450.prn$$

$$0$$

```
> readline(filenme);
```

Continue to read until the data appear.

"Camera Pic#   Bubble hPos   vPos   Camera Pic#   Bubble hPos   vPos"

```
> i := 1:  ex := 1:  while ex <> 0 do
> ex := readline(filenme):i:=i+1:  od:  i-1;
```

$$95$$

```
> fclose(filenme):
```

This action effectively throws away the first line

```
> readlib(readdata):fopen(filenme,READ):
> readline(filenme);
```

"Camera Pic#   Bubble hPos   vPos   Camera Pic#   Bubble hPos   vPos"

```
> format := [string,integer,integer,float,float,
> string,integer,integer,float,float];
```

```
> readdata(filenme,format);
```

format := [*string, integer, integer, float, float, string, integer, integer, float, float*]

```
[["A", 40, 1, 2.907, 9.017, "B", 40, 1, 4.752, 9.145], ["A", 41, 1, 2.833, 8.502, "B", 41, 1, 4.799, 8.703],
 ["A", 42, 1, 3.088, 9.534, "B", 42, 1, 6.425, 9.293], ["A", 43, 1, 3.129, 9.097, "B", 43, 1, 6.285, 8.852],
 ["A", 44, 1, 3.236, 8.692, "B", 44, 1, 6.175, 8.418], ["A", 45, 1, 3.329, 8.280, "B", 45, 1, 6.089, 7.953],
 ["A", 46, 1, 3.311, 7.842, "B", 46, 1, 6.079, 7.493], ["A", 47, 1, 3.281, 7.375, "B", 47, 1, 6.411, 7.076],
 ["A", 48, 1, 3.162, 6.955, "B", 48, 1, 6.205, 6.612], ["A", 49, 1, 3.007, 6.531, "B", 49, 1, 6.265, 6.186],
 ["A", 50, 1, 2.901, 6.086, "B", 50, 1, 6.314, 5.733], ["A", 51, 1, 2.822, 5.622, "B", 51, 1, 6.259, 5.295],
 ["A", 52, 1, 2.793, 5.156, "B", 52, 1, 6.178, 4.833], ["A", 53, 1, 2.862, 4.715, "B", 53, 1, 6.080, 4.409],
 ["A", 40, 2, 2.470, 4.132, "B", 40, 2, 3.030, 4.659], ["A", 41, 2, 2.458, 3.620, "B", 41, 2, 4.304, 9.355],
 ["A", 42, 2, 2.773, 8.042, "B", 42, 2, 6.651, 9.540], ["A", 43, 2, 2.747, 7.577, "B", 43, 2, 6.544, 9.075],
 ["A", 44, 2, 2.672, 7.090, "B", 44, 2, 6.268, 8.733], ["A", 45, 2, 2.636, 6.590, "B", 45, 2, 5.834, 8.801],
 ["A", 46, 2, 2.647, 6.097, "B", 46, 2, 5.698, 8.335], ["A", 47, 2, 2.620, 5.644, "B", 47, 2, 6.000, 7.056],
 ["A", 48, 2, 2.533, 5.185, "B", 48, 2, 5.483, 7.453], ["A", 49, 2, 2.486, 4.724, "B", 49, 2, 5.401, 7.004],
 ["A", 50, 2, 2.476, 4.244, "B", 50, 2, 5.393, 6.548], ["A", 51, 2, 2.441, 3.755, "B", 51, 2, 5.446, 6.078],
 ["A", 52, 2, 2.450, 3.277, "B", 52, 2, 5.513, 5.628], ["A", 53, 2, 2.484, 2.798, "B", 53, 2, 5.570, 5.174],
 ["A", 40, 3, 2.163, 9.330, "B", 40, 3, 2.790, 6.907], ["A", 41, 3, 1.952, 5.877, "B", 41, 3, 3.165, 4.142],
 ["A", 42, 3, 2.356, 3.105, "B", 42, 3, 4.898, 8.247], ["A", 43, 3, 2.214, 2.567, "B", 43, 3, 4.905, 7.771],
 ["A", 44, 3, 2.057, 2.051, "B", 44, 3, 5.824, 9.252], ["A", 45, 3, 1.946, 1.516, "B", 45, 3, 5.965, 8.392],
 ["A", 46, 3, 1.895, .982, "B", 46, 3, 5.782, 7.938], ["A", 47, 3, 1.934, .451, "B", 47, 3, 5.625, 7.872],
 ["A", 48, 3, 1.619, 5.351, "B", 48, 3, 4.531, 5.508], ["A", 49, 3, 1.511, 4.877, "B", 49, 3, 4.597, 5.019],
 ["A", 50, 3, 1.324, 4.332, "B", 50, 3, 4.621, 4.525], ["A", 51, 3, 1.540, .730, "B", 51, 3, 5.305, 6.435],
 ["A", 52, 3, 1.624, .218, "B", 52, 3, 5.247, 6.188], ["A", 53, 3, 1.199, 3.951, "B", 53, 3, 5.157, 5.892],
 ["A", 40, 4, 1.848, 6.384, "B", 40, 4, 2.452, 9.531], ["A", 41, 4, 1.889, 8.861, "B", 41, 4, 2.904, 6.415],
 ["A", 42, 4, 1.957, 5.335, "B", 42, 4, 4.237, 8.888], ["A", 43, 4, 1.882, 4.809, "B", 43, 4, 4.212, 8.377],
 ["A", 44, 4, 1.723, 4.284, "B", 44, 4, 4.816, 7.305], ["A", 45, 4, 1.526, 6.813, "B", 45, 4, 4.712, 6.847],
 ["A", 46, 4, 1.604, 6.323, "B", 46, 4, 4.630, 6.393], ["A", 47, 4, 1.624, 5.831, "B", 47, 4, 5.672, 7.553],
 ["A", 48, 4, 1.319, 6.765, "B", 48, 4, 4.522, 5.852], ["A", 49, 4, 1.005, 5.701, "B", 49, 4, 4.547, 5.352],
 ["A", 50, 4, 1.376, 1.213, "B", 50, 4, 4.506, 4.819], ["A", 51, 4, 1.159, 4.866, "B", 51, 4, 4.642, 4.053],
 ["A", 52, 4, 1.188, 4.423, "B", 52, 4, 4.746, 3.645], ["A", 53, 4, .881, 2.710, "B", 53, 4, 4.832, 3.218],
 ["A", 41, 5, 1.306, 8.158, "B", 41, 5, 2.332, 9.050], ["A", 42, 5, 1.646, 8.371, "B", 42, 5, 3.281, 3.622],
 ["A", 43, 5, 1.422, 8.331, "B", 43, 5, 3.317, 3.096], ["A", 44, 5, 1.754, 8.221, "B", 44, 5, 4.185, 7.832],
 ["A", 45, 5, 1.681, 7.882, "B", 45, 5, 4.238, 7.346], ["A", 46, 5, 1.526, 7.433, "B", 46, 5, 4.328, 6.835],
 ["A", 47, 5, 1.367, 7.059, "B", 47, 5, 4.538, 5.958], ["A", 48, 5, .946, 6.138, "B", 48, 5, 2.701, .453],
 ["A", 49, 5, 1.235, 1.721, "B", 49, 5, 2.520, 2.291], ["A", 50, 5, 1.075, 5.280, "B", 50, 5, 2.472, 1.784],
 ["A", 51, 5, 1.099, 3.758, "B", 51, 5, 4.446, 4.281], ["A", 52, 5, .916, 3.239, "B", 52, 5, 4.339, 3.723],
 ["A", 53, 5, .948, 5.393, "B", 53, 5, 4.144, 3.231], ["A", 42, 6, 1.411, 8.731, "B", 42, 6, 3.007, 5.883],
 ["A", 43, 6, 1.759, 8.536, "B", 43, 6, 3.086, 5.337], ["A", 44, 6, 1.488, 7.316, "B", 44, 6, 3.257, 2.560],
 ["A", 45, 6, 1.526, 3.775, "B", 45, 6, 3.122, 2.043], ["A", 46, 6, 1.327, 3.275, "B", 46, 6, 2.964, 3.823],
 ["A", 47, 6, 1.201, 2.762, "B", 47, 6, 4.423, 6.319], ["A", 48, 6, 1.161, 2.238, "B", 48, 6, 2.645, 2.810],
 ["A", 49, 6, 1.180, 6.461, "B", 49, 6, 2.325, 5.259], ["A", 50, 6, .833, 4.044, "B", 50, 6, 2.329, 4.765],
 ["A", 51, 6, .896, 3.594, "B", 51, 6, 2.505, 1.280], ["A", 52, 6, .949, 5.699, "B", 52, 6, 2.607, .766],
 ["A", 53, 6, .414, 2.750, "B", 53, 6, 2.719, .215], ["A", 42, 7, 1.606, 8.984, "B", 42, 7, 2.152, 8.593],
 ["A", 43, 7, 1.537, 7.876, "B", 43, 7, 2.014, 8.078], ["A", 44, 7, 1.294, 7.897, "B", 44, 7, 3.119, 4.809],
 ["A", 45, 7, 1.141, 7.459, "B", 45, 7, 3.077, 4.311], ["A", 46, 7, 1.026, 6.995, "B", 46, 7, 2.943, 1.523],
 ["A", 47, 7, .956, 6.568, "B", 47, 7, 2.804, 3.324], ["A", 48, 7, .909, 5.000, "B", 48, 7, 2.222, 5.727],
 ["A", 51, 7, .966, 5.947, "B", 51, 7, 2.240, 4.307], ["A", 52, 7, .534, 3.148, "B", 52, 7, 2.095, 3.827],
 ["A", 44, 8, .873, 6.828, "B", 44, 8, 1.932, 7.614], ["A", 45, 8, .812, 6.371, "B", 45, 8, 1.901, 7.133],
 ["A", 46, 8, .810, 5.892, "B", 46, 8, 1.970, 6.651], ["A", 47, 8, .873, 5.451, "B", 47, 8, 2.784, .992]]
```

```
> dd := %:
```

```
> dd[1];
```
$$["A", 40, 1, 2.907, 9.017, "B", 40, 1, 4.752, 9.145]$$

```
> op(2..4,%),op(9,%),(op(5,%)+op(10,%))/2;
> # bubble number, time, x, y, and average vertical position
```
$$40, 1, 2.907, 4.752, 9.081000000$$

Limits the
accuracy and the
display
requirements

```
>  Digits:=3:  # set Digits to 3 decimals
>  fix := proc(x); [op(2..4,x),op(9,x),(op(5,x)+op(10,x))/2]; end;
```

These data are the camera frame (time), notional bubble number and the x, y and average z positions

Collects just the required data

$$fix := \mathbf{proc}(x)\,[\text{op}(2..4,\, x),\, \text{op}(9,\, x),\, 1/2 \times \text{op}(5,\, x) + 1/2 \times \text{op}(10,\, x)]\,\mathbf{end}$$

```
>  data := map(fix,dd);
```

This output is massaged

```
data :=
[[40, 1, 2.907, 4.752, 9.09], [41, 1, 2.833, 4.799, 8.60], [42, 1, 3.088, 6.425, 9.42],
[43, 1, 3.129, 6.285, 8.98], [44, 1, 3.236, 6.175, 8.56], [45, 1, 3.329, 6.089, 8.12],
[46, 1, 3.311, 6.079, 7.67], [47, 1, 3.281, 6.411, 7.23], [48, 1, 3.162, 6.205, 6.79],
[49, 1, 3.007, 6.265, 6.37], [50, 1, 2.901, 6.314, 5.92], [51, 1, 2.822, 6.259, 5.46],
[52, 1, 2.793, 6.178, 5.00], [53, 1, 2.862, 6.080, 4.57], [40, 2, 2.470, 3.030, 4.40],
[41, 2, 2.458, 4.304, 6.49], [42, 2, 2.773, 6.651, 8.79], [43, 2, 2.747, 6.544, 8.33],
[44, 2, 2.672, 6.268, 7.92], [45, 2, 2.636, 5.834, 7.70], [46, 2, 2.647, 5.698, 7.22],
[47, 2, 2.620, 6.000, 6.35], [48, 2, 2.533, 5.483, 6.33], [49, 2, 2.486, 5.401, 5.86],
[50, 2, 2.476, 5.393, 5.40], [51, 2, 2.441, 5.446, 4.92], [52, 2, 2.450, 5.513, 4.46],
[53, 2, 2.484, 5.570, 3.99], [40, 3, 2.163, 2.790, 8.13], [41, 3, 1.952, 3.165, 5.01],
[42, 3, 2.356, 4.898, 5.69], [43, 3, 2.214, 4.905, 5.18], [44, 3, 2.057, 5.824, 5.66],
[45, 3, 1.946, 5.965, 4.96], [46, 3, 1.895, 5.782, 4.46], [47, 3, 1.934, 5.625, 4.17],
[48, 3, 1.619, 4.531, 5.44], [49, 3, 1.511, 4.597, 4.95], [50, 3, 1.324, 4.621, 4.44],
[51, 3, 1.540, 5.305, 3.59], [52, 3, 1.624, 5.247, 3.21], [53, 3, 1.199, 5.157, 4.93],
[40, 4, 1.848, 2.452, 7.96], [41, 4, 1.889, 2.904, 7.64], [42, 4, 1.957, 4.237, 7.12],
[43, 4, 1.882, 4.212, 6.60], [44, 4, 1.723, 4.816, 5.80], [45, 4, 1.526, 4.712, 6.84],
[46, 4, 1.604, 4.630, 6.36], [47, 4, 1.624, 5.672, 6.70], [48, 4, 1.319, 4.522, 6.32],
[49, 4, 1.005, 4.547, 5.53], [50, 4, 1.376, 4.506, 3.02], [51, 4, 1.159, 4.642, 4.47],
[52, 4, 1.188, 4.746, 4.04], [53, 4, 0.881, 4.832, 2.97], [41, 5, 1.306, 2.332, 8.61],
[42, 5, 1.646, 3.281, 6.00], [43, 5, 1.422, 3.317, 5.72], [44, 5, 1.754, 4.185, 8.03],
[45, 5, 1.681, 4.238, 7.62], [46, 5, 1.526, 4.328, 7.14], [47, 5, 1.367, 4.538, 6.51],
[48, 5, 0.946, 2.701, 3.30], [49, 5, 1.235, 2.520, 2.01], [50, 5, 1.075, 2.472, 3.53],
[51, 5, 1.099, 4.446, 4.02], [52, 5, 0.916, 4.339, 3.48], [53, 5, 0.948, 4.144, 4.32],
[42, 6, 1.411, 3.007, 7.31], [43, 6, 1.759, 3.086, 6.94], [44, 6, 1.488, 3.257, 4.94],
[45, 6, 1.526, 3.122, 2.91], [46, 6, 1.327, 2.964, 3.55], [47, 6, 1.201, 4.423, 4.54],
[48, 6, 1.161, 2.645, 2.53], [49, 6, 1.180, 2.325, 5.86], [50, 6, 0.833, 2.329, 4.41],
[51, 6, 0.896, 2.505, 2.44], [52, 6, 0.949, 2.607, 3.23], [53, 6, 0.414, 2.719, 1.49],
[42, 7, 1.606, 2.152, 8.79], [43, 7, 1.537, 2.014, 7.98], [44, 7, 1.294, 3.119, 6.36],
[45, 7, 1.141, 3.077, 5.89], [46, 7, 1.026, 2.943, 4.26], [47, 7, 0.956, 2.804, 4.95],
[48, 7, 0.909, 2.222, 5.37], [51, 7, 0.966, 2.240, 5.14], [52, 7, 0.534, 2.095, 3.50],
[44, 8, 0.873, 1.932, 7.23], [45, 8, 0.812, 1.901, 6.76], [46, 8, 0.810, 1.970, 6.28],
[47, 8, 0.873, 2.784, 3.23]]
```

```
>  nd := nops(data);
```

$$nd := 94$$

```
>  xyz := map(x->[op(3..5,x)],data):
>  xyz[10];
```

$$[3.007, 6.265, 6.37]$$

4.2.2 A lookup of Data

a 'lookup' is not conventional

Knowing the time and the bubble, the position can be found from a lookup, pt[time][bubble]. Without other knowledge cluttering up the analysis; the camera number, or the number of bubbles visible at any time; the position of a single bubble can be found.

NULL is an empty sequence

```
> pt := 'pt': 'pt = lookup';
> ttt := NULL:# seq of times
> for i from 1 to nd do;
> da := data[i]; dr:=op(2,da); tt:=op(1,da);
> pt[tt][dr]:=[op(3..5,da)];# point information
> ttt := ttt,tt; # times of camera shapshots
> od:
```

$$pt = lookup$$

```
> pt[46][1];
```

$$[3.311, 6.079, 7.67]$$

The sequence ttt was collected so that it contains all the times for which camera shots are available. Converting it into a set automatically makes sure there are no duplicates. It is then converted into a list and sorted to make the exact time sequence of times. There are 14 frames in the data set, starting at 40 and ending at 53.

```
>{ttt}: times := sort([op(%)]);
```

$$times := [40, 41, 42, 43, 44, 45, 46, 47, 48, 49, 50, 51, 52, 53]$$

```
> ntimes :=nops(times);
```

$$ntimes := 14$$

```
> tstart := op(1,times);
```

$$tstart := 40$$

```
> tend :=op(ntimes,times);
```

$$tend := 53$$

The following is one way of collecting the coordinates of all the bubbles at time 46. It is rather crude; it presumes there are less than 20, it gives no identity to the bubbles and, to be useful, could not allow for bubbles that were not seen by the camera. As this analysis evolves, it will be clear that the power of Maple allows that the identities of the bubbles in the data set can be maintained. The sequences can be cut and pasted, with 'dropout' and 'dropin' bubbles, and a whole animation can carry on.

```
> select(hastype,[seq(pt[46][i],i=1..20)],list);
```

$$[[3.311, 6.079, 7.67], [2.647, 5.698, 7.22], [1.895, 5.782, 4.46], [1.604, 4.63, 6.36],$$
$$[1.526, 4.328, 7.14], [1.327, 2.964, 3.55], [1.026, 2.943, 4.26], [.81, 1.97, 6.28]]$$

```
> i:='i';
```

$$i := i$$

4.2.3 Bubble Sequences

The routine below collects just the sequence in which the bubbles appear.
Some bubbles may be missing; bubble 5 may appear before bubble 1 or whatever; and the sequence of bubbles in every frame is preserved.

```
> for i from tstart to tend do
> bubbs.i := NULL: od:
> for i from 1 to nd do; data[i];
> ti := op(1,%);
> bubbs.ti:=bubbs.ti,op(2,%%);
> od:  i := 'i':
```

The bubbles that
appear in each
time shot

```
> bubbs.40;bubbs.41;bubbs.42;bubbs.43;
```
$$1, 2, 3, 4$$
$$1, 2, 3, 4, 5$$
$$1, 2, 3, 4, 5, 6, 7$$
$$1, 2, 3, 4, 5, 6, 7$$

```
> nops([bubbs.43]);bubbs.43[3];
```
$$7$$
$$3$$

```
> ms := max(seq(op(2,data[i]),i=1..nd));
```

The notional
bubbles with the
largest number

$$ms := 8$$

4.2.4 Identification with Colour

Each bubble is associated with a colour. To keep the colours individualistic and
as distinguishable as possible, eight are chosen, to correspond roughly with the
binary presentation of a octal number. A white background, the Red, Green,
Blue sequence RGB, 1, 1, 1 is useless and is converted into another colour.

yellow is also
nearly invisible

```
> col7 := proc(n) local N,R,G,B,S;
> N:=irem(n,8);
> if N=0 then 0,0,0; # makes black
> elif N=7 then 0,.7,.7;
> elif N=6 then .7,0,.7;# purple colour, yellow nearly invisible
> else;
> R := iquo(N,4); irem(%%,4);
> G := iquo(%,2);
> B := irem(%%,2);
> R,G,B;
> fi;
> end:
```

The following colours are used.

```
> one,col7(1),three,col7(3),seven,col7(7),ten,col7(10);
```

one, 0, 0, 1, *three*, 0, 1, 1, *seven*, 0, .7, .7, *ten*, 0, 1, 0

For each bubble, the formated COLOR part of the PLOT structure is set up in sequence co.

```
>  co := seq(COLOR(RGB,col7(i)),i=1..ms):
```

```
>  co[3], co[8];
```
$$\text{COLOR}(RGB, 0, 1, 1),\ \text{COLOR}(RGB, 0, 0, 0)$$

Exercise 4.3 Visibility of Colours

Review the colours of the bubbles, with a view to getting colours that are distinctly different to your eye. Use the four formats; colour=cyan, COLOR(RGB,.1,.4,.3), COLOR(HUE,0.77), and COLOR(HSV,.77,.9,1). See section 3.7.3, page 157

```
>  ?plot,color
>  ?plot,structure
```

4.2.5   A Simplest Plot

A single image is plotted, of all 7 bubbles present in frame 44.

```
>  seq(pt[44][i],i=1..7);
```

[3.236, 6.175, 8.56], [2.672, 6.268, 7.92], [2.057, 5.824, 5.66], [1.723, 4.816, 5.80], [1.754, 4.185, 8.03], [1.488, 3.257, 4.94], [1.294, 3.119, 6.36]

```
>  with(plots):
>  with(plottools):
```

```
>  P[44] := seq(pointplot3d([pt[44][i]],
>  style=POINT,symbol=CIRCLE,
>  color=co[i])
>  ,i=1..7):
```

```
>  display(P[44],axes=BOXED);
```

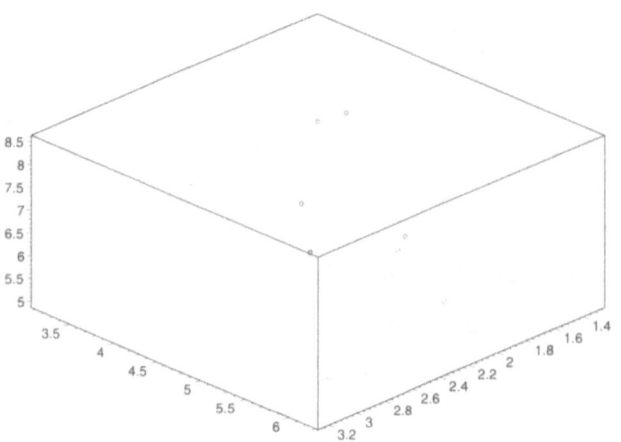

4.2.6　Bubble Spheres

for 2D, Kofler, pg
544, uses a sine
function

　The bubble images are too small. There are a number of ways to remedy this, including the manufacture of shaded spherical surfaces, large font TEXT and disk, as done in section 3.5. A routine from plottools is chosen; the bubbles are drawn as coloured spheres, to be part of a PLOT3D structure as POLYGONS in three dimensional points.

```
> with(plottools):
> sphere([5,5,5],3,numpoints=100,color=red):# radius=3
> PLOT3D(%);
```

We take out the lines with style=patchnogrid, lower its size, and it is a red bubble. Set the radius with parameter size.

```
> size := 0.5;
```

$$size := .5$$

4.2.7　Using Plot Structures

It seems that we can no longer ignore the detailed, baseline PLOT structure, the plotting functional that Maple always produces, which is hidden by the plot and display commands. In this case it is a PLOT3D structure and it allows an appropriate box to be drawn.

```
>   ?plot,structure
>   P[44] := seq(sphere(pt[44][i],size,
>   numpoints=100,
>   style=patchnogrid,
>   color=co[i])
>   ,i=1..7):

>   display(PLOT3D(P[44],AXESSTYLE(BOXED)));
```

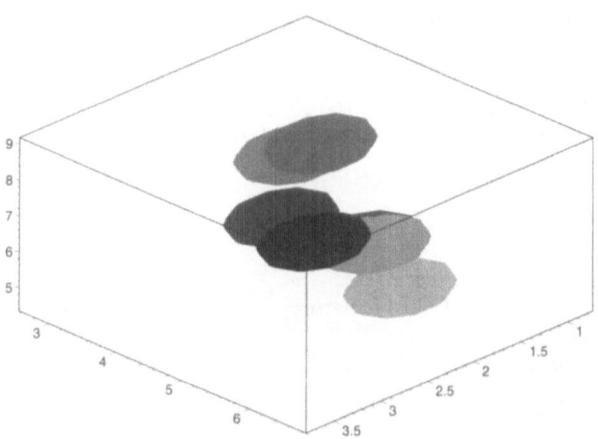

Then we constrain the axes and adjust the scales to give a reasonable view of the experiment.

```
>   opts := VIEW(1..5,3..8,5..10),
SCALING(CONSTRAINED),AXESSTYLE(BOXED):
display(PLOT3D(P[44],opts));
```

Strangely, one bubble doesn't want to be spherical

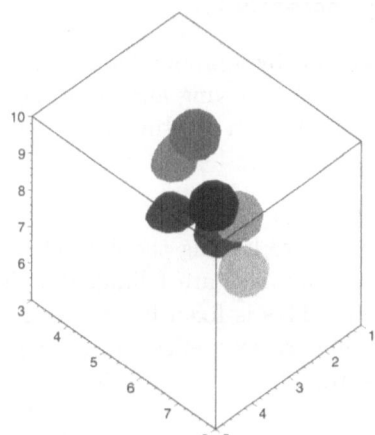

4.2.8   Animation of the Bubbles

Click on the
figure
We now put this into a sequence, in time. Note that when you click on
the figure, the menu bar changes. The buttons above, on the lower menu bar,
allow cycling through the PLOT3D list, the picture frames. One can proceed
frame by frame, stop and go, simply cycle through, speed up or slow the
animation. Use the three buttons on the left; the centre left button activates
the animated sequence, the far left stops it, the right one proceeds frame by
frame. Advice on the lower left of the screen tells what each button does.

Try the lower
menu buttons

The new, active plot is an animation, PPPP.

```
> #opts := VIEW(0..5,1..8,3..10),
> SCALING(CONSTRAINED),AXESSTYLE(BOXED):
```

Adjust the opts
after viewing

```
> for j from tstart to tend do;
> PP[j]:=seq(sphere(pt[j][i],size,
> numpoints=100,
> style=patchnogrid,
> color=co[i]),
> i=bubbs.j); od:
```

```
> PPP := [seq(PP[j],j=op(times))]:
```

```
> PPPP := map(PLOT3D,PPP,opts):
```

```
> display(PPPP,insequence=true);
```

Rotation of the figure is possible by 'grabbing' it with the mouse and manually turning it gently, adjusting the viewing angles displayed in the gadgets
on the left of the menu bar. This allows a scrutiny of the bubble locations and
what they are doing.

It is possible to
reverse the
VIEW range
The animation sequences through every bubble at every instant; not what
we wanted. Some of the bubbles are leaving the field of view; the ranges in
VIEW are increased. Also, the bubbles are falling, not rising; the vertical
direction needs to be reversed. This is fixed by reversing the z coordinate.
That is, the positions of the bubbles were, effectively, measured from the top
of the bubble column, downward.

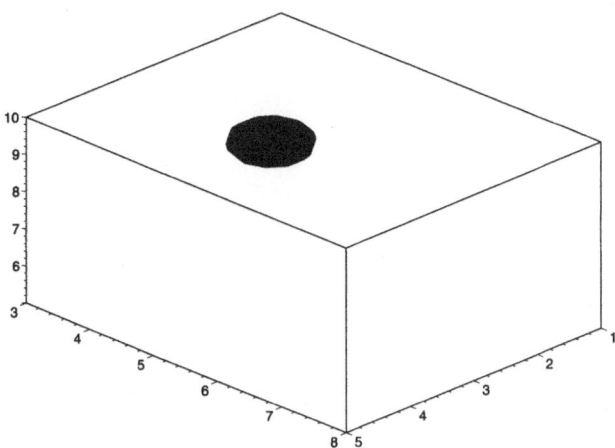

Back to the data set; reverse the z direction

```
>   i := 'i':  nd;
```
$$94$$
```
>   for i from 1 to nd do;
da := data[i]; db := op(2,da);
ptr[op(1,da)][db]:=
[op(3..4,da),-op(5,da)];# reverse of vertical, z
od:
>   data[2];pt[46][1];ptr[46][1];
```
$$[41, 1, 2.833, 4.799, 8.60]$$
$$[3.311, 6.079, 7.67]$$
$$[3.311, 6.079, -7.67]$$

The PLOT3D data structure makes it possible to combine the bubble images in a given frame, then sequence through the frames. This allows the sequencing we want for the animations. First we look at one frame, showing the bubbles.

```
>   opts := view=[0..5,1..8,-10..-3],
scaling=CONSTRAINED, axes=BOXED:
>   P[48] := PLOT3D(seq(sphere(ptr[48][i],size,
>   numpoints=100, style=patchnogrid,
>   color=co[i]), i=1..7)):
>   XX := display(P[48],opts):
```

Note that display is evaluated only once as the right side of an assignment so that using a ; on the statement makes a mess on the screen, a difficulty with most PLOT structures.

Edit/Undo result removes the mess

```
>   XX;
```

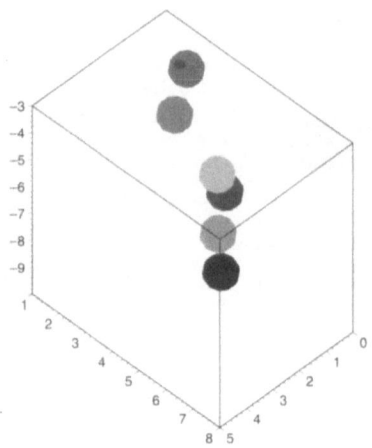

Exercise 4.4 Missing Bubble

Is one bubble missing? It seems so.

```
> bubbs.48;
```
$$1, 2, 3, 4, 5, 6, 7$$
```
> seq(co[i],i=bubbs.48);
```

COLOR(RGB, 0, 0, 1), COLOR(RGB, 0, 1, 0), COLOR(RGB, 0, 1, 1),
    COLOR(RGB, 1, 0, 0), COLOR(RGB, 1, 0, 1), COLOR(RGB, .7, 0, .7),
    COLOR(RGB, 0, .7, .7)

```
> 'time range  '.tstart, tend;
```
time range 40, 53
```
> op(times);
```
$$40, 41, 42, 43, 44, 45, 46, 47, 48, 49, 50, 51, 52, 53$$

4.2.9   Animated Swarms of Bubbles

In the following, a do loop and display plays the sequence; what the computer thinks the bubbles are doing.

```
> i:='i':   j:='j':k:='k':
> for j from tstart to tend do;
> P[j] := PLOT3D(seq(sphere(ptr[j][i],size,
> numpoints=100,
> style=patchnogrid,
> color=co[i]),
> i=bubbs.j)):od:
```

```
>  PP := [seq(P[k],k=op(times))]:

>  display(PP,insequence=true,lightmodel=light3,opts);
>  i:='i':  j:='j':  k:='k':
```

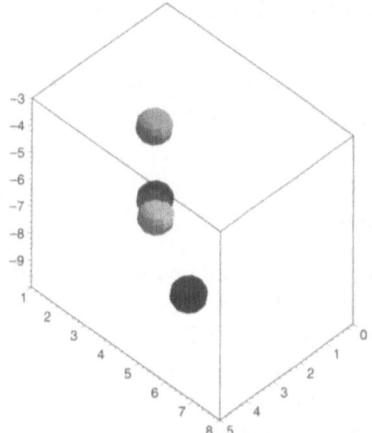

Grab the active 3D graph and rotate it to scrutinise the bubbles; you can see some of them leaving the box. About 20 degrees and 80 degrees are reasonable angles, though you should look at the bubbles from both the 'front' and the 'side'. Again, some of the bubbles seem to be missing. Also, notice that some of the bubbles seem confused. This is because the names, as found in the original data analysis, are confused. Is the black bubble that comes and goes really a separate bubble? The purple and light blue bubbles are quite confused. Further analysis may sort this out.

```
>  for j from tstart to tend do:
>  'frame '.j,' bubbles = ',bubbs.j;
>  od;
>  i:='i':  j :='j':
```

$$
\begin{aligned}
&\textit{frame 40, bubbles} = , 1, 2, 3, 4 \\
&\textit{frame 41, bubbles} = , 1, 2, 3, 4, 5 \\
&\textit{frame 42, bubbles} = , 1, 2, 3, 4, 5, 6, 7 \\
&\textit{frame 43, bubbles} = , 1, 2, 3, 4, 5, 6, 7 \\
&\textit{frame 44, bubbles} = , 1, 2, 3, 4, 5, 6, 7, 8 \\
&\textit{frame 45, bubbles} = , 1, 2, 3, 4, 5, 6, 7, 8 \\
&\textit{frame 46, bubbles} = , 1, 2, 3, 4, 5, 6, 7, 8 \\
&\textit{frame 47, bubbles} = , 1, 2, 3, 4, 5, 6, 7, 8 \\
&\textit{frame 48, bubbles} = , 1, 2, 3, 4, 5, 6, 7 \\
&\textit{frame 49, bubbles} = , 1, 2, 3, 4, 5, 6 \\
&\textit{frame 50, bubbles} = , 1, 2, 3, 4, 5, 6 \\
&\textit{frame 51, bubbles} = , 1, 2, 3, 4, 5, 6, 7 \\
&\textit{frame 52, bubbles} = , 1, 2, 3, 4, 5, 6, 7 \\
&\textit{frame 53, bubbles} = , 1, 2, 3, 4, 5, 6
\end{aligned}
$$

```
>  co;
```

COLOR(RGB, 0, 0, 1), COLOR(RGB, 0, 1, 0), COLOR(RGB, 0, 1, 1),
    COLOR(RGB, 1, 0, 0), COLOR(RGB, 1, 0, 1), COLOR(RGB, .7, 0, .7),
    COLOR(RGB, 0, .7, .7), COLOR(RGB, 0, 0, 0)

These are the bubbles that should be present, and their colours. Next, traces of the images are formed in streaks, to see the continuity of the tracks and identify the bubbles.

4.2.10 Streaks of Bubbles

It is clear that the rendering of each sphere image should have been saved. Without a 'save image' option in display, the earlier images of the bubbles must be redrawn to produce a streaking effect. This is done with a separate index k which 'collects' all the images from the start.

```
>   unassign('i','j','k','l'):
>   for j from tstart to tend do;
>   PS[j] := PLOT3D(seq(seq(sphere(ptr[k][i],size,
>   numpoints=100,
>   style=patchnogrid,
>   color=co[i]),
>   i=bubbs.j),k=tstart..j)):  od:

>   PPP := [seq(PS[l],l=op(times))]:
```

Twist it around a little to get a better perspective

```
>   display(PPP,insequence=true,lightmodel=light2,opts);
-.5cm
```

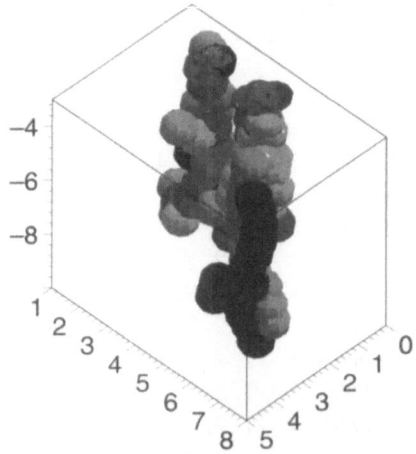

The 'worms' show that, indeed, some of the names are mixed. Perhaps you can sort them out.

Exercise* 4.5 POINT and TEXT

Repeat the above animation using style=POINT or STYLE(POINT) and symbol=CIRCLE or SYMBOL(CIRCLE). You may want to use plotpoint( ). It will be hard to see the points, unless they are streaked and become lines. Do the same with a large TEXT O; though the most economical way to do the drawing, the bubbles do not have a realistic 3D look.

Exercise 4.6 Light and Shading

Use ?plot3d,options, ?plot,color and/or ?plot3d[colorfunc] to arrange the ambientlight, and/or lightmodel to get pleasing spherical bubbles. You may want to increase numpoints to get more resolution and, of course, rotate the picture to get a best effect.

Exercise* 4.7 Identification of Bubbles

Look closely at the data – you have seen how the bubbles behave in ascent. Some of the bubbles jump around unreasonably. It is likely that the original identification of the bubbles is flawed. Adjust the data to care for this; reidentify individual bubbles and alter their names in the data listlist. When properly identified, the bubbles take an upward, cyclic path that is more or less continuous. Of course, a bubble or two may disappear from view and reappear, further up the column.

Bubbles moving up a water column were viewed through data lists of their positions. Bubble images were formed and viewed spatially; images that were variously corrupted. Maple lookup, reimaging, sorting and colouring has revealed a picture of bubbles in 3D. The bubbles were animated, their images rotated and streaked.

Analysis of data is in your hands. Now, consider that you want to further interpret data in a simplest way. Moving away from bubbles, intepretation through interpolation is presented in the next section.

4.3   Interpolation: An Application

Many a student and many consultants have asked the question: How can I present spatial data? A specific example is the groundwater problem, where one makes measurements of the water table in a region and wants to know the 'topography of the undergroundwater surface'. The main features that disturb the water table are lakes and rivers, wells and drains as well as transpiration from trees, evaporation and rainfall. Here we presume that a feature of lake, acting as a source or sink for groundwater, can be emulated and added to give a simulation of the water table contours.

In reading this, I ask your indulgence. Subsection 4.3.1 should not be executed at this point! Read through it and get a general feel for the data. Proceed through the following subsections (4.3.2 and 4.3.3) and, when you have assigned values to the data in section 4.3.4, page 213, you can activate subsection 4.3.1. It is ordinary in science and an important part of Maple; the real world does not read from top to bottom or from left to right. Trial and error, feedback and recursion are an important part of life and Maple.

4.3.1   A Rough Look at the Data

But first we look at the data in a traverse, through the lake.

This is cheating because the data are worked up later

```
>   sss := WL12,WL11,WL10,WL9,WL8,WL7,WL6,PL2;
>   data[sss[2]]; ntrav:= nops([sss]);
>   ipt := 'op(1,[xyc[sss[i]]]),op(1,[data[sss[i]]]) ':
```

First is a listing of the data points of interest; the cryptic notation refers to 'Water Level' WL and 'Pumping Level' PL measurements. All the data points are collected by measuring the water level in a hole in the ground, a bore, which is usually cased with a plastic or metal tube that is especially slotted to receive the water. Some bores are made just for this monitoring purpose and are generally called 'piezometers'; other bores are attached to pumps and, perhaps regularly, a large amount of water is pumped from them, they are 'pumping bores'. The second set of statements extracts the x value of the data point i, followed by the water table height of data point i. This sequence produces a traverse through WestLake, as initially presented in section 4.1; a simplest listing of distance along and water table height measured, in a rough straight line. A total of 8 points are chosen for the traverse, as substantiated in ntrav. The last statement, without output, presents the ordering of the datum point with index i. Notice how the level of evaluation is held back by the '. This is, in effect, any point i; a similar effect could be gained by making sure i is still unevaluated, unassign('i'). There is no real interest in i; it is simply a running index.

$$sss := WL12, \ WL11, \ WL10, \ WL9, \ WL8, \ WL7, \ WL6, \ PL2$$

$$2.864, \ 2.050$$

$$ntrav := 8$$

```
>   ROUGH := [seq([ipt],i=1..ntrav)];
```

$$ROUGH := [[\frac{141}{440}, 3.403], [\frac{149}{440}, 2.864], [\frac{79}{220}, 2.448], [\frac{3}{8}, 1.673], [\frac{199}{440}, .884], [\frac{207}{440}, .883],$$
$$[\frac{43}{88}, .745], [\frac{63}{110}, 2.282]]$$

```
>   plot(ROUGH,x=0.3..0.6,H=0..4,style=LINE);
```

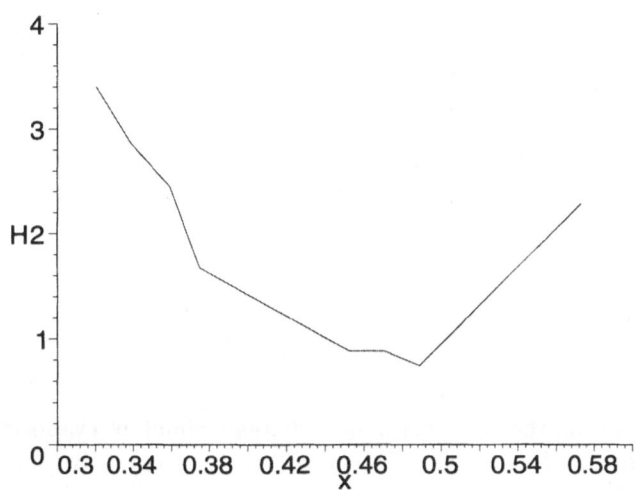

Groundwater Flows

Groundwater flows downhill following the hydraulic gradient, the slope in the water table. The lake generally appears as a low point; the trend is such that the lake is acting like a sink for water from the west and the east. This view is flawed but it does lead to an appropriate analytical form. Considering the lake as a pump, or rather a series of pumps, gives a pattern for the perimeter of the lake. This simplification is like 'Occam's Razor' and the 'Doctrine of Expository Simplicity'. A superposition of pumping drawdown cones, adding sinks, produces an analytical interpolation formula for the data.

See the plot in the last section

William of Occam, 1300-49

Page 478 of Forman S. Acton [1] ≈ Laplace's equation

4.3.2 Formulation

This simplest of formulations presumes steady state groundwater flows, without time dependence. And then only flow in a horizontal plane is allowed, following Darcy's law

$$\text{Darcy Velocity} = \text{K} * \text{fall/run}$$

where K is the 'hydraulic conductivity'. The Darcy Velocity is defined as the volumetric flow per unit crossectional area and is really a volumetric flux.

The fall/run is the simple statement that water flows downhill hydraulically, following the gradient. The sign is non-mathematical but this term is the simplest expression of the 'hydraulic gradient'. With a given depth of aquifer, d, the flow into a circular region around a sink is simply given by:

>   eq := G = 2*Pi*r*d*K*diff(H(r),r);

$$eq := G = 2\pi r d K \left(\tfrac{\partial}{\partial r} H(r)\right)$$

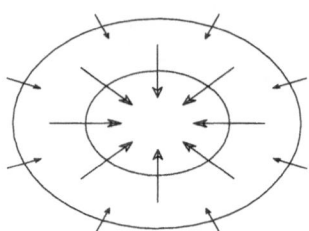

where r is the radius from the sink. Without recharge (rainfall or evapotransporation), G is constant for any r. This means that we can solve for the functional form for H(r):

>   assume(r>0); interface(showassumed=0): # removes the ~

>   dsolve({eq,H(rl)=Hl},H(r)): expand(rhs(%));

$$\frac{1}{2}\frac{G\ln(r)}{\pi d K} - \frac{1}{2}\frac{\ln(rl)\,G}{\pi d K} + Hl$$

This gives the general form we expect of the data, a logarithmic form with 2 parameters,

$$\textbf{A*ln(r) + B}$$

where the ln(rl) is absorbed into B. Let us explore the general shape of the constant height contours. First consider polar coordinates.

>   opts := -1..1,-1..1,coords=polar;

$$opts := -1..1,\ -1..1,\ coords = polar$$

>   plot({seq([ln(r)/2,t,t=0..2*Pi],r=1..5)},opts);

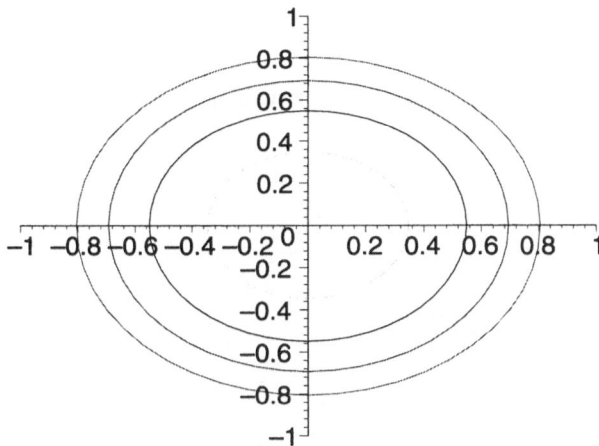

The simplest picture is of a lake acting as a single sink for underground water flow. The heights of the water table around the lake are ever increasing outward, following a logarithmic curve, as pictured. A more complex situation allows that different sinks and sources exist, roughly, where the lake is; the complex patterns of flows normally seen in the vicinity of a lake result from a adding of effects from all the sinks and sources. To locate each source or sink we need to use Cartesian coordinates; it immediately strikes us that the radial distance is made up of the sum of squares of x and y; these are circles and, in plotting, are implicit functions. Ordinary plot doesn't do surface contouring of implicit functions easily but a search of help files reveals that an appropriate contouring package is available.

When a lake supplies water to the aquifer, it acts as a source and recharges the aquifer

```
>  with(plots);
```

[*animate, animate3d, animatecurve, changecoords, complexplot, complexplot3d,*
   conformal, contourplot, contourplot3d, coordplot, coordplot3d, cylinderplot,
   densityplot, display, display3d, fieldplot, fieldplot3d, gradplot, gradplot3d,
   implicitplot, implicitplot3d, inequal, listcontplot, listcontplot3d, listdensityplot,
   listplot, listplot3d, loglogplot, logplot, matrixplot, odeplot, pareto, pointplot,
   pointplot3d, polarplot, polygonplot, polygonplot3d, polyhedra_supported,
   polyhedraplot, replot, rootlocus, semilogplot, setoptions, setoptions3d,
   spacecurve, sparsematrixplot, sphereplot, surfdata, textplot, textplot3d, tubeplot]

```
>  ln(x^2+y^2);
```

$$\ln(x^2 + y^2)$$

```
> contourplot(%,x=-1..1,y=-1..1,
> grid=[100,100], colour=blue,contours=[-3,-2,-1,0]);
```

A single sink has
a circular shape

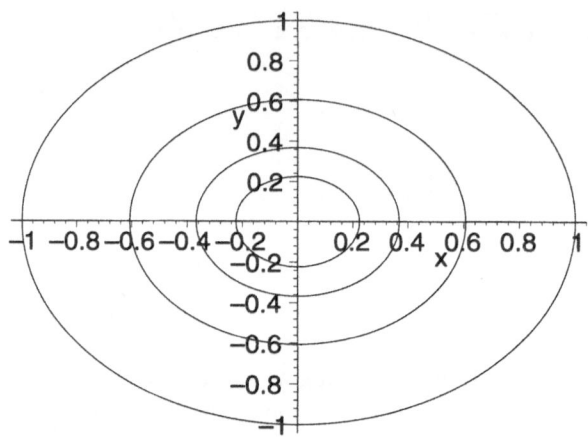

Composite Sinks

Next we consider the composite effect of more than one sink. These terms are weighted to keep the overall 'strength' of the sinks the same.

```
> 2/3*ln(x^2+y^2)+1/3*(ln((x-.25)^2+(y-.1)^2));
```

```
> opts:=grid=[100,100],colour=blue,contours=[-3,-2,-1,0]:
```

```
> contourplot(%%,x=-1..1,y=-1..1,opts);
```

Another sink
added, slightly
removed

$$\frac{2}{3}\ln(x^2 + y^2) + \frac{1}{3}\ln((x - .25)^2 + (y - .1)^2)$$

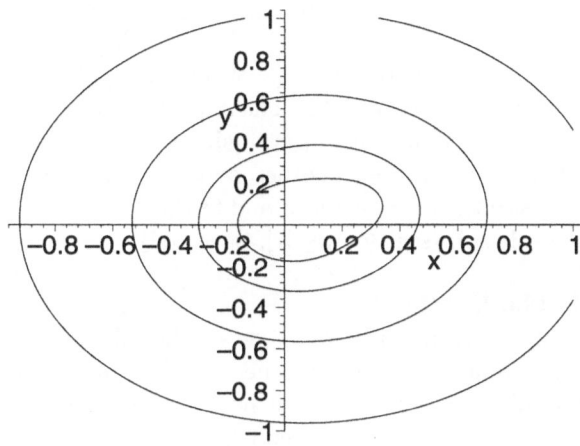

```
>   ln(x^2+y^2)/2+ln((x-.25)^2+(y-.1)^2)/4
>   +ln((x+.1)^2+(y-.25)^2)/4;
>   contourplot(%,x=-1..1,y=-1..1,opts);
```

$$\frac{1}{2}\ln(x^2 + y^2) + \frac{1}{4}\ln((x - .25)^2 + (y - .1)^2) + \frac{1}{4}\ln((x + .1)^2 + (y - .25)^2)$$

Another sink
added in the
upper left

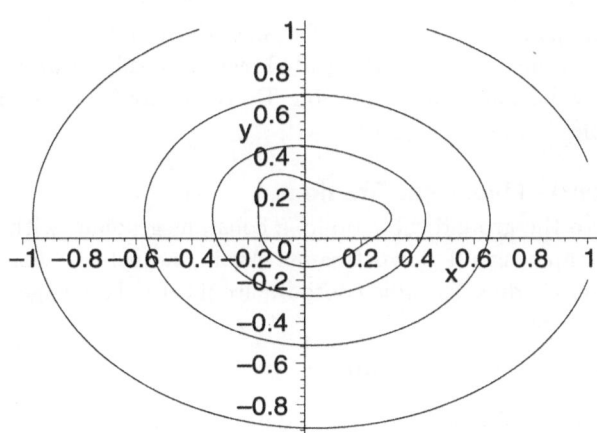

The shape of the center contour is expected to be, more or less, the shape
of the lake. With two or more 'sinks' or 'sources', it is possible to emulate

Lakes can be
sinks and sources

just about any combination; lakes, pumps or whatever. A 'flow-through' lake effectively has a source on one side and a sink on another. Most lakes are not symmetric. We simply combine 'sources' and 'sinks' to make whatever shape is required, or is necessary to simulate the shape of the water table at the lake. Here we have made a symmetric set of sinks. The important thing, other than **Darcy's Law**, page 205, is the use of **superposition**, where the solutions add. That is, we form one solution, then another, and so on. Then we multiply each by an arbitrary constant, add these terms and fit the boundary conditions

The shape of the real lake   (the data)–and have a solution (see Bouwer [7], page 204).

Exercise 4.8 Hydraulic Dipole

Draw a diagram of the case where one side of the lake acts as a sink (the side upstream hydraulically) and the other side acts as a source (the side downstream hydraulically). The 'model' has some resemblance to the 'flow-through' lakes on the Swan coastal plain in Western Australia. The lakes exist with a region of a sloping water table and water flows into the lakes on the east and out of the lakes on the west; The lakes are distorted expressions of the groundwater isopotentials. Remember to normalise the intensities (multiplication factors) so that the contours keep roughly the same scale; also, a lake always has the same height everywhere, though we may not know what the height is.

Exercise* 4.9 Continuum of Sinks

Consider a continuum of sinks, with intensities given by a simple function of position. Try a constant, a parabola, a sine or an exponential; it is expected that the function be zero everywhere except in the lake, or taper to zero at infinity; perhaps it would be piecewise smooth; its integral should be normalised to have a value of 1 when integrated over all radii. Adding up the effect of the sinks now becomes an integration. Do the analytical or numerical integrations in Maple.

Exercise 4.10 Gentle Hydraulic Gradient

Add a gradient to the groundwater table, a constant gradient with height equal to a*x + b, tapering off downstream. Again, adjust the heights or constants so that you produce sensible contours around the lake edges.

4.3.3 Choosing Appropriate Sinks

The points should be close to the lake   Looking at the problem from the side of the data, we read in the data, x,y,z; z is the height of the water table. For each point, we know the height of the water table and where the point is. The position enters into the formulas,

For each point, we calculate a distance to each sink   above, with different parameters for each sink and forms a set of equations in the parameters. Solving the set of equations gives a complete solution with the arbitrary and not-so-arbitrary parameters. With many data points, it is possible to do a least squares fit to get best values, see section 5.5, page 266.

The effect from each sink is associated with a radius, relative intensity and a position. Each massages the shape of the lake profile and the overall effect diminishes logarithmically out from the centre. The whole solution is composited from each contribution. Ultimately, this fitting gives an analytical function that fits the shape of the water table around the lake; it gives interpolated values for the water table between data points, and extrapolated values beyond.

With the Perry Lakes data, it seems that five sinks will do. The simulated lake is not perfect, but bears some semblance to the real lake. The positions are fixed at -0.2, -0.1, 0.0, +0.2, +0.3 of a kilometre from the lake centre, along an approximate SW/NW direction. The chosen values of relative intensities are 1/8, 1/7, 1/11, 1/7, 1/8; these will be recalculated. This procedure has minimal complication, yet it is meaningful and robust. The lake is about 20 m wide and 70 m long.

A trial and error could allow the positions to vary

Don't worry about the absolute scale; we will scale later

```
>  ln(x^2+y^2)/8+ln((x-.1)^2+(y-.1)^2)/7
>  +ln((x-.2)^2+(y-.2)^2)/11
>  +ln((x-.3)^2+(y-.3)^2)/7+ln((x-.4)^2+(y-.4)^2)/8;
>  contourplot(%,x=-1..1,y=-1..1,opts);
```

$$\frac{1}{8}\ln(x^2 + y^2) + \frac{1}{7}\ln((x - .1)^2 + (y - .1)^2) + \frac{1}{11}\ln((x - .2)^2 + (y - .2)^2)$$
$$+ \frac{1}{7}\ln((x - .3)^2 + (y - .3)^2) + \frac{1}{8}\ln((x - .4)^2 + (y - .4)^2)$$

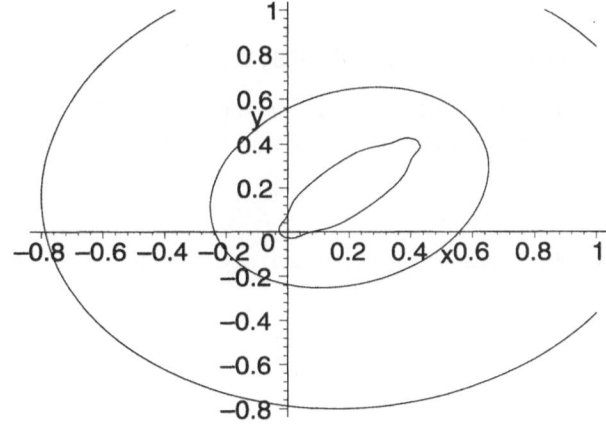

This 'matches' the actual lake contour reasonably, and is a start to fitting the water table data. Go back to section 4.1.1 and collect the data for the lake contours from the printed output at the end of the section. If the section is inactive, you can cut and paste the data; on Windows you can use the right mouse button to convert it to Maple input.

Pasted material
from page 184,
evalf(,4)

```
> opts:=THICKNESS(2),LINESTYLE(1):


> west_water :=[[.4068, .4636], [.4000, .4636],
> [.3955, .4636], [.3932, .4636], [.3977, .4705], [.4068, .4818],
> [.4182, .4909], [.4273, .4955], [.4364, .4977], [.4409, .4955],
> [.4409, .4909], [.4364, .4864], [.4296, .4818], [.4227, .4750],
> [.4136, .4705], [.4068, .4636]]:


> PPP := PLOT(CURVES(west_water,opts)):
> #PPP;


> opts:= STYLE(POINT),SYMBOL(BOX):


> bore_points := [[.6341, .7568], [.4523,
> .5045], [.3750, .4500], [.5727, .2409], [.3591, .4386], [.5159,
> .6705], [.2545, .4864], [.1886, .4409], [.4886, .5295], [.5705,
> .7114], [.4182, .4795], [.4705, .5159], [.5341, .6818], [.3386,
> .4273], [.3205, .4136], [.2386, .4727], [.5273, .5295]]:


> DP :=
> PLOT(CURVES(bore_points,COLOUR(RGB,1,0,0),opts)): #DP;
```

If the worksheet of section 4.1 is active or the above 2 sets of PLOT structures have been activated, the enlarged plot of WestLake is easy to manufacture.

```
> opts := axes=boxed,scaling=constrained:  # not for PLOT


> vu := view=[0.35..0.50,0.42..0.55]:


> tt := title='WestLake (Enlarged)':
```

The plot from
section 4.1 is
enlarged here

```
> TP := plots[display]({PPP,DP},tt,opts,vu):
> TP;
```

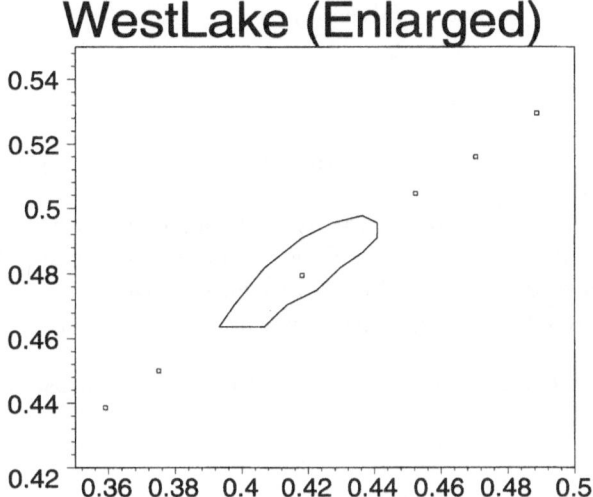

WestLake (Enlarged)

Indeed, the 'look' of the lake contours is similar to the artificial contours above.

4.3.4 Selecting Data

We carry on with a reading in of data points close to the lake. In this interpolation, and the number of equations that will be formed, we select a sufficient number to give the trends that might be expected. The number of data points must at least equal the number of unknown constants. We are adding together expressions in logarithms following the equation on page 206. For each sink there is a multiplier but also there is a $ln(rl)$ term. The rl terms combine to make an overall constant. Hence, with five sinks, at least five data points are required. When the equation set is finalised we will check the number of unknowns with indets( ). The data were contained in an Excel file; they were saved in ASCII text format, with spaces between, in file lakedata.prn.

We add the effect of gradient driven throughflow later and use more of the data

```
>  restart:
>  currentdir(); # listing of current directory structure
        "c:\\Program Files\ \Maple V Release 5.1\\BIN.WNT"
>  currentdir('c:/mplbook/v1'):  # changed directory
>  currentdir();
                    "c:\\mplbook\\v1"
>  filenme :='lakedata.prn';
>  fopen(filenme,READ);
```

$$\textit{filenme} := \textit{lakedata.prn}$$
$$0$$

```
>  readline(filenme); # returns line from file as a string
```

Recycle through the data as you please. It reads one line every time you activate the execution group. With a lot of data, and 'gobbledegook' at the start, just continue to read until something worthwhile appears.

"File cropped from Perry lakes file Murdoch.prn "

readline produces a 0 when no more data appears, so we can use it as a reader and a counter.

Back to the start of the file

```
>  fclose(filenme):  fopen(filenme,READ):
>  i := 0:  ex := 1:  while ex <> 0 do
>  ex := readline(filenme):i:=i+1:  od:  i;
```
$$53$$

```
>  fclose(filenme):
```

This is intentionally crude, to show the structure. There are 53 lines; we should check this out better in an editor, but since we read in 4 lines of heading, there should be 49 pieces of data. We are looking for points WL4, WL5, WL6, WL7, WL8, WL9, WL10, WL11, WL12, N1a, N2a, and, if needed,WL2, and WL3. The data are x and y position (WE direction and SN direction) and height in the borehole.

```
>  fopen(filenme,READ);
```
$$0$$

The x and y locations were scaled from a surveyed picture, on ghostview

```
>  ?readdata
```

```
>  readdata(filenme,[string,float,float,float]);
```

[["File"], ["measured"], ["name"], [""], ["PL1", 7.079, 2.880, 3.722],
   ["WL1", 5.722, 2.808, 3.645], ["WL2", 5.480, 2.776, 3.602],
   ["WL3", 5.308, 2.735, 3.546], ["N1a", 4.857, 2.710, 3.506],
   ["N1b", 4.945, 2.715, 3.515], ["N1c", 5.013, 2.711, 3.510],
   ["N2a", 5.083, 2.381, 3.119], ["N2b", 5.138, 2.399, 3.144],
   ["N2c", 5.214, 2.405, 3.142], ["WL4", 5.806, 2.341, 3.013],
   ["WL5", 7.426, 2.260, 2.817], ["WL6", 3.427, 2.682, 3.447],
   ["WL7", 3.507, 2.624, 3.442], ["WL8", 3.489, 2.605, 3.436],
   ["WL9", 4.221, 2.548, 3.395], ["WL10", 4.955, 2.507, 3.334],
   ["WL11", 5.345, 2.481, 3.295], ["WL12", 5.836, 2.433, 3.230],
   ["PL2", 4.953, 2.671, 3.477], ["WL13", 5.228, 2.702, 3.510],
   ["WL14", 5.212, 2.729, 3.530], ["WL15", 5.327, 2.752, 3.545],
   ["WL16", 5.288, 2.785, 3.562], ["WL17", 5.088, 2.823, 3.577], ["W26", 3.812],
   ["WL18", 5.746, 2.991, 3.797], ["WL19", 5.726, 2.948, 3.721],
   ["WL20", 5.439, 2.914, 3.661], ["N3a", 4.982, 2.864, 3.620],
   ["N3b", 5.106, 2.888, 3.629], ["N3c", 5.254, 2.887, 3.629],
   ["N4a", 4.655, 2.773, 3.463], ["N4b", 4.800, 2.770, 3.458],
   ["N4c", 4.905, 2.664, 3.377], ["WL21", 5.071, 2.720, 3.431],
   ["WL22", 5.067, 2.656, 3.390], ["WL23", 5.096, 2.482, 3.263],
   ["WL24", 5.170, 2.366, 3.187], ["WL25", 2.817, 1.978, 2.346],
   ["N5a", 11.075, 2.059, 2.584], ["N5b", 11.111, 2.078, 2.600],
   ["N5c", 11.130, 2.071, 2.592], ["EastLake", 3.070, 3.552],
   ["WestLake", 2.547, 3.435]]

See the detail in
Heck pp 110

Somehow, Maple seems to have got it right. One expects to read each line as a character and use a formatted decode to get it right with readline or fscanf and sscanf. The data consist of a name for the borehole, a Maple string, usually followed by 3 height measurements: the first is the TOC or 'top of casing', notionally ground height, measured relative to AHD 'Australian Height Datum-sea level'; the second is the measured distance from the TOC to the groundwater level on 10Mar97, the end of summer; and the last is the same measurement made on 15Sep97, the end of winter. All measured values are in metres.

We now convert this into a table with a structure such that the data is associated with the appropriate name.

```
>   AA := %:
```

```
> whattype(AA);
> nops(AA); AA[7]; op(1,%);
```

$$list$$
$$49$$
$$[\text{``WL2''}, 5.480, 2.776, 3.602]$$
$$\text{``WL2''}$$

```
> Data := table();
```

$$Data := \text{table}([$$
$$])$$

convert(,symbol)
suits Maple V
Release 4

```
> for i from 5 to 49 do
> op(1,AA[i]);
> convert(%,symbol);
> Data[%]:=op(2..4,AA[i]);
> od:
```

`Error, improper op or subscript selector`

Maple objects here because not all of the data have 4 operands–and the data are unread after item 29, when the Maple Error stopped the conversion. Items 29 and 30 in AA correspond to "WL17" and "W26" in Data.

```
> AA[29]; AA[30];
```

$$[\text{``WL17''}, 5.088, 2.823, 3.577]$$
$$[\text{``W26''}, 3.812]$$

```
> Data[WL17]; Data[W26];
```

$$5.088, 2.823, 3.577$$
$$Data_{W26}$$

The remedy is to use a correct range for each particular item.

```
> for i from 5 to 49 do
> op(1,AA[i]);
> convert(%,symbol);
> Data[%]:=op(2..nops(AA[i]),AA[i]);
> od:

> Data[WL17]; Data[W26];
```

$$5.088, 2.823, 3.577$$
$$3.812$$

This is a picture of data points read in as a list and converted into a table called Data. The fifth item in the list AA contains the measured height of water in Pumping bore PL1; the 49th item is the height of water in WestLake; with the table Data, we no longer need to know the list item number.

```
> AA[5]; AA[49];
```

$$[\text{``PL1''}, 7.079, 2.880, 3.722]$$

["WestLake", 2.547, 3.435]

The above may seem nonconventional but the table is indexed with a name we can relate to. If Joe Bloggs asks "What is the height of water in WestLake?", all one does is ask the computer the same question.

```
>  Data[WestLake];
```

2.547, 3.435

The height of the water in West Lake

The names of the locations we want are WL4, WL5, WL6, WL7, WL8, WL9, WL10, WL11, WL12, PL2, WL13, N1a, N2a, in a list.

```
>  ss := [WL4, WL5, WL6, WL7, WL8, WL9, WL10, WL11, WL12, PL2,
WL13, N1a, N2a];
>  ss[1];
```

$ss := [WL4, WL5, WL6, WL7, WL8, WL9, WL10, WL11, WL12, PL2, WL13, N1a, N2a]$

WL4

The water table values are the second and third numbers in the listing of the original data; they are relative to the top of casing TOC, the first number in the sequence. Reorganising, we scan through the table and the data of interest are adjusted to make a single table of the two water table values relative to sea level (AHD).

```
>  adjust := proc(ref,x,y) ref-x,ref-y end;
```

$adjust := \mathbf{proc}(ref, x, y)\, ref - x,\, ref - y\, \mathbf{end}$

```
>  data := table();
```

$$data := \text{table}([$$
$$])$$

The loop operation cleverly does the operation for every operand in ss.

```
>  for i in ss do ;
>  Data[i];adjust(%);data[i]:=%; od:
```

Repeating, data is now our table of measurements: the first value is the autumn level of the water table above sea level in metres; the 2th value is the spring level in metres. Here we will only deal with the autumn water table levels. But first, we note that the height of West Lake is a simple measurement and required no adjustment. A correction is necessary.

```
>  Data[WestLake];
```

2.547, 3.435

```
>  data[WestLake]:=op(1,[%]); # has no height adjustment
```

$data_{WestLake} := 2.547$

The AHD of West Lake

Horizontal Positions

The boreholes all need to be located properly in the horizontal positions x and y as well. In this case, the x, y data were digitised from a planar survey of the lakes, similar to the graph in section 4.1.

```
> xy[WL1] :=279,333:  xy[WL2] :=251,313:  xy[WL3]:=235,300:
> xy[WL4] :=105,208:  xy[WL5] := 83,194:  xy[WL6]:=215,233:
> xy[WL7] :=207,227:  xy[WL8] :=199,222:  xy[WL9]:=165,198:
> xy[WL10]:=158,193:  xy[WL11]:=149,188:  xy[WL12]:=141,182:
> xy[WL13]:=232,233:  xy[N1a] :=227,295:  xy[N2a]:=112,214:
> xy[PL2] :=252,106:  xy[lake]:=184,211:
> xy[WL9];
```

$$165, 198$$

The axes are aligned with the EW and SN directions, as normal; the units are arbitrary with 100 metres corresponding to 44 units. The angle of the lake, relative to WE, is arctan(79/118) or 0.59 radians; (184, 211) is the centre of the lake. Since we work in kilometres each set of coordinate data is multiplied by 0.1/44.

```
> xyc := map(proc(x,y) x/440,y/440 end, xy):
> op(op(1,xyc)); nops(op(%%));
```

$$[WL2 = (\frac{251}{440}, \frac{313}{440}), PL2 = (\frac{63}{110}, \frac{53}{220}), WL4 = (\frac{21}{88}, \frac{26}{55}), WL3 = (\frac{47}{88}, \frac{15}{22}),$$

$$WL5 = (\frac{83}{440}, \frac{97}{220}), N1a = (\frac{227}{440}, \frac{59}{88}), WL6 = (\frac{43}{88}, \frac{233}{440}), WL11 = (\frac{149}{440}, \frac{47}{110}),$$

$$WL10 = (\frac{79}{220}, \frac{193}{440}), WL12 = (\frac{141}{440}, \frac{91}{220}), WL7 = (\frac{207}{440}, \frac{227}{440}),$$

$$WL13 = (\frac{29}{55}, \frac{233}{440}), WL8 = (\frac{199}{440}, \frac{111}{220}), lake = (\frac{23}{55}, \frac{211}{440}),$$

$$WL1 = (\frac{279}{440}, \frac{333}{440}), N2a = (\frac{14}{55}, \frac{107}{220}), WL9 = (\frac{3}{8}, \frac{9}{20})]$$

$$17$$

4.3.5   Interpolation Equation

Now we develop our interpolation formula, a single, composed function for a small region around the lake. There are five sink terms, the logarithmic forms, a constant is added to adjust for absolute height above sea level, and a gentle, gradient term for regional groundwater flow through the region. The formula is a superposition of effects using mathematical linearity, a technique which is applicable in groundwater systems (see Bouwer [7], page 204). We have used Occam's Razor and left things out; the function contains the important mechanisms and should produce realistic trends around the lake.

```
>  i:='i':j := 'j':
>  eqn:=H[j]=Sum(A[i]*ln(r[i,j]),i=1..5) + C*x[j] + D*y[j] + E;
```

$$eqn := H_j = \left(\sum_{i=1}^{5} A_i \ln(r_{i,j})\right) + C\,x_j + D\,y_j + E$$

For each data point j, we calculate the logarithmic distance to each sink i, $\ln(r_{i,j})$. With the x_j and y_j locations for the same set of data, an equation can be written in the constants; a condition is set; 8 such conditions allows us to solve for the values of the 8 coefficients. Values for the coefficients make this equation an interpolation equation for the water table around the lake.

Excess data can improve the solution (page 266)

Of these data, we use only the x, y, r and H values for autumn. We select them now.

```
>  nops(ss);data[ss[2]];# this is a sequence
```

$$13$$

$$5.166, 4.609$$

[ ]'s make the type correct for op( )

```
>  autumn := [seq(op(1,[data[ss[i]]]),i=1..nops(ss))];
```

autumn := [3.465, 5.166, .745, .883, .884, 1.673, 2.448, 2.864, 3.403, 2.282, 2.526, 2.147, 2.702]

The x and y values are next:
```
>  xyvalues := seq([xyc[ss[i]]],i=1..nops(ss));
>  npts := nops([%]);
```

$$xyvalues := [\frac{21}{88}, \frac{26}{55}], [\frac{83}{440}, \frac{97}{220}], [\frac{43}{88}, \frac{233}{440}], [\frac{207}{440}, \frac{227}{440}], [\frac{199}{440}, \frac{111}{220}], [\frac{3}{8}, \frac{9}{20}], [\frac{79}{220}, \frac{193}{440}],$$
$$[\frac{149}{440}, \frac{47}{110}], [\frac{141}{440}, \frac{91}{220}], [\frac{63}{110}, \frac{53}{220}], [\frac{29}{55}, \frac{233}{440}], [\frac{227}{440}, \frac{59}{88}], [\frac{14}{55}, \frac{107}{220}]$$

$$npts := 13$$

It seems this approach is tedious; we could simply cut and paste or type the values into a spreadsheet. But this is also burdensome, especially with large amounts of data; it also promotes error and is usually flawed when there are missing data. Real world data tends to be sparse, disparate and/or quite different in kind. Maple can allow for this and reorganise the data; it can meld such data together to produce information, and knowledge.

Our venture proceeds to putting the positions and water table values together. Again, we use vectors (see Arrow, section 3.5) to locate positions and get distances.

```
>  with(linalg);
```

```
Warning, new definition for norm
```

```
Warning, new definition for trace
```

[*BlockDiagonal*, *GramSchmidt*, *JordanBlock*, *LUdecomp*, *QRdecomp*, *Wronskian*, *addcol*,
addrow, *adj*, *adjoint*, *angle*, *augment*, *backsub*, *band*, *basis*, *bezout*, *blockmatrix*,
charmat, *charpoly*, *cholesky*, *col*, *coldim*, *colspace*, *colspan*, *companion*, *concat*,
cond, *copyinto*, *crossprod*, *curl*, *definite*, *delcols*, *delrows*, *det*, *diag*, *diverge*,
dotprod, *eigenvals*, *eigenvalues*, *eigenvectors*, *eigenvects*, *entermatrix*, *equal*,
exponential, *extend*, *ffgausselim*, *fibonacci*, *forwardsub*, *frobenius*, *gausselim*,
gaussjord, *geneqns*, *genmatrix*, *grad*, *hadamard*, *hermite*, *hessian*, *hilbert*,
htranspose, *ihermite*, *indexfunc*, *innerprod*, *intbasis*, *inverse*, *ismith*, *issimilar*,
iszero, *jacobian*, *jordan*, *kernel*, *laplacian*, *leastsqrs*, *linsolve*, *matadd*, *matrix*,
minor, *minpoly*, *mulcol*, *mulrow*, *multiply*, *norm*, *normalize*, *nullspace*, *orthog*,
permanent, *pivot*, *potential*, *randmatrix*, *randvector*, *rank*, *ratform*, *row*, *rowdim*,
rowspace, *rowspan*, *rref*, *scalarmul*, *singularvals*, *smith*, *stackmatrix*, *submatrix*,
subvector, *sumbasis*, *swapcol*, *swaprow*, *sylvester*, *toeplitz*, *trace*, *transpose*,
vandermonde, *vecpotent*, *vectdim*, *vector*, *wronskian*]

```
> lakev := vector([xyc[lake]]);
```
The location of the lake

$$lakev := \left[\frac{23}{55}, \frac{211}{440} \right]$$

The other 'virtual sinks' add to simulate the total effect of the lake as determined by the initial 'fitting' of the lake shape. These sinks were effectively
$\sqrt{2}$ times 100 metres apart (0.141 km); two on either side of the lake centre.
With this in mind, and that the angle of the lake, relative to the WE line, is
0.59 radians, the x,y coordinates of the sinks are:

```
> dx:= 0.01*sqrt(2)*cos(0.59):
> dy := 0.01*sqrt(2)*sin(0.59):
> dd := vector([dx,dy]);
> xysinks :=
> seq(evalf(matadd(lakev,dd,1,i)),i=-2..2);
> nsinks :=nops([%]);
```

$$dd := \left[.008309406791 \sqrt{2}, .005563610229 \sqrt{2} \right]$$

$$xysinks := [.3946792667, .4638091884], [.4064305424, .4716773215],$$
$$[.4181818182, .4795454545], [.4299330940, .4874135875],$$
$$[.4416843698, .4952817206]$$

$$nsinks := 5$$

The distances $r_{i,j}$ are now calculated, for sink i and point j:

```
>   for i to nsinks do;
>   for j to npts do;
>   (matadd(xyvalues[j],xysinks[i],1,-1));
>   RR[i,j]:= sqrt(dotprod(%,%));od; od;
```

```
>   RR[2,3];
```

dotprod is a grotty way to find the magnitude of a vector

Exercise 4.11 Magnitude of a Vector

Look up the magnitude of a vector and find and use at least 2 commands in Maple to calculate the magnitude of a vector. Use op( ) to pull apart the vector and calculate the magnitude.

Exercise* 4.12 Distance along a traverse

Use the data above and/or your own calculated values to obtain the distances of each of the points in the above analysis from both WestLake and EastLake. Plot the data on a graph that shows the bore hole positions relative to the lakes.

4.3.6 Expanded Equations

Presently the xyvalues are available as a sequence of lists, the autumn water table data are in a list, and the RR values are the array defined above. The equation set is:

```
>   i := 'i':  j := 'j':
```

```
>   eqns := {seq(eqn,j=1..npts)};
```

$$eqns := \left\{ H_1 = \left(\sum_{i=1}^{5} A_i \ln(r_{i,1}) \right) + C\,x_1 + D\,y_1 + E, \right.$$

$$H_2 = \left(\sum_{i=1}^{5} A_i \ln(r_{i,2}) \right) + C\,x_2 + D\,y_2 + E,$$

$$H_3 = \left(\sum_{i=1}^{5} A_i \ln(r_{i,3}) \right) + C\,x_3 + D\,y_3 + E,$$

$$H_4 = \left(\sum_{i=1}^{5} A_i \ln(r_{i,4}) \right) + C\,x_4 + D\,y_4 + E,$$

$$H_5 = \left(\sum_{i=1}^{5} A_i \ln(r_{i,5}) \right) + C\,x_5 + D\,y_5 + E,$$

$$H_6 = \left(\sum_{i=1}^{5} A_i \ln(r_{i,6}) \right) + C\,x_6 + D\,y_6 + E,$$

$$H_7 = \left(\sum_{i=1}^{5} A_i \ln(r_{i,7}) \right) + C\,x_7 + D\,y_7 + E,$$

$$H_8 = \left(\sum_{i=1}^{5} A_i \ln(r_{i,8}) \right) + C\,x_8 + D\,y_8 + E,$$

$$H_9 = \left(\sum_{i=1}^{5} A_i \ln(r_{i,9}) \right) + C\,x_9 + D\,y_9 + E,$$

$$H_{10} = \left(\sum_{i=1}^{5} A_i \ln(r_{i,10}) \right) + C\,x_{10} + D\,y_{10} + E,$$

$$H_{11} = \left(\sum_{i=1}^{5} A_i \ln(r_{i,11}) \right) + C\,x_{11} + D\,y_{11} + E,$$

$$H_{12} = \left(\sum_{i=1}^{5} A_i \ln(r_{i,12}) \right) + C\,x_{12} + D\,y_{12} + E,$$

$$\left. H_{13} = \left(\sum_{i=1}^{5} A_i \ln(r_{i,13}) \right) + C\,x_{13} + D\,y_{13} + E \right\}$$

The x and y values for data point 5

In a list to keep the order

To keep this expression small, the substitution for the x and y values should be done before expanding. Back up a bit.

```
>   evalf(xyvalues[5][1]), evalf(xyvalues[5][2]);
                .4522727273, .5045454545
```

```
>   subs(x[j]=xyvalues[j][1],y[j]=xyvalues[j][2],H=autumn,r=RR,eqn):
>   [seq(%,j=1..npts)];
```

$$\left[3.465 = \left(\sum_{i=1}^{5} A_i \ln(RR_{i,1}) \right) + \frac{21}{88} C + \frac{26}{55} D + E, \right.$$

$$5.166 = \left(\sum_{i=1}^{5} A_i \ln(RR_{i,2}) \right) + \frac{83}{440} C + \frac{97}{220} D + E,$$

$$.745 = \left(\sum_{i=1}^{5} A_i \ln(RR_{i,3}) \right) + \frac{43}{88} C + \frac{233}{440} D + E,$$

$$.883 = \left(\sum_{i=1}^{5} A_i \ln(RR_{i,4}) \right) + \frac{207}{440} C + \frac{227}{440} D + E,$$

$$.884 = \left(\sum_{i=1}^{5} A_i \ln(RR_{i,5}) \right) + \frac{199}{440} C + \frac{111}{220} D + E,$$

$$1.673 = \left(\sum_{i=1}^{5} A_i \ln(RR_{i,6}) \right) + \frac{3}{8} C + \frac{9}{20} D + E,$$

$$2.448 = \left(\sum_{i=1}^{5} A_i \ln(RR_{i,7}) \right) + \frac{79}{220} C + \frac{193}{440} D + E,$$

$$2.864 = \left(\sum_{i=1}^{5} A_i \ln(RR_{i,8}) \right) + \frac{149}{440} C + \frac{47}{110} D + E,$$

$$3.403 = \left(\sum_{i=1}^{5} A_i \ln(RR_{i,9}) \right) + \frac{141}{440} C + \frac{91}{220} D + E,$$

$$2.282 = \left(\sum_{i=1}^{5} A_i \ln(RR_{i,10}) \right) + \frac{63}{110} C + \frac{53}{220} D + E,$$

$$2.526 = \left(\sum_{i=1}^{5} A_i \ln(RR_{i,11}) \right) + \frac{29}{55} C + \frac{233}{440} D + E,$$

$$2.147 = \left(\sum_{i=1}^{5} A_i \ln(RR_{i,12}) \right) + \frac{227}{440} C + \frac{59}{88} D + E,$$

$$\left. 2.702 = \left(\sum_{i=1}^{5} A_i \ln(RR_{i,13}) \right) + \frac{14}{55} C + \frac{107}{220} D + E \right]$$

Don't look now!

```
>  eqns := value(%):
```
Value activates Sum and automatic substitution for the RR values occurs.
This expands all the sums to give the whole equation set.
```
>  nops(eqns); unkns :=indets(eqns);nindets :=
>  nops(%);
```

Maple
automatically
substitutes in the
values

13

$$unkns := \{C, E, \mathrm{D}, A_1, A_2, A_3, A_4, A_5\}$$
$$nindets := 8$$

We can only use 8 of these equations, unless we intend to do a least squares, best fit (see section 5.5, page 266).

The first 8
equations are
selected

```
>   constants :=
>   solve({op(1..nindets,eqns)},unkns);
```

$$constants := \{\mathrm{D} = -14.57115043,\ E = 19.02189993,\ C = 4.194708982,\ A_1 = 84.21942789,$$
$$A_4 = -360.6197741.\ A_5 = 56.14408738,\ A_3 = 647.9987452,\ A_2 = -421.7298143\}$$

Exercise* 4.13 Other Data

Use some of the other data points, of the 13 above, to fit the data. Plot the results. Considering that one wants a 'smooth' fit, which set of data gives a 'best' value?

4.3.7   A Single Equation in x and y

We now explore the nature of the solution. We get rid of all the 'bits' and put everything in terms of x and y.

The 'i' removes
the i definition
from the loop so
it doesn't cause
trouble

```
>   for i to nsinks do;
>   xs[i]  := xysinks[i][1];
>   ys[i]  := xysinks[i][2];
>   od:    i := 'i':
>   xs, seq(xs[i], i=1..nsinks);
>   ys, seq(ys[i], i=1..nsinks);
```
xs, .3946792667, .4064305424, .4181818182, .4299330940, .4416843698
ys, .4638091884, .4716773215, .4795454545, .4874135875, .4952817206

```
>   Sum(A[i]*ln(sqrt((x-xs[i])^2+(y-ys[i])^2)),i=1..nsinks);
```

$$\sum_{i=1}^{5} A_i \ln(\sqrt{x^2 - 2\,x\,xs_i + xs_i{}^2 + y^2 - 2\,y\,ys_i + ys_i{}^2})$$

Sum may not
evaluate fully if
included inside a
complex equation

```
>   value(%)+C*x+D*y+E: #includes the steady gradient terms
>   soln := evalf(subs(constants,%));
```

$soln := 84.21942789 \ln(\sqrt{x^2 - .7893585334\,x + .3708906868 + y^2 - .9276183768\,y})$

$\quad - 421.7298143 \ln(\sqrt{x^2 - .8128610848\,x + .3876652814 + y^2 - .9433546430\,y})$

$\quad + 647.9987452 \ln(\sqrt{x^2 - .8363636364\,x + .4048398760 + y^2 - .9590909090\,y})$

$\quad - 360.6197741 \ln(\sqrt{x^2 - .8598661880\,x + .4224144706 + y^2 - .9748271750\,y})$

$\quad + 56.14408738 \ln(\sqrt{x^2 - .8833687396\,x + .4403890653 + y^2 - .9905634412\,y})$

$\quad + 4.194708982\,x - 14.57115043\,y + 19.02189993$

```
>  evalf(xyvalues[5][1]); evalf(xyvalues[5][2]);
```

$$.4522727273, .5045454545$$

```
>  subs({x=0.45,y=0.50},soln); evalf(%);
```

$84.21942789 \ln(.06610717359) - 421.7298143 \ln(.05196606296)$

$\quad + 647.9987452 \ln(.03782572009) - 360.6197741 \ln(.02368751781)$

$\quad + 56.14408738 \ln(.009560957065) + 13.62394376$

$$-1.41969224$$

```
>  autumn[5];
```

$$.884$$

```
>  data[WestLake];
```

$$2.547$$

The height of
WestLake

```
>  with(plots):
>  contourplot(soln,x=0..1,y=0..1, grid=[100,100],
>  contours=[2.,2.547,3.,3.5,4.,4.5,5,5.5,6.,6.5,7.,7.5]);
```

The Inferred Water Table Heights

The graph, next page, is certainly not what was expected and is not a careful fitting of the data. It seems to show that the flow pattern is fully 3D and that there is a source or sink of water to the south and west. To get the pattern right, it is necessary to combine bore hole data with lake data and use all the data as a statistical base, see page 266, section 5.5. Of particular interest are the data in the traverse through West Lake–the rather dry part of it. WL6, WL7 and WL8 are to the northwest of the small pool and WL9, WL10,WL11 and WL12 are to the southeast. The exercises attempt to take the analysis further.

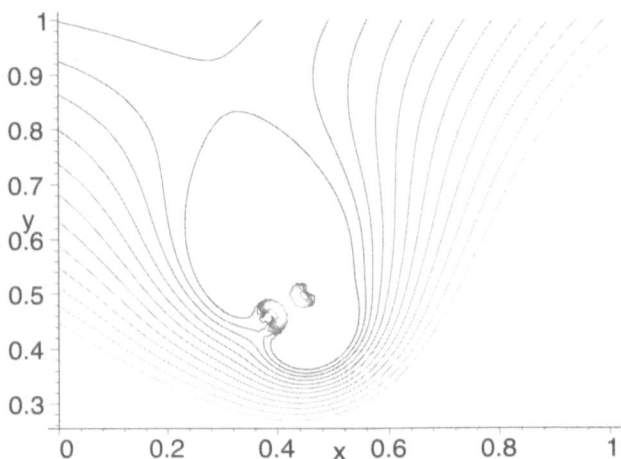

Exercise 4.14 Lake and Groundwater Isopotentials

Plot the shape of the lake on the same scale as above. Establish and mark (with numbers) the isohyets so that they are clear. Is the solution wrong?

Exercise 4.15 Compare with the Data

Plot the data for the traverse through the lake on the same plot with the 'analytical' solution. Hopefully there is agreement; the points were used in the fitting of the interpolation equation.

Exercise* 4.16 Gradient and Flow

Find the interpolation equation, above, and explore its properties. What is the groundwater gradient through the the Perry Lake system? Look at the terms in the original formulation; if the hydraulic conductivity K is 10 m/day and the depth d of the aquifer is 20 m, what is the overall flow into the lake? You have to add the effects of all the flow terms.

Exercise* 4.17 Line Integrals

Use Darcy's law, as given under Formulation, section 4.3.2, page 205 and work up a line integral form for the total flow into West Lake. Put the interpolation formula into the expression and integrate to get the total flow into the lake. Does it agree with the value obtained above? This is not a trivial exercise; you will need to know about integrals and fluxes.

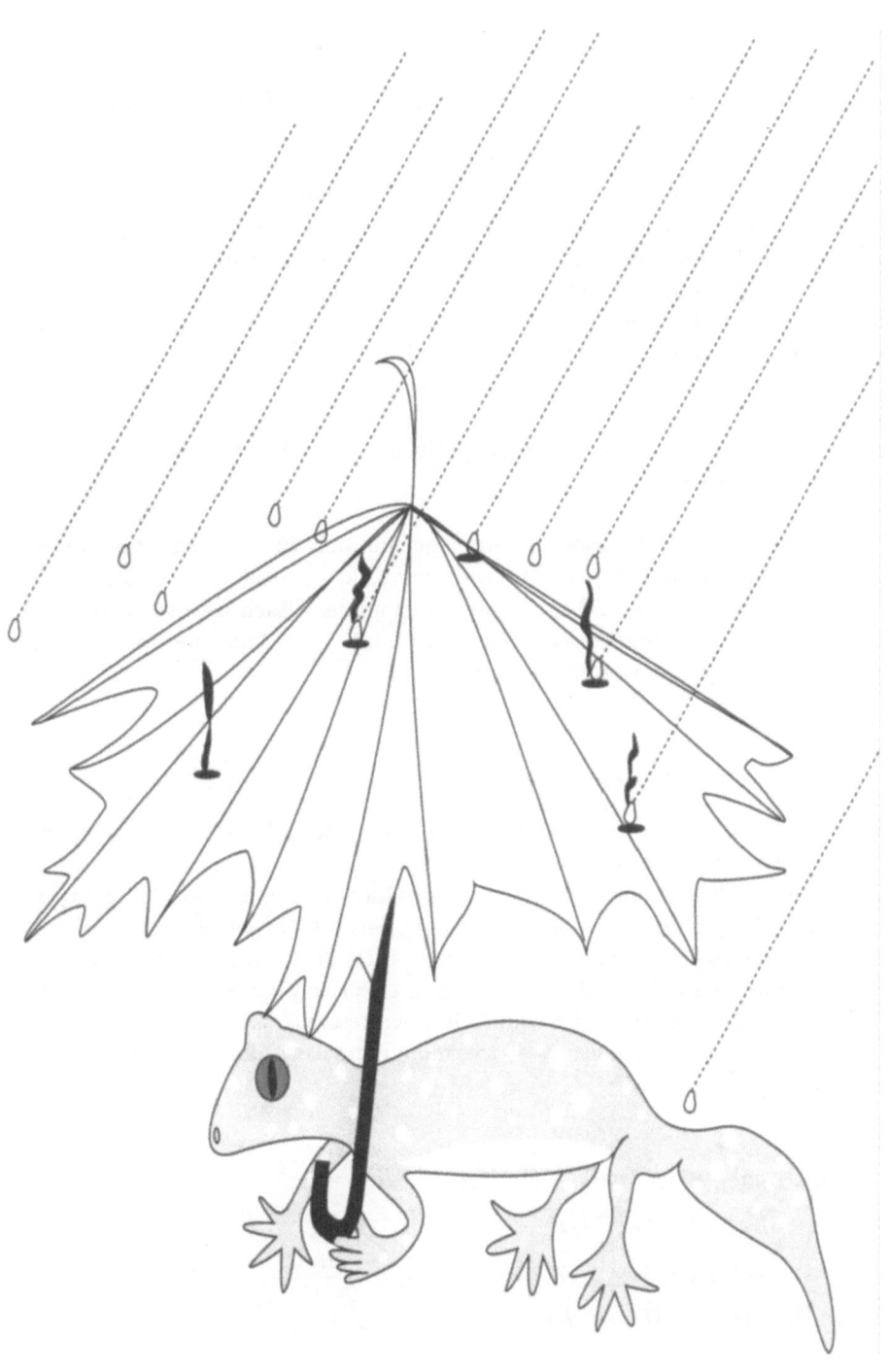

interface(longdelim=false);

4.4   Acid Rain: A Look at Solution Chemistry

The power of Maple is apparent when tackling the solution of multivariate polynomials. Here the equilibrium chemistry of droplets is considered, in examples with three levels of complexity. Considered first is the simplest case of ionic equilibrium with species derived from Carbon Dioxide CO_2. Then, the case of species generated from three interacting sources, including Sulfur Dioxide SO_2 and Ammonia NH_3. Lastly, chemical kinetics are included with oxidation to sulfate ion. The results are all applicable to the problem of acid rain, resulting from excess sulfurous emissions and carbon dioxide. With effort from the user, many more interactive species may be included, even organics and solids. Maple can solve these problems, provided the user evolves an appropriate structure.

These three specific cases are important not only in cloud and rain physics and air pollution but also in our waterways, in our soils and in our foods. The technique is presented as a series of examples. Each uses three types of conditions; 'equilibrium mass action constants', 'mass balances' and a single electroneutrality equation.

4.4.1   Carbon Dioxide

The pH of a solution of Carbon Dioxide CO_2 in water H_2O, as may be appropriate near the turn of the century.

Simply put, CO_2 dissolves in water to produce 'carbonic acid'. This breaks up into hydrogen ions H^+ and bicarbonate ions HCO_3^- which, in turn, break up into more hydrogen ions and carbonate ions $CO_3^=$. Water is an interloper that supplies a small number of ions and effectively links the acid H^+ and the basic OH^- ion. Here the equilibrium expressions or 'mass action ratios', the K values, are expressions of thermodynamic equilibrium. In a steady, equilibrium state

$$CO_2(gas) + H_2O(lqd) \rightleftharpoons H_2CO_3 \qquad K_{hc} = \frac{[H_2CO_3]}{p_{co_2}} = 34/1000$$

$$H^+ + HCO_3^- \rightleftharpoons H_2CO_3 \qquad K_{1c} = \frac{[H_2CO_3]}{[H^+] \cdot [HCO_3^-]} = 2250000$$

$$H^+ + CO_3^= \rightleftharpoons HCO_3^- \qquad K_{1c} = \frac{[HCO_3^-]}{[H^+] \cdot [CO_3^=]} = 21400000000$$

$$H^+ + OH^- \rightleftharpoons H_2O \qquad K_w = \frac{1}{[H^+] \cdot [OH^-]} = 10^{14}$$

they must apply, together with the electroneutrality equation

$$[H^+] = [OH^-] + \left[HCO_3^-\right] + 2 \cdot [CO_3^=]$$

The [ ]'s represent 'concentration of' in moles per litre. The ratios given are 'mass action' ratios, not activities; it is presumed that the concentrations are small and the activity coefficients can be assumed to be unity. A feature here is that all the constants (the K values given for 25C), are formation constants so that the numbers used are not small fractions. This makes it easier to put into Maple though it is not essential. As Maple is a symbolic manipulator, it might be better to use the symbolic values of the constants. Later, using another language like Fortran, the values are inserted. This does not change the essential structure of the solution but generally will fill pages with symbolic code; it is not particularly enlightening during testing of an idea. Here we only use floating point numbers when they can not be avoided; this makes the work easier for Maple and removes roundoff error.

The question asked is: If the concentration of CO_2 in the atmosphere is ~sometime early in the 21st century~ 400 parts per million by mass, what is the expected pH of rainwater? This corresponds to a partial pressure of CO_2 (p_{co_2}) of $\frac{400 \cdot 29/44}{1000000}$ moles of CO_2 per mole of air or about $264/1000000$ atmospheres of CO_2. All we do now is combine the above equations, using Maple. It is noteworthy that there are 5 equations and 5 unknowns in this equation set.

The Maple codes omit the square brackets and the subscripts.

```
>   restart:
>   #pCO2 := 205/1000000; # 311 parts per million by mass
>   pCO2 := 264/1000000; # 400 parts per million by mass
```

Above, two estimates values of CO_2 can be assigned. The first is a low value, most likely appropriate early this century. The second is a value predicted in the first few decades of the 21st century.

$$pCO2 := \frac{33}{125000}$$

```
>   H2CO3 := 34/1000*pCO2;
```

A variation of Henry's law for CO_2 states that the partial pressure of CO_2 is proportional to the concentration of dissolved CO_2 , or carbonic acid.

$$H2CO3 := \frac{561}{62500000}$$

Note that this is just a number; the way we have done the assignment follows a natural progression towards a solution. The units of the carbonic acid concentration are moles/litre.

Watch the output as we proceed.

```
>   HCO3 := H2CO3/(2250000*H);
```

A statement of equilibrium between bicarbonate ion and carbonic acid. Again, the equilibrium constant can be variously derived and is purposely put into an integer form.

$$HCO3 := \frac{187}{46875000000000} \frac{1}{H}$$

Now we have an unknown, the hydrogen ion concentration. This is the driving argument, or 'joining argument' for the activity in the solution. And we want to calculate the acidity or pH of the drops, which is, directly, the negative log10 of H.

```
>   CO3 := HCO3/(214*10**8*H);
```

A last ionization of carbonic acid; the carbonate ion is formed from the ionization of bicarbonate ion.

$$CO3 := \frac{187}{10031250000000000000000000} \frac{1}{H^2}$$

The scene becomes more adventuresome. The hydrogen ion comes in as an inverse square. Again, if we know the H concentration, we know the bicarbonate ion concentration.

The last equilibrium equation, and the least, relates more directly to everyday experience.

```
>   OH := 1/(10**14*H);
```

The classic equilibrium relation of water, in which it ionises to form hydrogen ion and hydroxyl ion. At a standard temperature of 25 degrees, this relation establishes the 0-14 pH scale we know

$$OH := \frac{1}{100000000000000} \frac{1}{H}$$

Importantly, we have reformed to help the structure of the solution. In this we simply use the fact that, with each assignment, the variable on the left no longer needs to be considered as a variable. The same could be accomplished by putting the equations as written above into a set and methodically using the subs command. The results are the same but the above method is best for the Maple novice; this helps with learning as it keeps the expressions simple. Of course, it would also be better to leave the K values in the solution as parameters (though known) so that the values can be easily altered. This would allow one to investigate the effect of error (which can be gross), temperature or whatever – play 'what if' games.

The final expression is automatically formed as an indirect substitution into the electroneutrality equation. Again, this simply states that there can be no excess electric charge in the solution, the sum of all the negative charges must equal the sum of all the positive charges. That means we multiply the doubly charged species concentrations by a factor of two and ignore the uncharged species. The only positive ion in this pure CO_2 solution is H.

The method disobeys Corless' first rule

```
>   eqn := H-OH-HCO3-2*CO3;
```

$$eqn := H - \frac{5999}{1500000000000000} \frac{1}{H} - \frac{187}{50156250000000000000000} \frac{1}{H^2}$$

```
>  collect(%,H):
```
This cleans up the expression, but is unnecessary in this simple case.
```
>  fsolve(eqn,H=1E-4);
```

$$.1999879937\,10^{-5}$$

```
>  H := %; # the hydrogen ion is assigned a value
```

$$H := .1999879937\,10^{-5}$$

Remember that Maple, by convention, takes an expression as being equal to zero in fsolve( ). The guess value, 1E-4, could just as well be presented as a range; say, 0..1, but 0 is not possible.

Range is a separate argument

It is interesting that we expect that this 'pure rainwater' will have a pH of potatoes or sour milk,
```
>  pH := evalf(-log10(H),3);
```

$$pH := 5.70$$

which is just slightly acid, despite the large amount of CO_2.

Exercise 4.18 Present day values

Now, in the year 2000, the measured levels of CO_2 are around 370 parts per million by mass. Calculate the expected pH value. If the CO_2 levels have linearly increased from values in the year 1900 of about 311 ppm, what is the trend in pH levels? Plot the pH values as a function of time.

Exercise 4.19 Insert K values

Redo the last equation set inserting the K values as algebraic quantities. Then substitute for the K values. The result (the pH) will be the same, unless you erred. The Arrhenius Equation predicts the variation of K values with temperature; with global warming, one expects additional variation in the pH.

Exercise 4.20 Multiply by H^3

Multiply eqn by H^3 and expand; this is not good practice because it adds solutions. However, the expression then clearly becomes a cubic. With the unknown in the numerator, range is effective in fsolve. Solve for the pH.

Exercise 4.21 Find the Other Solutions

The above equation for the H ion concentration should have 3 solutions, two of which are imaginary. Find the complex solutions.

4.4.2 Carbon Dioxide, Sulfur Dioxide and Ammonia

Here we include a few more species, from sulfur dioxide SO_2 and ammonia NH_3. Both of these constituents may have natural sources but man contributes a significant quantity of the atmospheric sulfur dioxide with smelting of ores and the burning of sulfurous coals. Ammonia, along with other gases is exuded from swamps and tip sites; animals contribute ammonia through their urine. Sulfur dioxide is an acid material and ammonia is a basic material; it is not clear in any one case whether the rainwater will be acid or basic but the acidity usually wins. The additional equilibrium constants are:

$$SO_2(gas) + H_2O(lqd) \rightleftharpoons H_2SO_3 \qquad K_{hs} = \frac{[H_2SO_3]}{p_{CO_2}} = 124/100$$

$$H^+ + HSO_3^- \rightleftharpoons H_2SO_3 \qquad K_{1s} = \frac{[H_2SO_3]}{[H^+]\cdot[HSO_3^-]} = 79$$

$$H^+ + SO_3^= \rightleftharpoons HSO_3^- \qquad K_{1s} = \frac{[HSO_3^-]}{[H^+]\cdot[SO_3^=]} = 1600000$$

$$NH_3(gas) + H_2O(lqd) \rightleftharpoons NH_4OH \qquad K_{ha} = \frac{[NH_4OH]}{p_{NH_3}} = 57$$

$$NH_4^+ + OH^- = NH_4OH \qquad K_{1a} = \frac{[NH_4OH]}{[NH_4^+][OH^-]} = 56470$$

and the electroneutrality condition contains additional ionic species:

$$\left[H^+\right] + \left[NH_4^+\right] = [OH^-] + \left[HCO_3^-\right] + 2\cdot[CO_3^=] + \left[HSO_3^-\right] + 2\cdot[SO_3^=]$$

The calculation requires additional Maple statements. But first we must make H a variable again. Here we start to see the real reason we have been unevaluating variables relentlessly in the past chapters. The whole physical or chemical problem has changed; we introduced new variables; it must happen that the assigned variables change, and be allowed to do so.

unassign could be
used

> H := 'H': # H becomes unevaluated

This 'releases' H from the 'value' above and it becomes a name again. Really, this means we have redefined H, to be ready for the next calculation.

> pSO2 := 14/10^9; # see Scott and Hobbs

An ordinary, unpolluted level of 14 billionths of an atmosphere. This corresponds to 14 moles $SO_2/10\hat{\ }9$ moles air or 64/29*14/10^9 grams SO_2 per gram of air.

$$pSO2 := \frac{7}{500000000}$$

We follow the same substitution plan we used for CO_2 , structurally replacing it with SO_2 . The sulfurous acid concentration is calculated first.

> H2SO3 := 124/100*pSO2;

$$H2SO3 := \frac{217}{12500000000}$$

Then the bisulfite ion concentration.

```
>  HSO3 := H2SO3/(H*79);
```

$$HSO3 := \frac{217}{987500000000} \frac{1}{H}$$

and, lastly, the sulfite ion concentration.

```
>  SO3 := HSO3/(H*16*10**6);
```

This species is of importance in calculating the oxidation to sulfate.

$$SO3 := \frac{217}{15800000000000000000} \frac{1}{H^2}$$

The ammonia equilibria are very similar. We chose an ordinary level of NH_3, as might be found in a paddock in England.

```
>  pNH3 := 7/10^9;
```

$$pNH3 := \frac{7}{1000000000}$$

```
>  NH4OH := 57*pNH3;
```

$$NH4OH := \frac{399}{1000000000}$$

```
>  NH4 := NH4OH/(OH*56470);
```

$$NH4 := \frac{3990000}{5647} H$$

```
>  eq := H+NH4-OH-HCO3-2*CO3-HSO3-2*SO3;
```

$$eq := \frac{3995647}{5647} H - \frac{26513921}{118500000000000000} \frac{1}{H} - \frac{544202699}{198117187500000000000000000} \frac{1}{H^2}$$

```
>  HH := fsolve(eq,H=1E-5);
```

$$HH := .6158375349 \, 10^{-6}$$

```
>  pH := evalf(-log10(HH),3);
```

$$pH := 6.21$$

And the pH increases in this instance, the ammonia overcoming the weak acidity of the sulfur dioxide.

Exercise* 4.22 Mass Balance in a Box

Consider an alternative situation, with a given amount of CO_2, SO_2 and NH_3 in a box of volume 1 m^3. The pressures of the two gases is not kept constant but is controlled by the amount initially in the box . The initial concentration is the same as given above; all other conditions are the same. Allow that the water droplets with a volume of 0.01litres are placed in the box; they are initially free of any of the gases. What is the final pH of the droplets? What are the final vapour pressures of the two gases? Remember that the equilibrium constants (mass action constants) are derived such that the species concentrations in the liquid are in moles/litre and the gas pressures are in atmospheres. The molecular weights are, respectively, 44, 64, and 17 for CO_2, SO_2 and NH_3; that of air is effectively 29 grams/mole; the gases obey the physical rules that (pressure %) equals (volume %) equals (mole %) and there are 22.4 litres of a gas in a mole at standard conditions. This situation simulates the atmosphere in a more enclosed situation, including the loss of gas on absorption into the droplets.

4.4.3 Kinetics of Sulfate Formation

Now we add an extra whammy and allow oxidation of the SO_2 in the water droplets. This produces both acidity and sulfate ion $SO_4^=$, effectively sulfuric acid, which is a strong acid; continuity of mass (mass balance) requires that, as the sulfate is formed, the other S-containing species are depleted. The sulfate ion adds to the electroneutrality equation on the right side. Following Scott and Hobbs[25], the oxidized species is thought to be the sulfite ion $SO_3^=$. The kinetic equation for $SO_4^=$ is a first-order Differential Equation with a right hand side that has some complexity.

$$\tfrac{\partial}{\partial t} SO_4 = k\, SO_3$$

The solution follows from the lead above and an additional term in the electroneutrality equation.

```
>   eqn := eq - 2*SO4;
```

$$eqn := \frac{3995647}{5647} H - \frac{26513921}{118500000000000000} \frac{1}{H} - \frac{544202699}{198117187500000000000000} \frac{1}{H^2} - 2\,SO4$$

but solve still copes well with the task. It produces three solutions to the cubic expression, only one of which is realistic:

```
>   assume(H>0,SO4>0); HH:='HH':
```

```
>   HH := [solve(eqn,H)]:
```

```
>   nops(HH); # solution is a sequence of 3
                        3
```

Check the values with a small amount of SO_4, to see which solution is correct.

```
>   subs(SO4=1/10^6,HH):
```

```
>   evalf(%);
```

$$[.6171440811\,10^{-6}, \ -.4844772929\,10^{-6} + .12\,10^{-15}\,I, \ -.1298402123\,10^{-6} - .10\,10^{-15}\,I]$$

```
>   is(%[1]>0), is(%[2]>0), is(%[3]>0);
```

true, false, false

```
>   H := HH[1];
```

assume is necessary to get the roots out

Horrible thing to print
Not a trivial solution

$$H := \frac{1}{32318454190682318411500467736395000000}(502860406954256961830\backslash$$

$1246043014267589917120890075722673976492216652180779284843 4\backslash$

$770212759822500000000 SO_4 + 6552257498938251651449828253366 48\backslash$

$6629081883758219844414704947620260902758734571604981216703 0\backslash$

$2320 + 28233842466499364210172216528941026413902343971390789\backslash$

$0845481773550032023877237373570000000000000000000000000000 SO_4{}^3 +$

$29380660514071209810143697041327204728907400126353370891915 5\text{sqrt}(-$

$104853264283606811726728042562650429146220644481398682296812\backslash$

$604709026254819 - 9764511476512806643453841002287555744524 59\backslash$

$5839835366865672158306750000000000000000 SO_4{}^2 + 763387922301899\backslash$

$63298200005611296023830184753965372777372454835778129600000 \backslash$

$000 SO_4 + 428615457511267302067509075440542085313862290432 77\backslash$

$929890702720000000000000000000000000 SO_4{}^3))^{(1/3)} -$

$$\frac{5170952670509170945840074837823 2}{15625}($$

$$-\frac{149724111887}{14545433687040} - \frac{12456487890625000000}{143686754537481} SO_4{}^2\Big) \Big/ (502860406954\backslash$$

$256961830124604301426758991712089007572267397649221665218 07\backslash$

$79284843477021275982250000000000 SO_4 + 655225749893825165144982\backslash$

$82533664866290818837582198444147049476202609027587345716049\backslash$

$8121670302320 + 28233842466499364210172216528941026413902343\backslash$

$97139078908454817735500320238772373735700000000000000000000 \backslash$

$0000 SO_4{}^3 +$

$29380660514071209810143697041327204728907400126353370891915 5\text{sqrt}(-$

$104853264283606811726728042562650429146220644481398682296812\backslash$

$604709026254819 - 9764511476512806643453841002287555744524 59\backslash$

$5839835366865672158306750000000000000000 SO_4{}^2 + 763387922301899\backslash$

$63298200005611296023830184753965372777372454835778129600000 \backslash$

$000 SO_4 + 428615457511267302067509075440542085313862290432 77\backslash$

$929890702720000000000000000000000000 SO_4{}^3))^{(1/3)} + \frac{11294}{11986941} SO_4$

The kinetic constant (turnover constant) is altered to give the fractional turnover per day. The rate of depletion of $SO_3{}^=$ ion or the rate of increase of $SO_4{}^=$ ion is given by a complex expression with squared terms, cubic terms, square roots and cubic roots.

```
>  k := 1/10*60*24:# turnover constant per day
>  rate := k*SO3;
```

Continuing, we plot the rate of reaction and find that it is an ever-decreasing, hyperbolic function of the amount of $SO_4^=$ ion. The rate of production eventually goes to zero, presumably because the rise in acidity decreases the $SO_3^=$ concentration nearly to zero.

```
>  opt1 := title='Rate of Sulfate Production':
>  plot(rate,SO4=0..0.001,opt1);
```

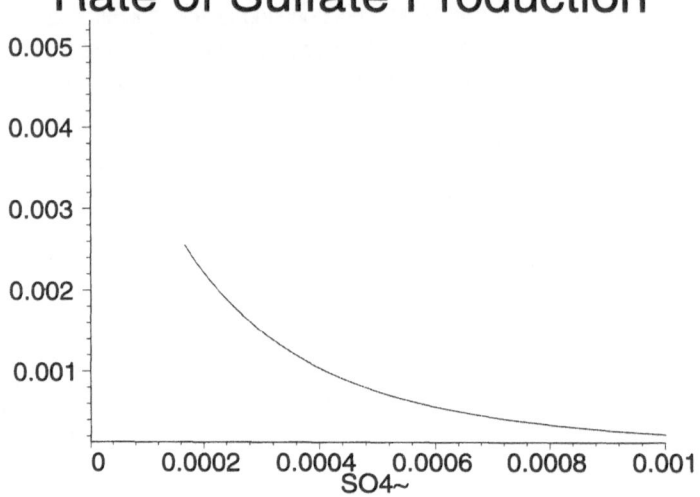

Rate of Sulfate Production

The plot doesn't continue to SO4=0. We have lost the function at concentrations below about 0.00015 moles/litre SO4 ! The rate is well-known near SO4=0 and at higher values, where the rate goes to zero.

```
>  subs(SO4=0,rate):  evalf(%);
                      .005214742022
>  limit(rate,SO4=infinity);
                          0
```

Fortunately, it isn't critical here. We will deal with the inverse of the rate, the time per unit change in SO4, irate.

Maple 6 doesn't show the discontinuity

Exercise* 4.23 Discontinuities in Rate

Exercise your ability to seek out a discontinuity on a real function. The function that predicts the rate of $SO_4^=$ formation as a function of the amount of $SO_3^=$ seems flawed in some way. It only is well-behaved between SO4 = 0 and 0.0000039 and above SO4=0.00015. Why? When imaginary values are obtained, use evalc( ), and RootOf ( ) structures can be removed with allvalues( ), provided the numbers are without unknowns. Use discont( ) to get some idea of the discontinuities in the function. Plot the detail one sees at low and high concentrations. Consider the possibility that the results are produced by roundoff error or, that Maple has not collected the terms properly, perhaps unreasonably splitting off imaginary terms. It is even possible that another root should have been chosen in the calculation of H.

Exercise 4.24 Number of Real Solutions

Look at the general mathematical qualities of the electroneutrality balance, given above, and the solutions HH. How many real solutions does it have? Use tools like realroot( ). Try not to approximate the solutions unnecessarily.

4.4.4 Sulfate Change with Time

We pursue the relation between sulfate and time. There are various ways of looking at this; a simplest one is to consider small changes, particularly a small change in sulfate. With infinitesimals, quantities so small they are near to zero, it is rather easy to show the relationship. Then, we let Maple add all these small quantities up, integrate. The approach recognises that the rate is a function of the amount of SO4. Reversing our minds, we solve for the time at which there is a given value of SO4.

The calculus

t is a function of SO4

The formulation is nothing more than a statement of $dt = dt$ written $dt = \frac{dt}{dSO4}\ dSO4$. In Maple it is written as a partial derivative which is the inverse rate.

```
>  'rate' = Diff(SO4,t); 'irate' = Diff(t,SO4);
```
$$rate = \tfrac{\partial}{\partial t}\ SO4$$
$$irate = \tfrac{\partial}{\partial SO4}\ t$$

```
>  irate := simplify(1/rate);
```

$$irate := \frac{5}{10328477397116593096657915257908165550400107440418264003343 97487}\Big($$

$$\%1^{(2/3)}+$$
$$11009464304836640362623141942627827893224519282560271317062706755$$
$$+\ 9272140235876077711117470890707046047673849148104262649 00\backslash$$
$$00000000000000SO_4{}^2$$
$$+\ 3045018922088346844616039093000000000\ SO_4\ \%1^{(1/3)})^2 \Big/ \%1^{(2/3)}$$

$$\%1 := 502860406954256961830124604301426758991712089007572267397649\backslash$$
$$221665218077928484347702127598225000000000SO_4 + 6552257498938\backslash$$
$$251651449828253366486629081883758219844414704947620260902 75\backslash$$
$$873457160498121670302320 + 2823384246649936421017221652894 10\backslash$$
$$264139023439713907890845481773550032023877237373570000000 00\backslash$$
$$00000000000000SO_4{}^3+$$
$$29380660514071209810143697041327204728907400126353370891915 5\text{sqrt}(-$$
$$10485326428360681172672804256265042914622064481398682296812\backslash$$
$$604709026254819 - 9764511476512806643453841002287555744524 59\backslash$$
$$58398353668656721583067500000000000000000SO_4{}^2 + 7633879 22301899\backslash$$
$$632982000056112960238301847539653727773724548357781296000 00\backslash$$
$$000SO_4 + 4286154575112673020675090754405420853138622904327 7\backslash$$
$$9298907027200000000000000000000000000SO_4{}^3)$$

irate is in
Typeset Notation
from
Options/Output
Display

```
>   plot(irate);
```

Plotting error, empty plot

The plotting is unsuccessful, as expected. Explore the nature of irate.

```
>   ee := subs(SO4=1/10000,irate):
>   evalf(ee); hastype(%,I);
```
$$294.7146617 - .8682688581\ 10^{-8}\ I$$
true

The imaginary bit seems to come from a conversion error and relates to the number of digits used.

```
>   evalf(ee); evalf(ee,5); evalf(ee,20); evalf(ee,30);
```
$$294.7146617 - .8682688581\ 10^{-8}\ I$$
$$294.72 - .0017818\ I$$
$$294.71466183533908067 + .47403574773358943957\ 10^{-20}\ I$$
$$294.714661835339080667363462266 + .34478190887635011800606032003 9\ 10^{-27}\ I$$

A 'brute force' decision here is that there is an algebraic or a numerical difficulty, inherent in the evaluation of rate, irate and, probably, HH, the originally calculated hydrogen ion calculation. What is clear is that the imaginary part gets ever smaller the larger the number of digits used in the calculation. This also shows the need to do calculations with infinite precision as far as possible, before any floating point conversions.

The solution is clear, though; take the real part of the solution. The imaginary solutions are small. In fact, an investigation of the nature of the solutions would show that all the solutions are real.

```
> Re(evalf(ee));
```
$$294.7146617$$
```
> opt1 := thickness=2,title='Inverse rate':
> plot(Re(evalf(irate)),SO4=0..1/1000,opt1);
```

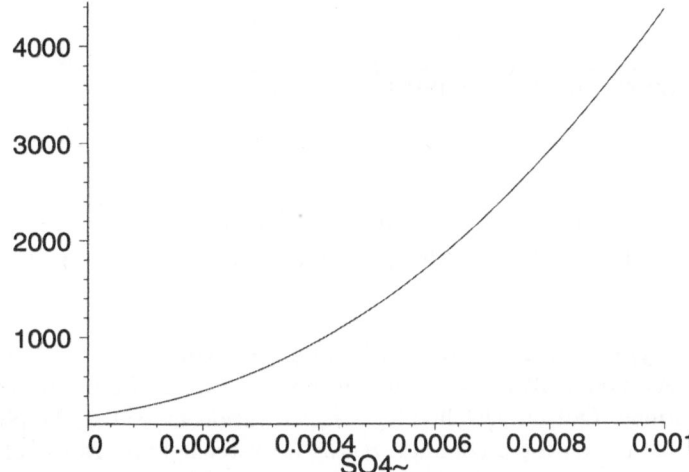

```
> Refloat := proc(x) evalf(x); Re(%) end;
```
$$Refloat := \mathbf{proc}(x)\,\mathrm{evalf}(x)\,;\,\Re(\%)\,\mathbf{end}$$
```
> TP := proc(f,x) global SO4, Refloat;
```

evalhf( ) could be
effective here

```
> evalf(Int(Refloat(f),SO4=0..x)); # numerically integrate
> end;
```
$$TP := \mathbf{proc}(f,\,x)\,\mathbf{global}\,SO4,\,Refloat;\,\mathrm{evalf}(\mathrm{Int}(\mathrm{Refloat}(f),\,SO4=0..x))\,\mathbf{end}$$
```
> TP(irate,0), TP(irate,1/10000), TP(irate,1/1000), TP(irate,1/100);
```
$$0,\, .02394181577,\, 1.652415553,\, 1349.798407$$
```
> opt := thickness=2,title='Sulfate formation with time':
```

```
>  rng := x=0..1/1000:
>  plot(TP(irate,x),rng,opt);
```

plot takes more
than 10 minutes

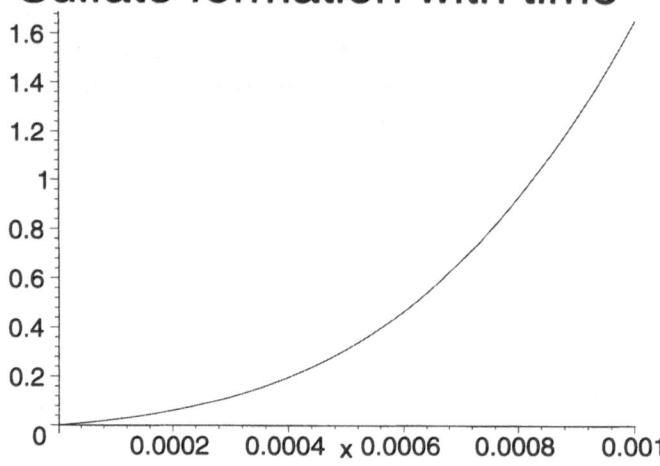

Sulfate formation with time

This calculation has little efficiency, but shows the effect; in a day and a half the concentration grows to something like 0.001 moles per litre SO4 in the water.

Exercise* 4.25 Typeset Notation and Maple Notation

Consider the detailed makeup of irate, the inverse rate of formation of sulfate. With Options, Output Display, Typeset Notation activated; Maple should make a substitution for a common factor in irate; it appears as %1. Change to an alternative for output, Maple Notation, which, like lprint( ) is easily copied. Copy the factor as fctr.

Output in Maple
Notation is easily
copied

```
>  #fctr := %1:  # collect fctr for further exploration
```

Carry on and break apart fctr until you find the offending part using

```
>  #whattype(fctr);
>  #nops(fctr);
>  #op(1..4,fctr);
```

Substitute values for the argument and try to get an idea of where the imaginary part comes from. Perhaps you can solve the problem. It might involve assume( ); see the appendix AAArgh, page 358. When you are all done, it is a good idea to return to Editable Maths as an output option.

Return to
Editable Maths

Exercise 4.26 Reorganise the Plot

Redo the plot above and clean it up. Put in appropriate labels with AXESLABEL( ) and an appropriate title with TITLE( ) and TEXT( ). Use the title " The time required for a given concentration of Sulfate to Evolve"; make it a large size. Reverse the abscissa and ordinate with a procedure. The alternative is to use implicit plot and redo the calculation, which is very inefficient with the method used.

4.4.5 Fitting the Data

Of course, we have not tried straightforward integration of irate.

```
>   int(irate,SO4=0..x):   int(irate,SO4=1..1/1000):
```

The 'suck it and see' doesn't work because Maple has some difficulty integrating the formula with all the mixed roots and powers. Probably the best approach is to use dsolve(irate,SO4,numeric) with a Runga-Kutta integration routine. However, such a smooth function lends itself to the use of a simple fit. Hence the solution was completed using a quadratic fit to the inverse of the rate in the range of interest. It is a more than adequate fit.

```
>   ?fit

>   ?fit[leastsquare]

>   with(stats[fit]);
```
$$[leastmediansquare, \ leastsquare]$$

```
>   xdat := []:   ydat := []:

>   for i from 0 to 10 do

>   i/10000:

>   eval(Refloat(subs(SO4=%,irate))):

>   xdat := [op(xdat), %%]; # SO4 becomes abscissa

>   ydat := [op(ydat), %%]; # irate is the ordinate

>   od:   i:='i':

>   eqn := y=a*x^2+b*x+c;
```
$$eqn := y = a\,x^2 + b\,x + c$$

```
>   leastsquare[[x,y],eqn,{a,b,c}]([xdat,ydat]);
```
$$y = .3790755189\,10^{10}\,x^2 + 351022.1765\,x + 212.6237077$$

```
>   irate1 := rhs(%);
```
$$irate1 := .3790755189\,10^{10}\,x^2 + 351022.1765\,x + 212.6237077$$

4.4.6 The Sulfate Concentrations

The inverse rate is equivalent to $\frac{dt}{dSO4}$ so $dt = \frac{dt}{dSO4}\ dSO4$ or $dt = irate\ dSO4$

The time at which the sulfate concentration reaches a given level x is given by

```
>   t=Int('irate',SO4=0..x);
```

$$t = \int_0^x irate\ dSO4$$

```
>   opt := thickness=2,
>   labels=["time in days","SO4"],
>   title="Sulfate formation with time",
>   titlefont=[TIMES,BOLD,16]:
>   t=subs(x=SO4,int(irate1,x));   eee := %:
>   plots[implicitplot](eee,t=0..2,SO4=0..0.001,opt);
```

$$t = .1263585063\ 10^{10}\ SO4^3 + 175511.0883\ SO4^2 + 212.6237077\ SO4$$

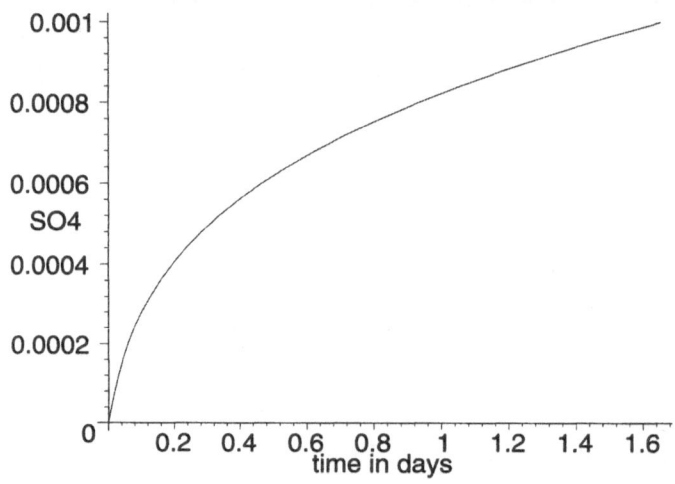

Sulfate formation with time

4.4.7 The pH

A last calculation gives the pH variation with time. This is an indication of the time required to create acid rain in the given polluted, wet environment. The analytical forms, above, have SO4 as an implicit parameter in the calculation; the implicit nature of the data persists. A robust way around this is to simply use a listing of the required data points.

```
>   #H; # H still is defined
```

```
>  dat := NULL:
>  ee := rhs(eee);
>  for i from 0 to 100 do
>  i/100000;
>  evalf(subs(SO4=%,ee));
>  -log10(evalf(subs(SO4=%%,H)));  # Hydrogen ion concentration
>  dat := dat, [%%,%];
>  od:
>  dat := [dat]:

>  opt := thickness=2,
>  labels=["time in days","pH"],
>  title="Increasing Acidity of Rainwater",
>  titlefont=[TIMES,BOLD,16]:
>  rng :=view =[0..1.6,5.4..6.1]:
>  plot(dat,rng,opt);
```

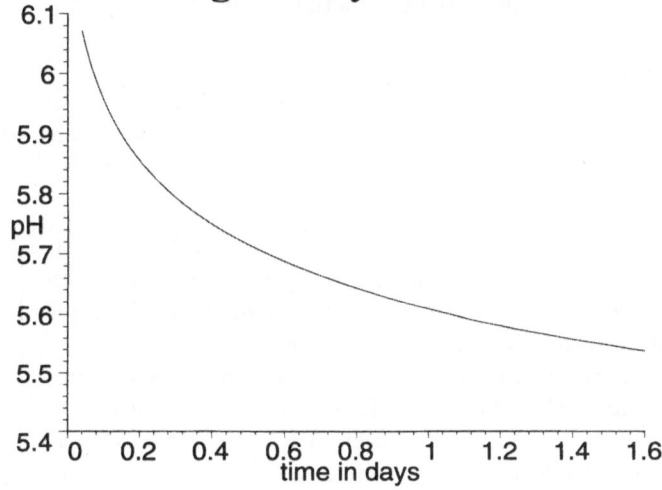

The *pH* is now approaching the value for beer.

Exercise 4.27 Using a Parametric Plot

Complete the above pH calculation using a parametric plot. Ordinary plot suffices, see section 3.5. The two equations are guided by the parameter SO4. Think of it like the location of a boat on the ocean; at any one traveltime the boat has a given position in x and y; giving the traveltime automatically generates the position coordinates x and y. SO4 simply takes the place of the traveltime. The expected format is

```
> ?plot,parametric
> plot([ee,'H'=H,SO4=0..0.001],op(rhs(rng)),opt);
```

Exercise* 4.28 Comparison with Data

There was no attempt to prove that the quadratic fitting of the irate data was tolerable. It is. Anyway, compare the data from the numerical integration with the quadratic regression line. One way is to simply take the PLOT structure and paste on the curve for the quadratic, with linestyle = 2 (dots), perhaps. More importantly, compare the final curve with that of Scott and Hobbs[25]. If one uses precisely the environment of CO_2, SO_2 and NH_3 used by these authors, the results should overlay the curves in the paper. Note that multiplying the data by a conversion factor of 96000 will convert the SO4 data from moles per litre to micrograms per cm^3.

Exercise* 4.29 Effect of Ammonia

Redo the above calculations for a range of NH_3 concentrations, perhaps as high as 1E-9. Make composite plots to display the data. Show that, indeed, the presence of ammonia enhances the production of $SO_4^=$.

Exercise* 4.30 Limiting pH

There is disagreement as to whether or not there is a limiting value of pH in this model. It seems that such a limiting form should exist. Use the algebraic forms of the solutions to find the limiting pH, with and without NH_3.

> The mathematics of practical solution chemistry produces a self-limiting acidity in droplets. The chemistry involves interwoven and relatively complicated expressions. The numerical solution is tractable and smooth when done with Maple.
>
> In the next Chapter we will add expressions to care for large numbers of species and solids, and deal with chemistry/solution effects in soils and kidney stones [26]; see section 5.7, page 289. Generally, the use of multiple assignments and Maple loses the tedium of such analysis. Also, the analysis is complete and not marginalised by limiting assumptions.

Acknowledgements for Chapter 4

The Bubble data is from Soeren Scheid of the University of Bremen, Germany. The material on Perry Lakes derives from John Rich and the Laboratory Manual, Water and Earth Sciences, Murdoch University. The general acid rain model is presented in the book by Lyons and Scott, *Principles of Air Pollution Meteorology*; the solutions evolved while I was on study leave at the Department of Computing and Mathematics at Deakin University, with Professor Peter Kloeden.

Chapter 5

Simultaneous Solutions: Matrices

Chemical Balances: The Breakfast Table

Air Quality Index: Least Squares

Turbulence: Rotating Statistics

Struvite: Fertilizer and Kidney Stone

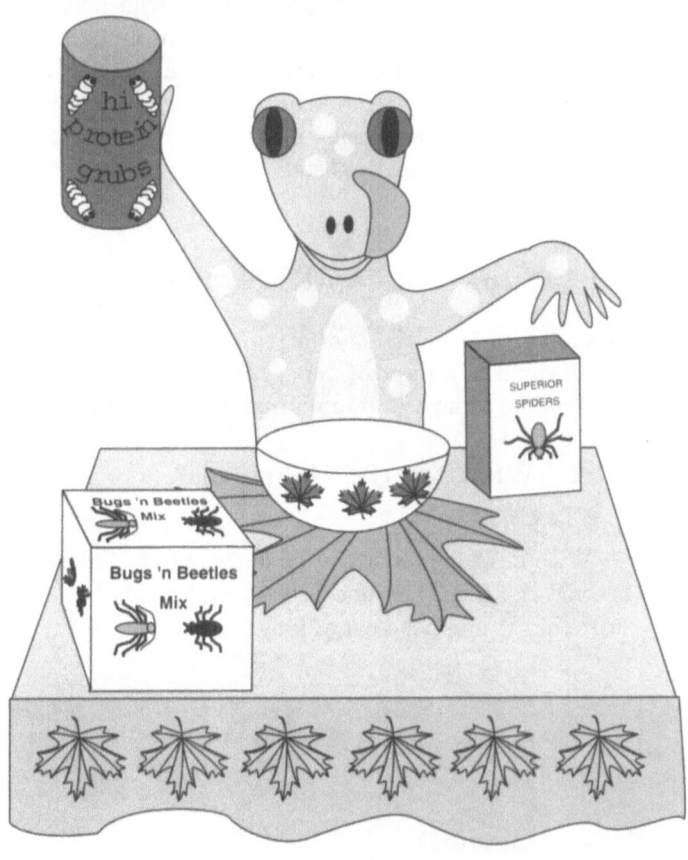

Simultaneous Solutions: Matrices

Übersicht

The manufacture and solving of equations in many variables, including polynomials, matrices, and statistics. Details of finding, substitution and manipulation. Mass balances, in chemistry, our rivers and our bodies. Correction of possible errors in measurements of the force of the wind. Completing a difficult solution, complex because of the number of operations, manufacturing a kidney stone.

Guidewords:

polynomials, indets( ), solve( ), remove( ), select( ), anames( ), unames( ); matrices, &*, inverse, transpose( ), evalm( ), det( ); breakfast foods, diet, units; chemical balance, copper, aspirin, isolve( ); dilution, river, concentration; turbulence, wind, transformation, linear, rotation, projection, statistics, diad; struvite, phosphorus, ammonia, magnesium, precipitation, ions, electroneutrality; Newton-Raphson

5.1   Equations and Matrices

Developing a formulation or attempting a solution should no longer be limited by our mind or a piece of paper. Maple can allow for extreme variation in the problem, an almost unlimited number of variables and equations.

5.1.1   What are the Unknowns?

Large Expressions

With large expressions, the names of the unknowns or the indeterminates become clouded. The one below could result from a horrid calculation; consider that it might be pages long.

```
>   restart;
>   e1 := x*y*(z^q + (r-s)^10 + 3*ln(zzz+m)/(z+c));
```

$$e1 := x\,y\,(z^q + (r - s)^{10} + 3\,\frac{\ln(zzz + m)}{z + c})$$

```
>   indets(e1);
```

$$\{x,\, y,\, z,\, q,\, r,\, s,\, zzz,\, m,\, c,\, z^q,\, \ln(zzz+m)\}$$

Make this into a list so the order means something, using op( )

```
>  [op(%)];
```

$$[x,\, y,\, z,\, q,\, r,\, s,\, zzz,\, m,\, c,\, z^q,\, \ln(zzz+m)]$$

The list has more variables than expected, but the symbols are easily selected.

```
>  op(1,%);whattype(%);
```

$$x$$
$$symbol$$

```
>  select(type,%%%,symbol);
```

$$[x,\, y,\, z,\, q,\, r,\, s,\, zzz,\, m,\, c]$$

Presuming these are the required variables, proceed with a solution. For instance, it may be desirable for c to be written as an expression in z and zz, with the values of other indeterminates known:

```
>  solve(e1,c);
```

$$
\begin{aligned}
&-(45\,r^8\,s^2\,z - 120\,r^7\,s^3\,z + 210\,r^6\,s^4\,z - 252\,r^5\,s^5\,z + 210\,r^4\,s^6\,z - 120\,r^3\,s^7\,z + 45\,r^2\,s^8\,z\\
&- 10\,r\,s^9\,z + r^{10}\,z + s^{10}\,z + z^q\,z - 10\,r^9\,s\,z + 3\ln(zzz+m))/(45\,r^2\,s^8 - 10\,r^9\,s\\
&+ 45\,r^8\,s^2 + 210\,r^6\,s^4 - 252\,r^5\,s^5 + 210\,r^4\,s^6 - 120\,r^3\,s^7 - 10\,r\,s^9 + r^{10} + s^{10} + z^q\\
&- 120\,r^7\,s^3)
\end{aligned}
$$

```
>  subs({r=2,s=1,m=0.5},%);
```

$$-\frac{z + z^q\,z + 3\ln(zzz + .5)}{1 + z^q}$$

The desired
solution

Multiple Equations

As we have seen, solve is also useful in solving simultaneous equations.

```
>  e1 := 10*x1 + 3*x2 + 5/10*x3 +x4 -x5 = 20;
>  e2 := x1 + x2 + 6*x3 + x5 = 0;
>  e3 := convert([seq(x.i,i=1..5)],'+') = 23;

>  e4 := convert([seq(i*x.i,i=1..4)],'+') = 50;
>  e5 := x4 + x5 + c = -10;
>  e6 := 16*x1 + c = 3;
```

catenation is
awkward in sum

$$e1 := 10\,x1 + 3\,x2 + \frac{1}{2}\,x3 + x4 - x5 = 20$$
$$e2 := x1 + x2 + 6\,x3 + x5 = 0$$
$$e3 := x1 + x2 + x3 + x4 + x5 = 23$$
$$e4 := x1 + 2\,x2 + 3\,x3 + 4\,x4 = 50$$
$$e5 := x4 + x5 + c = -10$$

$$e6 := 16\,x1 + c = 3$$

It may appear that there are only numbered variables

```
> seq(x.i, i=1..5);
```
$$x1,\ x2,\ x3,\ x4,\ x5$$

But c is present in the last two equations. It is good practice to check out how many unknowns actually exist in a given equation or set of equations. This can show up errors in substitution or assignment.

```
> eqns := {seq(e.i,i=1..6)}:
> vars := indets(eqns);
```
$$vars := \{c,\ x5,\ x4,\ x3,\ x2,\ x1\}$$

```
> solve(eqns,vars);
```
$$\{x1 = \frac{496}{211},\ x2 = \frac{-470}{211},\ x3 = \frac{-366}{211},\ x4 = \frac{3023}{211},\ x5 = \frac{2170}{211},\ c = \frac{-7303}{211}\}$$

Maple 6 uses || for catenation, see upgrades page 38

Catenation .

As already noted, the catenation operator   .   can be used to make names of variables.

```
> x.(1..4);
```
$$x1,\ x2,\ x3,\ x4$$

```
> (1..4);
```
$$1..4$$

```
> $ (1..4);
```

The $ is the evaluation function

$$1,\ 2,\ 3,\ 4$$

```
> x.%;
```
$$x1,\ x2,\ x3,\ x4$$

```
> x[$ (1..4)]; # indexed
```
$$x_{1,2,3,4}$$

```
> 1,2,5,6; # number missing
```
$$1,\ 2,\ 5,\ 6$$

```
> x.%;
```
$$x1,\ x2,\ x5,\ x6$$

```
> 3,2,1,5; # wrong order
```
$$3,\ 2,\ 1,\ 5$$

```
> x.%;
```

$$x3,\ x2,\ x1,\ x5$$

```
>    1,2,3,'rat',a,b,c; # mixed
```

$$1,\ 2,\ 3,\ rat,\ a,\ b,\ c$$

```
>    y.%;
```

$$y1,\ y2,\ y3,\ yrat,\ ya,\ yb,\ yc$$

```
>    a,b,c,d,e;
```

$$a,\ b,\ c,\ d,\ e$$

```
>    x.%;
```

$$xa,\ xb,\ xc,\ xd,\ xe$$

```
>    x[%%]; # indexed
```

$$x_{a,\,b,\,c,\,d,\,e}$$

```
>    f(x.(1..5));
```

$$f(x1,\ x2,\ x3,\ x4,\ x5)$$

```
>    diff(%,x.(1..3));
```

$$\frac{\partial^3}{\partial x3\,\partial x2\,\partial x1}\,f(x1,\ x2,\ x3,\ x4,\ x5)$$

Strings

The operator works with strings; it will change symbols into strings, following the lead. This is equivalent to the cat( ) command.

```
>    "".'the fat lady who takes the X-ray--Peter Connery';
```

"the fat lady who takes the X-ray–Peter Connery"

```
>    "We are done but for ".%;
```

"We are done but for the fat lady who takes the X-ray–Peter Connery"

```
>    ''."If you want something badly enough".
>    'be careful or you'll get it.--Gary Player';
```

If you want something badly enough be careful or you'll get it.–Gary Player

Remember strings cannot be evaluated

Again, Maple 6 uses || for catenation, see pages 38 and 52

5.1.2   Properties of Polynomials

coeff

```
>    ss := 5*(R+6)^4 + x^7 - 3*u*w^8*x^4 + Pi;
```

$$ss := 5\,(R+6)^4 + x^7 - 3\,u\,w^8\,x^4 + \pi$$

The leading coefficient is

```
>    lcoeff(ss);
```

$$1$$

The trailing coefficient is

```
>    tcoeff(ss);
```

Maple does some simplification to separate the constant part as trailing.

$$6480 + \pi$$

```
>  op(4,ss);
```
$$\pi$$

The coefficients of special terms are collected with another form of leading coefficient.

```
>  lcoeff(ss, [u,w], 't');
```
$$-3\,x^4$$

Then t takes the value of the u,w term, with the appropriate powers.

```
>  t;
```
$$u\,w^8$$

```
>  coeffs(ss);
```

This doesn't work because there is no reference

```
Error, invalid arguments to coeffs
>  coeffs(ss,[x,w],'t');
```

In turn, these are the coefficients of the power combination in t.
$$-3\,u,\, 5\,(R+6)^4 + \pi,\, 1$$

```
>  t;
```
$$w^8\,x^4,\, 1,\, x^7$$

Collect

It is a bit more orderly to collect coefficients with respect to x and then look for the coefficients.

```
>  sss := collect(ss,x); coeffs(%,x,'t');
```
$$sss := 5\,(R+6)^4 + x^7 - 3\,u\,w^8\,x^4 + \pi$$
$$-3\,u\,w^8,\, 5\,(R+6)^4 + \pi,\, 1$$

```
>  t;
```
$$x^4,\, 1,\, x^7$$

The coefficient of R to the power 3 is

```
>  collect(ss,R); coeff(%,R,3);
```
$$5\,R^4 + 120\,R^3 + 1080\,R^2 + 4320\,R + 6480 + x^7 - 3\,u\,w^8\,x^4 + \pi$$
$$120$$

The expansion is of different degree in each variable.

```
>  degree(ss), degree(ss,x), degree(sss,R);
```
$$13,\, 7,\, 4$$

Poor name, don't confuse it with 'leading'

The 'lowest' degree is the same for both variables.

```
>  ldegree(sss,x), ldegree(sss,R);
```
$$0, 0$$

```
>  x^7 - 3*u*w^8*x^4;
```
$$x^7 - 3\,u\,w^8\,x^4$$

```
>  ldegree(%,x);
```

4

allvalues and subs

The command subs is handy for checking answers.

```
>   p := 3*y^5 - 20*y^4 + 3*y^3 - 3*y = 0;
```
$$p := 3\,y^5 - 20\,y^4 + 3\,y^3 - 3\,y = 0$$
```
>   ans := [solve(p,y)];
```

The solution is a bit of a mess; the five answers can be seen more clearly by forcing a numerical result. This is done with allvalues( ) which converts RootOf( ) forms to floating point numbers.

This RootOf form can be used directly by Maple

$$ans := [0, \text{RootOf}(3\,\_Z^4 - 20\,\_Z^3 + 3\,\_Z^2 - 3)]$$
```
>   remove(has,%,0); # removes the zeros
>   exsol := allvalues(%):
>   remove(has,[%],I);
```

$$\left[\frac{5}{3} + \frac{1}{6}\sqrt{\%2} + \frac{1}{6}\sqrt{\frac{188\,\%1^{(1/3)}\sqrt{\%2} + 3\sqrt{\%2}\,\%1^{(2/3)} - 33\sqrt{\%2} + 1820\,\%1^{(1/3)}}{\%1^{(1/3)}\sqrt{\%2}}},\right.$$
$$\left.\frac{5}{3} + \frac{1}{6}\sqrt{\%2} - \frac{1}{6}\sqrt{\frac{188\,\%1^{(1/3)}\sqrt{\%2} + 3\sqrt{\%2}\,\%1^{(2/3)} - 33\sqrt{\%2} + 1820\,\%1^{(1/3)}}{\%1^{(1/3)}\sqrt{\%2}}}\right]$$

$$\%1 := 563 + 10\sqrt{3183}$$
$$\%2 := \frac{94\,\%1^{(1/3)} - 3\,\%1^{(2/3)} + 33}{\%1^{(1/3)}}$$

Two solutions are not imaginary
```
>   nops(%);
```
2

```
>   ansf := evalf(%%);
```
$$ansf := [6.516831042, -.476521210]$$
```
>   select(hastype,%,positive);
```
$$[6.516831042]$$
```
>   ansf := op(%);
```

$$ansf := 6.516831042$$
```
>   subs(y=ansf,p);
```

But, even so, there is error.

$$.00002707 = 0$$

To prove the solutions are correct, we back substitute the exact solutions, but it produces a mess.

```
>  subs(y=exsol[1],p), subs(y=exsol[2],p):
>  simplify([%]);
```

$$[0 = 0, \, 0 = 0]$$

This is convincing; different algorithms are used in evaluation and simplification, so we can be confident with the result.

Try another polynominal.

```
>  restart:
>  p := 7*y^4 - 59*y^3 + 19*y^2 + 166*y -1008 = 0;
```

$$p := 7\,y^4 - 59\,y^3 + 19\,y^2 + 166\,y - 1008 = 0$$

```
>  ans :=solve(p,y);
```

$$ans := 8, \, \frac{-18}{7}, \, \frac{3}{2} + \frac{1}{2}\,I\,\sqrt{19}, \, \frac{3}{2} - \frac{1}{2}\,I\,\sqrt{19}$$

```
>  subs(y=ans[2],p);
```

easy proof, partly because the polynominal could be factored

$$0 = 0$$

```
>  factor(p);
```

$$(y - 8)\,(7\,y + 18)\,(y^2 - 3\,y + 7) = 0$$

subs and assignment

Making assignments is like traveling a thin road with an abrupt cliff on one side. On the other side, an alternative, using Ditto % is a dangerous gorge. Stick with subs, if at all possible, as a vehicle to get you through.

Consider the assignment of a value to the symbol y, above.

```
>  y := -18/7;
```

$$y := \frac{-18}{7}$$

This makes the polynominal useless.

```
>  p;
```

$$0 = 0$$

Indeed, the solution is correct for the polynominal, but it is likely you want to make further use of y as a running variable and p as a polynominal. Expect to keep track of your assigned variables and use subs as much as possible. If you are curious, or need information on assigned names, have a look at anames( ) or unames( ).

```
>  an := [anames()]:
>  nops([anames()]);
```
$$364$$

Adjust the range
to find y

```
>  sort([anames()])[(%-20)..%]; # sort assigned names, last 21
```

[*solve/rec/rename, solve/rec/series, solve/rec/signum, solve/rec/trig, solve/rec/unify,*
solve/rec2, solve/reduce, solve/rootunit, solve/split, solve/twotwo, sqrt/nested,
sqrt/primes, sqrt/product, testCBZ, type/AndProp, type/EvalfableProp,
type/JustLinear, type/JustOnce, type/JustPoly, type/LinearProp, type/OrProp,
type/ParamProp, type/PropRange, type/RealRange, type/RootOf, type/algext,
type/algnum, type/linear, type/polynom, type/property, y]

```
>  member(evaln(y), [anames()]); # y must be a name
```
true
```
>  member('y',sort([anames()]),'k');
```
true
```
>  k;
```
$$364$$
```
>  has([anames()],'p');
```
true

5.1.3 Matrix Operations

Matrices are derived from lists of linear simultaneous equations, stacked vertically. Take the case with two unknowns, x and y, and two predicting equations.
The operations of matrix multiplication and addition are defined to preserve
the operations over the equations. The operation of multiplication &* consists of selecting a row from the first matrix and a column from the second.
Multiply the elements in pairs; row along, column down. Add the result and
place it in the position of the original row and column. Do this for all rows
and columns. You should do this with the example below, with Maple and by
hand.

```
>  eqns := {2*x + 3*y = 12, x - y = 2};
```
$$eqns := \{2\,x + 3\,y = 12,\ x - y = 2\}$$
```
>  solve(eqns,{x,y});
```
$$\{y = \frac{8}{5},\ x = \frac{18}{5}\}$$
```
>  A := matrix(2,2,[[2,3],[1,-1]]);
```
$$A := \begin{bmatrix} 2 & 3 \\ 1 & -1 \end{bmatrix}$$

```
>  X := matrix(2,1,[[x],[y]]);
```

$$X := \begin{bmatrix} x \\ y \end{bmatrix}$$

```
>  B := matrix(2,1,[[12],[2]]);
```

$$B := \begin{bmatrix} 12 \\ 2 \end{bmatrix}$$

The matrix equation is written $AX = B$; the inverse of matrix A is expected to be matrix AI.

Maple 6 uses
a . for
non-commutative
multiplication

```
>  'A'&*'X'='B',' so we expect ', 'X'='AI'&*'B';
```

$$A \& * X = B, \text{ so we expect }, X = AI \& * B$$

```
>  with(linalg):
>  AI := inverse(A);
```

$$AI := \begin{bmatrix} \dfrac{1}{5} & \dfrac{3}{5} \\ \dfrac{1}{5} & \dfrac{-2}{5} \end{bmatrix}$$

```
>  AI&*B;
```

$$AI \& * B$$

```
>  evalm(%);
```

$$\begin{bmatrix} \dfrac{18}{5} \\ \dfrac{8}{5} \end{bmatrix}$$

The answer is the same; a matrix equation is just another way of solving linear simultaneous equations, useful with a large number of unknowns.

5.2   Breakfast Foods

Listed below are some analyses of cold breakfast foods. These are percents
and should be listed as fractions. In pondering over breakfast, consider the
situation where your breakfast satisfies all your daily needs for *fibre*, *kjoules*,
protein and *water*.

The energy row is
in units of kjoules
per gm

```
> restart:
> Breakfast :=
> [[ FOODS , Granose, Muesli, Sultanas, Milk],
> [fibre  ,    13   ,   10  ,     6   ,   0 ],
> [kjoules,    15   ,   25  ,    30   ,   1 ],
> [protein,    12   ,   25  ,     0   ,   2 ],
> [water  ,    12   ,   12  ,    40   ,  90 ]]:
> op(1,%):  op(1,%); op(2..5,%%); foods := %:
```
$$FOODS$$
$$Granose,\ Muesli,\ Sultanas,\ Milk$$

```
> 'Percentage of';
```

The second
argument of
map2 feeds op

```
> op(2..5, map2(op,1,Breakfast));
```
$$Percentage\ of$$
$$fibre,\ kjoules,\ protein,\ water$$

The daily requirements are given on the cereal box. The tables are put
into a matrix form, including some water.

```
> with(linalg):
```

Warning, new definition for norm

Warning, new definition for trace

```
> matrix(1,4,[10,4000,50,1500]);
```
$$\begin{bmatrix} 10 & 4000 & 50 & 1500 \end{bmatrix}$$

```
> Requirements;
```
$$Requirements$$

```
> B :=transpose(%%);
```
$$B := \begin{bmatrix} 10 \\ 4000 \\ 50 \\ 1500 \end{bmatrix}$$

That is, we require 50 gms of fibre, 4000Kjoules of energy, 30 gms of
protein and 1500gms of water each day. This, of course, is dependent on the
individual.

The matrix equation is of form $AX = B$. The three matrices; A, X, and B represent the 4 balance equations. For each foodstuff, there is a certain requirement of total expected intake in a single day, made up from the expected percentages of the amount in each foodstuff. For instance, the total gms of fibre comes from the the sum of the 15% in the Granose, the 12% in the muesli, and the 6% in the sultanas; there is none in the milk. This total should equal 50 gms. Matrix multiplication carries through with the 'row-across', 'column-down' companion multiplication and addition sequence.

The matrix equation is set up and checked to make sure it represents the appropriate equations. Solution follows immediately, with inversion of the A matrix.

>   `[op(2..5,Breakfast)];`

    $[[\textit{fibre}, 13, 10, 6, 0], [\textit{kjoules}, 15, 25, 30, 1], [\textit{protein}, 12, 25, 0, 2],$
    $[\textit{water}, 12, 12, 40, 90]]$

>   `aa := map2(remove,hastype,%,symbol);`

    $aa := [[13, 10, 6, 0], [15, 25, 30, 1], [12, 25, 0, 2], [12, 12, 40, 90]]$

>   `d100 := x->x/100;[map(d100,aa[1]),aa[2],map(d100,aa[3]),map(d100,aa[4])];`

conversion of some rows to fractions

$$d100 := x \to \frac{1}{100}\, x$$

$$[[\frac{13}{100}, \frac{1}{10}, \frac{3}{50}, 0], [15, 25, 30, 1], [\frac{3}{25}, \frac{1}{4}, 0, \frac{1}{50}], [\frac{3}{25}, \frac{3}{25}, \frac{2}{5}, \frac{9}{10}]]$$

These statements are worth a little analysis: map2 takes hastype, and the extra argument symbol, checks the inner list and applies remove. The lower list is a simple pasting together of our A matrix. This conversion might be done with map(map( )).

>   `A := convert(%,matrix);`

$$A := \begin{bmatrix} \dfrac{13}{100} & \dfrac{1}{10} & \dfrac{3}{50} & 0 \\[2mm] 15 & 25 & 30 & 1 \\[2mm] \dfrac{3}{25} & \dfrac{1}{4} & 0 & \dfrac{1}{50} \\[2mm] \dfrac{3}{25} & \dfrac{3}{25} & \dfrac{2}{5} & \dfrac{9}{10} \end{bmatrix}$$

Then, the matrix equation is

>   `evalm(A) &* matrix(4,1,[x.(1..4)])=evalm(B);`

$$\begin{bmatrix} \frac{13}{100} & \frac{1}{10} & \frac{3}{50} & 0 \\ 15 & 25 & 30 & 1 \\ \frac{3}{25} & \frac{1}{4} & 0 & \frac{1}{50} \\ \frac{3}{25} & \frac{3}{25} & \frac{2}{5} & \frac{9}{10} \end{bmatrix} \&* \begin{bmatrix} x1 \\ x2 \\ x3 \\ x4 \end{bmatrix} = \begin{bmatrix} 10 \\ 4000 \\ 50 \\ 1500 \end{bmatrix}$$

or, the full list of balance equations

```
>  evalm(%);
```

$$\begin{bmatrix} \frac{13}{100} x1 + \frac{1}{10} x2 + \frac{3}{50} x3 \\ 15\,x1 + 25\,x2 + 30\,x3 + x4 \\ \frac{3}{25} x1 + \frac{1}{4} x2 + \frac{1}{50} x4 \\ \frac{3}{25} x1 + \frac{3}{25} x2 + \frac{2}{5} x3 + \frac{9}{10} x4 \end{bmatrix} = \begin{bmatrix} 10 \\ 4000 \\ 50 \\ 1500 \end{bmatrix}$$

linsolve( ) from linalg will solve this in a single operation; this shows the detail.

```
>  AI := inverse(A);
```

$$AI := \begin{bmatrix} \frac{847250}{64517} & \frac{-3539}{129034} & \frac{-164650}{64517} & \frac{5625}{64517} \\ \frac{-405000}{64517} & \frac{922}{64517} & \frac{331900}{64517} & \frac{-8400}{64517} \\ \frac{-85425}{64517} & \frac{9189}{258068} & \frac{-196425}{64517} & \frac{3625}{129034} \\ \frac{-21000}{64517} & \frac{-908}{64517} & \frac{65000}{64517} & \frac{71250}{64517} \end{bmatrix}$$

```
>  evalm(AI&*B): X := evalf(%,3);
```

$$X := \begin{bmatrix} 24.8 \\ 56.3 \\ 19.1 \\ 1650. \end{bmatrix}$$

```
>  evalm(matrix(4,1,[foods])= X);
```

$$\begin{bmatrix} Granose \\ Muesli \\ Sultanas \\ Milk \end{bmatrix} = \begin{bmatrix} 24.8 \\ 56.3 \\ 19.1 \\ 1650. \end{bmatrix}$$

One can argue with these answers. The combination isn't necessarily pleasant. If you raise the protein content of the milk or change the constitution of the muesli very much, you end up eating negative amounts of Muesli , Granose, or Sultanas. It probably means that the muesli or the sultanas have a less than desirable constitution. Anyway, this is a very sparing diet with a very low protein milk, normally difficult to get in Australia. Ordinary diets must consider different fats, calcium, vitamins, and minerals as well–and the body must deal with it. The maths, here, may give a clue to the problems the body has in organising food. This mix is not very good, but, if it is your diet, you should think of adjusting your brands so that the constitutions are different, and drink more water.

Exercise 5.1 No Granose

Complete the above for a man with a protein requirement of 70 gms per day, who eats no Granose and drinks full milk with a protein content of 4% and an energy content of 2 kjoules per gram. His Muesli has a lot of sultanas, an energy content of 15KJ per gm and is 30% water. The package says that his sultanas have an energy content of 1200 kjoules in every 100 grams.

Exercise 5.2 Unlimited water

Complete the original calculation leaving out the water balance. Complete Exercise 5.1 with the water balance left out.

Exercise* 5.3 With Fat

Look up some appropriate values for fat in the foods suggested. One might expect about 1% for the Granose, 5% for the Muesli, 0% for the Sultanas and 1% in the milk. Calculate the amount of fat consumed with the above diet. Further, estimate how much fat one should eat in a day. Complete the calculation with a 5x5 matrix.

Exercise* 5.4 A Working Diet

Consider three different muelis in the market. One is low in protein, 12% (value for wheat); another is high in honey with a kjoule content of 35 kjoules per gm; a third is high in nuts, with a fat content of 50%. Complete the above balances and see if you can get realistic results. If you can not, adjust the constitutions to make the numbers work. Remember mass can not generally be lost and the body must cope with what you give it.

5.3   Balancing a Chemical Equation

An ordinary chemistry problem that can be perplexing, especially if there are complex chemical species involved. It often seems there is not enough information but the requirement for integer values for a, b, c, d, e, f and g allow for a simple, integer solution with isolve( ) in Maple.

Copper Half reaction

a Cu_2S + b H^+ + NO_3^- -> a Cu^{++} + eNO + f S_8 + g H_2O

```
> ceq := a*Cu2S + b*Hp + NO3m = d*Cu2p + e*NO + f*S8 + g*H2O;
```
$$ceq := a\,Cu2S + b\,Hp + NO3m = d\,Cu2p + e\,NO + f\,S8 + H2O$$

Consider the balance on each atom.

```
> Cu := 2*a = d;
```
$$Cu := 2\,a = d$$

```
> S := a = 8*f;
```
$$S := a = 8\,f$$

```
> H := b = 2*g;
```
$$H := b = 2\,g$$

```
> N := c = e;
```
$$N := c = e$$

```
> Ox := 3*c = e + g;
```
$$Ox := 3\,c = e + g$$

The requirement of charge neutrality adds one additional equation.   O is protected

```
> Q := b - c = 2*d;
```
$$Q := b - c = 2\,d$$

```
> isolve({Cu,S,H,N,Ox,Q});
```
$$\{a = 24\,\_N1,\, b = 128\,\_N1,\, f = 3\,\_N1,\, g = 64\,\_N1,\, c = 32\,\_N1,\, e = 32\,\_N1,\, d = 48\,\_N1\}$$

_N1 refers to an unknown parameter. The lowest possible whole numbers are required.

```
> subs(_N1=1,%);
```
$$\{a = 24,\, b = 128,\, f = 3,\, g = 64,\, c = 32,\, e = 32,\, d = 48\}$$

```
> subs(%,ceq);
```
$$24\,Cu2S + 128\,Hp + NO3m = 48\,Cu2p + 32\,NO + 3\,S8 + H2O$$

Preparation of Aspirin

In reality several reactions are involved; water enters the reaction but it may be a reactant or a product.

```
> a*C6H5OH + b*NaOH + c*CO2 + d*CH3OH  = e*NaOOCCOCH3C6H4 + f*H2O;
> ceq:=%:
```
$$a\,C6H5OH + b\,NaOH + c\,CO2 + d\,CH3OH = e\,NaOOCCOCH3C6H4 + f\,H2O$$

```
> C := 6*a + c + d = 9*e;
```
$$C := 6\,a + c + d = 9\,e$$

```
> H := 6*a + b + 4*d = 7*e +2*f;
```

errors here
$$H := 5\,a + b + c + 4\,d = 8\,e + 2\,f$$

```
> Q := a+b+2*c + d =3*e + f ;
```
$$Q := a + b + 2\,c + d = 3\,e + f$$

```
> Na := b = e;
```
$$Na := b = e$$

error here

```
> isolve({C,H,Ox,Na});
```

$$\{a = 10\_N1 - 25\_N2,\ b = 7\_N1 - 17\_N2,\ f = 2\_N1,\ c = 3\_N1 - 6\_N2,$$
$$e = 7\_N1 - 17\_N2,\ d = 3\_N2\}$$

There are two unknown constants. Since they can have any value and all the coefficients should be real and as small as possible, we choose one of them to be 1 and adjust the other one so as to make both of them as small as possible.

```
> subs(_N2=1,%);
```

$$\{f = 2\_N1,\ a = 10\_N1 - 25,\ b = 7\_N1 - 17,\ c = 3\_N1 - 6,\ e = 7\_N1 - 17,\ d = 3\}$$

```
> subs(_N1=3,%);
```

$$\{d = 3,\ f = 6,\ a = 5,\ b = 4,\ c = 3,\ e = 4\}$$

It is clear that we put the water on the appropriate side of the equation.

```
> subs(%,ceq);
```

$$5\,C6H5OH + 4\,NaOH + 3\,CO2 + 3\,CH3OH = 4\,NaOOCCOCH3C6H4 + 6\,H2O$$

And our aspirin, sodium acetylsalicylate, is ready! No!! There is an error in the analysis. The correct answer is

$$C6H5OH + 5\,NaOH + 12\,CO2 + 27\,CH3OH = 5\,NaOOCCOCH3C6H4 + 42\,H2O$$

Exercise 5.5 Precipitation of Sulfur

As in the above, the interaction of sodium thiosulfate with sulfuric acid produces small sulfur particles in the water. This is a rather simple balance. Work out the coefficients with Maple.

```
> ceq := a*NaS2O5 + b*H2SO4  = c*S8 + d*NaSO4 + e*H2O;
```

Exercise 5.6 Quantitative Reaction of Molybdenum

```
> ceq := a*Mo2O3 + b*KMnO4 + c*H2SO4 = d*MoO3 + e*MnSO4 + f*H2O
```

Exercise* 5.7 Biochemical Carbohydrate Oxidation

Adenosine triphosphate, ATP, and adenosine diphosphate, ADP, are agents for energy storage and recovery in living systems. Both have very complex structures; it suffices here to simply give the stoichiometric chemical formulas; $ATP = C_{10}H_{16}N_5O_{13}P_3$ and $ADP = C_{10}H_{15}N_5O_{10}P_2$.

```
>   ceq := a*C3H4O3 + b*O2 + c*ADP + d*H3PO4 = e*CO2 + f*H2O + g*ATP
```

5.4 Dilution: Rivers and Solutions

Dilutions can come from mixing solutions in a beaker or mixing rivers. Multiple equations and matrices can make these 'solutions' smooth and revealing.

River Dilution

Matrices are methods to madness in the analysis of river flows. The rivers Har, Ino, Jel and Kee; H, I , J and K; meet in a relatively short reach and form the major river Confusion. The flow in Confusion is measured to be 50 million m^3/day. The flows in H, I, J and K are unknown. However, the salt levels, the nitrogen levels and the phosphorus levels have been measured in all the rivers; they are as shown below. The question is, what are reasonable estimates of the flows in rivers H, I, J, and K?

```
>   restart;
>   [['Water Flow in Millions of m3 per Day'],
>   ['Salt Flow in Kilograms per day'],
>   ['Nitrogen Flow in grams per day'],
>   ['Phosphorus Flow in grams per day']]:
>   Balances := convert(%,array);
```

$$Balances := \begin{bmatrix} \textit{Water Flow in Millions of m3 per Day} \\ \textit{Salt Flow in Kilograms per day} \\ \textit{Nitrogen Flow in grams per day} \\ \textit{Phosphorus Flow in grams per day} \end{bmatrix}$$

```
>   'In the river Confusion the flows are';
>   B := matrix(4,1,[[50],[1500],[300],[250]]);
```

In the river Confusion the flows are

$$B := \begin{bmatrix} 50 \\ 1500 \\ 300 \\ 250 \end{bmatrix}$$

```
>   'The concentration Matrix for the constituents is';
>   [[1,1,1,1],[40,30,40,20],[15,4,5,6],[4,3,9,3]]:
>   A := convert(%,matrix);
```

The concentration Matrix for the constituents is

$$A := \begin{bmatrix} 1 & 1 & 1 & 1 \\ 40 & 30 & 40 & 20 \\ 15 & 4 & 5 & 6 \\ 4 & 3 & 9 & 3 \end{bmatrix}$$

Look at the elements in rows: The balance for water flow in the first row simply says that the flows of water should add up to make the total flow in the river Confusion. The second row balances the flows of salt. Look at the units and how they cancel; a flow of salt is a product of a concentration and a water flow. The concentrations are in kilograms of salt per million m^3 of water, for each particular river. The third and fourth rows, similarly, are balances of Nitrogen and Phosphorus flows with concentration units of grams of Nitrogen or Phosphorus per million m^3 of water.

If the determinant is non-zero, the equations are independent and there is a solution.

```
> with(linalg):  det(A);
```

There is a solution

```
Warning, new definition for norm

Warning, new definition for trace
```
$$750$$

```
> IA := inverse(A);
```

$$IA := \begin{bmatrix} \dfrac{-17}{25} & \dfrac{2}{125} & \dfrac{2}{25} & \dfrac{-1}{25} \\[2mm] \dfrac{2}{15} & \dfrac{11}{150} & \dfrac{-2}{15} & \dfrac{-4}{15} \\[2mm] \dfrac{-29}{75} & \dfrac{-1}{375} & \dfrac{-1}{75} & \dfrac{13}{75} \\[2mm] \dfrac{29}{15} & \dfrac{-13}{150} & \dfrac{1}{15} & \dfrac{2}{15} \end{bmatrix}$$

```
> X := evalm(IA&*B);
```

$$X := \begin{bmatrix} 4 \\ 10 \\ 16 \\ 20 \end{bmatrix}$$

```
> 'The river flows in millions of m3 per day';
> [[H],[I],[J],[K]]:
> convert(%,array)=evalm(X);
```

The river flows in millions of m3 per day

$$\begin{bmatrix} H \\ I \\ J \\ K \end{bmatrix} = \begin{bmatrix} 4 \\ 10 \\ 16 \\ 20 \end{bmatrix}$$

Note the mix of matrices and arrays. The matrix command produces arrays with a restricted ordering, starting with row and column numbers at 1 and in a numerical progression. There is no real difference between matrices and arrays the way they are used here.

Problems in dilution are conceptually not unlike the chemical balance of section 5.3. The mathematics are the basis of tracer techniques that can determine flows and sources of pollution.

Mixing Solutions

Two separate solutions of chemicals are 0.1 Molar in $CaCl_2$ and 0.05 Molar in KCl, respectively. How much of each solution must be added to a litre of water to make the final solution 0.02 Molar in Ca^{++} and 0.05 M in Cl^- ? The scheme water used in the preparation has a Ca^{++} concentration of 0.002 Molar and a Cl^- concentration of 0.01 Molar.

Here we let:

x = volume of $CaCl_2$ solution (litres)

y = volume of KCl solution (litres)

z = volume of scheme water (litres)

Then:

```
>   eq1  :=     x +       y +         z  = 1.00:  # water mass balance
>   eq2  := 0.1*x              + 0.002*z = 0.02:  # Ca mass balance
>   eq3  := 0.2*x + 0.05*y + 0.010*z     = 0.05:  # Cl mass balance
>   eq1;eq2;eq3;
```

$$x + y + z = 1.00$$
$$.1\,x + .002\,z = .02$$
$$.2\,x + .05\,y + .010\,z = .05$$

```
>   eqs  := {eq1,eq2,eq3}; vars := {x,y,z};
```
$$eqs := \{.2\,x + .05\,y + .010\,z = .05, .1\,x + .002\,z = .02, x + y + z = 1.00\}$$
$$vars := \{z, x, y\}$$

```
>   solve(eqs,vars);
```
$$\{x = .1860465116, y = .1162790698, z = .6976744186\}$$

Matrices and Arrays

The same solution can be obtained with matrices. We have already used the matrix command. Here we emphasise that the matrix command is a variant of the array command; it creates arrays with a restricted ordering, starting with row and column numbers at 1 and in a numerical progression. The package linalg is a source of many matrix operations.

```
>   with(linalg);
```

Have a look at
the commands
available

[*BlockDiagonal, GramSchmidt, JordanBlock, LUdecomp, QRdecomp, Wronskian, addcol, addrow, adj, adjoint, angle, augment, backsub, band, basis, bezout, blockmatrix, charmat, charpoly, cholesky, col, coldim, colspace, colspan, companion, concat, cond, copyinto, crossprod, curl, definite, delcols, delrows, det, diag, diverge, dotprod, eigenvals, eigenvalues, eigenvectors, eigenvects, entermatrix, equal, exponential, extend, ffgausselim, fibonacci, forwardsub, frobenius, gausselim, gaussjord, geneqns, genmatrix, grad, hadamard, hermite, hessian, hilbert, htranspose, ihermite, indexfunc, innerprod, intbasis, inverse, ismith, issimilar, iszero, jacobian, jordan, kernel, laplacian, leastsqrs, linsolve, matadd, matrix, minor, minpoly, mulcol, mulrow, multiply, norm, normalize, nullspace, orthog, permanent, pivot, potential, randmatrix, randvector, rank, ratform, row, rowdim, rowspace, rowspan, rref, scalarmul, singularvals, smith, stackmatrix, submatrix, subvector, sumbasis, swapcol, swaprow, sylvester, toeplitz, trace, transpose, vandermonde, vecpotent, vectdim, vector, wronskian*]

```
>   A := matrix([[1,1,1],[0.1,0.,0.002],[0.2,0.05,0.010]]);
>   whattype(%); lprint(%%);
```

$$A := \begin{bmatrix} 1 & 1 & 1 \\ .1 & 0. & .002 \\ .2 & .05 & .010 \end{bmatrix}$$

array

```
array(1 .. 3, 1 .. 3,[(2, 2)=0.,(1, 2)=1,(1, 1)=1,(2, 3)=.2e-2,(3,
1)=.2,(3, 2)=.5e-1,(3, 3)=.10e-1,(2, 1)=.1,(1, 3)=1])
>   B := matrix(3,3,[1,1,1,0.1,0.,0.002,0.2,0.05,0.010]);
>   whattype(%); lprint(%%);
```

A second way of
using matrix

$$B := \begin{bmatrix} 1 & 1 & 1 \\ .1 & 0. & .002 \\ .2 & .05 & .010 \end{bmatrix}$$

array

```
array(1 .. 3, 1 .. 3,[(2, 2)=0.,(1, 2)=1,(1, 1)=1,(2, 3)=.2e-2,(3,
1)=.2,(3, 2)=.5e-1,(3, 3)=.10e-1,(2, 1)=.1,(1, 3)=1])
>   C := array([[1,1,1],[0.1,0.,0.002],[0.2,0.05,0.010]]);
>   whattype(%); lprint(%%);
```

$$C := \begin{bmatrix} 1 & 1 & 1 \\ .1 & 0. & .002 \\ .2 & .05 & .010 \end{bmatrix}$$

array

```
array(1 .. 3, 1 .. 3,[(2, 2)=0.,(1, 2)=1,(1, 1)=1,(2, 3)=.2e-2,(3,
1)=.2,(3, 2)=.5e-1,(3, 3)=.10e-1,(2, 1)=.1,(1, 3)=1])

>   equal(A,B),equal(B,C),equal(C,A);
```

true, true, true

A matrix is an specialised 2D array with indices starting with 1. For the simple way we are using array, all of the last three forms are essentially identical, in structure and type. However, Maple may not know these forms are nearly identical and may require conversions.

```
> matadd(A,B,1,-1), matadd(B,C,-2,+2); # A-B, -2*B+2*C
```

$$\begin{bmatrix} 0 & 0 & 0 \\ 0. & 0. & 0. \\ 0. & 0. & 0. \end{bmatrix}, \begin{bmatrix} 0 & 0 & 0 \\ 0. & 0. & 0. \\ 0. & 0. & 0. \end{bmatrix}$$

```
> matadd(A,C,-1,+2); # -C+2*A
```

$$\begin{bmatrix} 1 & 1 & 1 \\ .1 & 0. & .002 \\ .2 & .05 & .010 \end{bmatrix}$$

An array doesn't have to start its indices at 1. Errors may occur when the data are used in further calculations and the alignment is mixed. Take care when using convert(,matrix).

```
> array(0 ..  2, -1 ..+1,
> [[1,1,1],[0.1,0.,0.002],[0.2,0.05,0.010]]): lprint(%);

array(0 .. 2, -1 .. 1,[(1, 0)=0.,(1, 1)=.2e-2,(2, -1)=.2,(0, -1)=1,(2, 0)=.5e-1,(0, 0)=1,(0, 1)=1,(1, -1)=.1,(2, 1)=.10e-1])

> convert(%%,matrix); lprint(%);
```

The indices now start with 1

$$\begin{bmatrix} 1 & 1 & 1 \\ .1 & 0. & .002 \\ .2 & .05 & .010 \end{bmatrix}$$

```
array(1 .. 3, 1 .. 3,[(2, 2)=0.,(1, 2)=1,(1, 1)=1,(2, 3)=.2e-2,(3, 1)=.2,(3, 2)=.5e-1,(3, 3)=.10e-1,(2, 1)=.1,(1, 3)=1])
```

Solution with Matrices

Back to our initial problem. The solution is rather straightforward with matrix operations.

```
> X := matrix(3,1,[x,y,z]);
```

$$X := \begin{bmatrix} x \\ y \\ z \end{bmatrix}$$

```
> B := matrix(3,1,[1.,.02,.05]);
```

$$B := \begin{bmatrix} 1. \\ .02 \\ .05 \end{bmatrix}$$

```
> multiply(A,X);
```

$$\begin{bmatrix} x + y + z \\ .1\,x + .002\,z \\ .2\,x + .05\,y + .010\,z \end{bmatrix}$$

```
> AI := inverse(A);
```

$$AI := \begin{bmatrix} -.02325581395 & 9.302325581 & .4651162791 \\ -.1395348837 & -44.18604651 & 22.79069767 \\ 1.162790698 & 34.88372093 & -23.25581395 \end{bmatrix}$$

```
> X = multiply(AI,B);
```

The answers, in a
column matrix

$$X = \begin{bmatrix} .1860465117 \\ .116279070 \\ .697674419 \end{bmatrix}$$

Exercise 5.8 Linsolv

Use linalg[linsolv]( ) to complete solutions to the last two dilution problems.

5.5   An Air Quality Index: Least Squares

The matrix form of the least-squares calculation

$$(Xt\&*X)\&*C=Xt\&*Y$$

is a 'least squares' solution to a regression calculation. Here the X matrix is simply the table of air pollution data, having far more rows of data than there are parameters that may be fitted to the data. The multiplication does the 'maximum likelihood' calculation and produces the 'best' values of the parameters necessary to fit the known data. This forms an 'Air Quality Index', the 'y-index'. The 'y-index' is considered as a measure of the quality of life in Metro City. It presumes that a person is positively affected by sunshine (0-1000 $watts/m^2$), variable x1; % of area covered by trees and vegetation, x2; the negative charge density in the air (0-100 million $charges/m^3$), x3; and negatively affected by air pollution (0-100 $micrograms/m^3$), x4. Using the data given below and the Linear Model:

$$C1\ x1 + C2\ x2 + C3\ x3 + C4\ x4 = y$$

The calculation seeks the most likely values of the coefficients C_i, i=1..4. Nine sets of data for, respectively, x1, x2, x3, x4 and y are supplied.

```
>  d1 :=     0,  50,  10,  20, 19:
>  d2 :=    50,  10,  50,  80, 13:
>  d3 := 1000,  70,  60,  75, 72:
>  d4 :=  720,  30,  80,   0, 66:
>  d5 :=   16,   0,   0,  90, -9:
>  d6 :=  500, 100,  75,  10, 74:
>  d7 :=   60,  70, 100, 100, 45:
>  d8 :=  800,  15,   5, 100, 28:
>  d9 :=  650,  20,  90,  60, 55:
```

```
>  X := array([seq([seq(d.i[j],j=1..4)],i=1..9)]);
```

$$X := \begin{bmatrix} 0 & 50 & 10 & 20 \\ 50 & 10 & 50 & 80 \\ 1000 & 70 & 60 & 75 \\ 720 & 30 & 80 & 0 \\ 16 & 0 & 0 & 90 \\ 500 & 100 & 75 & 10 \\ 60 & 70 & 100 & 100 \\ 800 & 15 & 5 & 100 \\ 650 & 20 & 90 & 60 \end{bmatrix}$$

```
>  Y := array([seq([d.i[5]],i=1..9)]);
```

$$Y := \begin{bmatrix} 19 \\ 13 \\ 72 \\ 66 \\ -9 \\ 74 \\ 45 \\ 28 \\ 55 \end{bmatrix}$$

```
>  Xt := transpose(X);
```

$$Xt := \begin{bmatrix} 0 & 50 & 1000 & 720 & 16 & 500 & 60 & 800 & 650 \\ 50 & 10 & 70 & 30 & 0 & 100 & 70 & 15 & 20 \\ 10 & 50 & 60 & 80 & 0 & 75 & 100 & 5 & 90 \\ 20 & 80 & 75 & 0 & 90 & 10 & 100 & 100 & 60 \end{bmatrix}$$

```
>  multiply(Xt,X);
```

$$\begin{bmatrix} 2837256 & 171300 & 226100 & 210440 \\ 171300 & 23925 & 23975 & 17750 \\ 226100 & 23975 & 36350 & 25350 \\ 210440 & 17750 & 25350 & 44225 \end{bmatrix}$$

```
>   det(%);
```
The determinant exists

$$9937869737998875000$$

```
>   inverse(%%);
```

$$
\begin{bmatrix}
\dfrac{498508841}{636023663231928} & \dfrac{-1553662387}{795029579039910} & \dfrac{-338260979}{132504929839985} & \dfrac{-196366114}{132504929839985} \\[2ex]
\dfrac{-1553662387}{795029579039910} & \dfrac{255912683848}{1987573947599775} & \dfrac{-9539193263}{132504929839985} & \dfrac{-737065146}{662524649199925} \\[2ex]
\dfrac{-338260979}{132504929839985} & \dfrac{-9539193263}{132504929839985} & \dfrac{13741516937}{132504929839985} & \dfrac{-12192553252}{662524649199925} \\[2ex]
\dfrac{-196366114}{132504929839985} & \dfrac{-737065146}{662524649199925} & \dfrac{-12192553252}{662524649199925} & \dfrac{26937370393}{662524649199925}
\end{bmatrix}
$$

```
>   multiply(Xt,Y);
```

$$
\begin{bmatrix}
217876 \\
20170 \\
25580 \\
17350
\end{bmatrix}
$$

```
>   multiply(%%,%):

>   C := evalf(%);
```

$$
C := \begin{bmatrix}
.04033939526 \\
.3103994055 \\
.3252359466 \\
-.1106465355
\end{bmatrix}
$$

```
>   [['sunshine in watts/m2'],
>   ['% cover by vegetation'],
>   ['millions of negative charges/m3'],
>   ['micrograms air pollution /m3']];

>   'y-Index' := convert(%,array);
```

$$
y - Index := \begin{bmatrix}
sunshine\ in\ watts/m2 \\
\%\ cover\ by\ vegetation \\
millions\ of\ negative\ charges/m3 \\
micrograms\ air\ pollution\ /m3
\end{bmatrix}
$$

```
>   'The y-value Quality Index Coefficients';

>   eval('y-Index')=eval(C);
```

The y − value Quality Index Coefficients

$$
\begin{bmatrix}
sunshine\ in\ watts/m2 \\
\%\ cover\ by\ vegetation \\
millions\ of\ negative\ charges/m3 \\
micrograms\ air\ pollution\ /m3
\end{bmatrix}
=
\begin{bmatrix}
.04033939526 \\
.3103994055 \\
.3252359466 \\
-.1106465355
\end{bmatrix}
$$

These are the best, maximum likelihood, values of the constants needed to calculate the 'y-index'. As an example of its use, consider a day on which there is moderate sunshine, 500 watts/m2, in a cleared area, 0 % vegetation, with no negative charge and high air pollution, 100 micrograms/m3:

```
>  dd := matrix(1,4,[500, 0, 0, 100]);
```
$$dd := \begin{bmatrix} 500 & 0 & 0 & 100 \end{bmatrix}$$

The matrix command is needed here because arrays are not tolerated by multiply.

```
>  yindex := multiply(dd,C);
```
$$yindex := \begin{bmatrix} 9.10504408 \end{bmatrix}$$

This suggests that, on a scale 0 to 100%, this is not a very pleasant day.

Using Least Squares

Of course, the same calculation could have been done using leastsqrs( ). Done properly, leastsqrs( ) simply solves the regression equation directly and produces the values of the coefficients, either directly or in the equation form.

```
>  YY := convert(Y,vector);
```
$$YY := [19, 13, 72, 66, -9, 74, 45, 28, 55]$$

```
>  leastsqrs(X,YY);
```
$$\begin{bmatrix} \dfrac{2138067495469}{53001971935994}, & \dfrac{41129451451782}{132504929839985}, & \dfrac{43095366288706}{132504929839985}, & \dfrac{-2932242285270}{26500985967997} \end{bmatrix}$$

```
>  evalf(%);
```

same answer

$$[.04033939526, .3103994055, .3252359466, -.1106465355]$$

Another way is to simply set up a set of equations.

```
>  eqns :=
>{seq(convert([seq(d.i[j]*C.j,j=1..4)],'+')=d.i[5],i=1..9)};
```
This is the equivalent, over determined equation set.

```
>  consts := indets(eqns);
```

```
>  leastsqrs(eqns,consts);
```

```
>  evalf(%);
```

$eqns := \{16\,C1 + 90\,C4 = -9,\ 500\,C1 + 100\,C2 + 75\,C3 + 10\,C4 = 74,$
$60\,C1 + 70\,C2 + 100\,C3 + 100\,C4 = 45,\ 800\,C1 + 15\,C2 + 5\,C3 + 100\,C4 = 28,$
$650\,C1 + 20\,C2 + 90\,C3 + 60\,C4 = 55,\ 50\,C2 + 10\,C3 + 20\,C4 = 19,$
$50\,C1 + 10\,C2 + 50\,C3 + 80\,C4 = 13,\ 1000\,C1 + 70\,C2 + 60\,C3 + 75\,C4 = 72,$
$720\,C1 + 30\,C2 + 80\,C3 = 66\}$

$$consts := \{C3,\ C2,\ C4,\ C1\}$$

$$\{C3 = \frac{43095366288706}{132504929839985},\ C2 = \frac{41129451451782}{132504929839985},\ C4 = \frac{-2932242285270}{26500985967997},$$

$$C1 = \frac{2138067495469}{53001971935994}\}$$

$$\{C2 = .3103994055,\ C3 = .3252359466,\ C4 = -.1106465355,\ C1 = .04033939526\}$$

5.6 Turbulence: Rotating Statistics

The measurement of turbulence statistics in the wind usually uses the cross-correlations between the different wind components. This depends on the coordinate system and can easily be corrupted by an error in alignment. In the normal atmospheric condition, the vertical wind is small, perhaps only a few tenths of a metre/sec; the horizontal wind is much larger, perhaps 10 metres/sec. It is usual to try to align the measuring system into the wind; the measuring systems do not respond properly when the wind is at a large angle to the sensor. An alignment error of 0.1/10 or 0.01 radians, about 0.5 degrees, can produce a 100% error in the correlation between the horizontal and the vertical wind. A strong horizontal wind, with a small vertical misalignment, allows that a portion of this horizontal wind is seen as a component of the vertical wind and this vertical component is most certainly correlated with the horizontal wind. The correlation between the horizontal and the vertical wind may be totally corrupted by this effect. And boundary layer studies rely on this correlation; the correlation between the vertical and the horizontal wind is a measure of the force of the turbulent wind on the ground. Wind erosion studies and air pollution studies rely on these measurements.

5.6.1 Transformation Matrices

After the measurements are taken, and the statistics are calculated, transformation matrices can be applied to the data to correct for misalignment. The direction of the mean wind is known after the measurements. The transformation matrix is calculated from the known values of the mean wind; the diad or crosscorrelation matrix is subjected to the transformation and all the statistics can be accurately corrected to align with the mean wind.

The three matrices below rotate each of the three wind components through angles a , b , and c. A rotation of the coordinates about the x axis changes components in the y and z direction, given by cos and sin functions in an antisymmetric way. This is all quantified by a TA transformation matrix; multiplying a wind vector by TA on the left gives the transformed wind vector in the new coordinate system. Corresponding to the other two rotations there are transformation matrices TB and TC which show similar effects in rotation around the y and z axes. This is a linear, additive effect, so the combination of three rotations, through three separate angles, is the same as three independent rotations; the effect is given by a single transformation matrix T.

The angle a is the rotation, in a right hand sense, about the x-axis. That means that if the thumb of the right hand is pointed along the x-axis, the rotation follows the direction of the fingers. The x-axis is, approximately, located along the direction of the wind. The y-axis is located in the horizontal, perpendicular to both the x-axis and the vertical; it points to the right when one is looking along the x-axis. The z-axis is perpendicular to both the x-axis and the y-axis and points upward. /bin/bash: matrix: command not found

y-axis is opposite to some conventions

```
> restart: with(plots): with(plottools);
```

[*arc, arrow, circle, cone, cuboid, curve, cutin, cutout, cylinder, disk, dodecahedron, ellipse, ellipticArc, hemisphere, hexahedron, homothety, hyperbola, icosahedron, line, octahedron, pieslice, point, polygon, project, rectangle, reflect, rotate, scale, semitorus, sphere, stellate, tetrahedron, torus, transform, translate, vrml*]

```
> ?arrow
> ax := arrow([0,0,0], [1,0,0], .1, .2, .2,color=red):
> ay := arrow([0,0,0], [0,-1,0], .1, .2, .2,color=blue):
> az := arrow([0,0,0], [0,0,1], .1, .2, .2,color=green):
> tx := TEXT([1,0,0],"x",COLOUR(RGB,1,0,0)):
> ty := TEXT([0,-1,0],"y",COLOUR(RGB,0,0,1)):
> tz := TEXT([0,0,1],"z",COLOUR(RGB,0,1,0)):
> rng := view=[0..1,-1..0,0..1]:
> opts := projection=1,orientation=[-115,60]:
> plots[display]([ax,ay,az],tx,ty,tz,rng,opts);
```

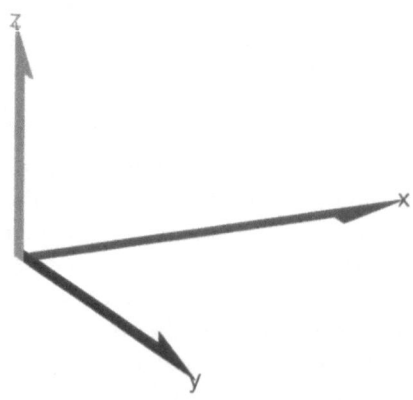

Standing at the origin and looking along the direction of the x-axis, a right-handed rotation of the axes through an angle a appears as a redistribution of the vectors, in components along the new axes. The breaking up of the vector arrows is presented below; starting out with a test_arrow and using plottools and editing (cut-and-paste) to scale, rotate, translate and colour the arrows. Note that the inclusion of axes in arrow( ) seems to prevent the full PLOT structure from rotating.

Maple bug?

```
>  opts := scaling=constrained, xtickmarks=0, ytickmarks=0:

>  test_arrow := arrow([0,0], [0,10], .2, .4, .1, color=green);
>  plots[display](test_arrow, axes=none, opts);
```

test_arrow := POLYGONS([[.1000000000, 0], [−.1000000000, 0], [−.1000000000, 9.000000000], [.1000000000, 9.000000000]], [[−.2000000000, 9.000000000], [0, 10.], [.2000000000, 9.000000000]], STYLE(*PATCHNOGRID*), COLOUR(*RGB*, 0, 1.00000000, 0)), CURVES([[.1000000000, 0], [−.1000000000, 0], [−.1000000000, 9.000000000], [−.2000000000, 9.000000000], [0, 10.], [.2000000000, 9.000000000], [.1000000000, 9.000000000], [.1000000000, 0]])

The arrow is copied directly from the top of the plot structure. Try to follow the rotations, scaling and translations; the statements may be tested using display( ).

```
>  az :=POLYGONS([[.1,0],[-.1,0],[-.1,9],[.1,9]],[[-.2,9],[0,10],
>  [.2,9]],STYLE(PATCHNOGRID),COLOUR(RGB,0,1,0)):

>  rotate(az,-Pi/6,[0,0]):

>  az1 := scale(%,sqrt(3)/2,sqrt(3)/2,[0,0]):

>  rotate(az,Pi/3,[0,0]):   scale(%,1/2,1/2,[0,0]):

>  az2 := translate(%,10*sqrt(3)/4,30/4):

>  arrowz := [az,az1,az2]:   # vertical arrow set

>  map(rotate,arrowz,-Pi/2):
```

The above PLOT structure is hand edited to alter the COLOUR.

```
>   arrowy:=[POLYGONS([[0,-.1],[0,.1],[9.,.1],[9.,-.1]],[[9.,.2],
>   [10.,0],[9.,-.2]],STYLE(PATCHNOGRID),COLOUR(RGB,0,0,1)),
>   POLYGONS([[-.0433012702,-.075],[.04330127020,.075],[6.793301274,
>   -3.822114318],[6.706698733,-3.972114318]],[[6.836602544,
>   -3.747114318],[7.5,-4.33012702],[6.663397463,-4.047114318]],
>   STYLE(PATCHNOGRID),COLOUR(RGB,0,0,1)),POLYGONS([[7.54330127,
>   -4.35512702],[7.45669873,-4.30512702],[9.70669873,-.408012702],
>   [9.79330127,-.458012702]],[[9.66339746,-.383012702],[10,0],
>   [9.836602541,-.483012702]],STYLE(PATCHNOGRID),COLOUR(RGB,0,0,1))]:


>   az := TEXT([1,4],"a",COLOUR(RGB,0,1,0)):

>   ay := TEXT([4,-1],"a",COLOUR(RGB,0,0,1)):
>   tt1 := TEXT([4,12],
>   "Rotation about the x-axis through angle a",
>   FONT(TIMES,BOLD,14),COLOUR(RGB,1,0,0)) :
>   tt2 := TEXT([4,11],
>   "2D plot in the plane of the y & z-axes",
>   FONT(TIMES,BOLD,12),COLOUR(RGB,0,0,0)) :

>   display(arrowz,arrowy,az,ay,tt1,tt2,opts);
```

Rotation about the x-axis through angle a
2D plot in the plane of the y & z-axes

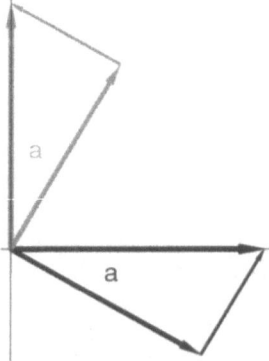

The new axes are in the directions of the medium sized arrows. A right-handed rotation about the x-axis diminishes the y vector by the factor of cos(a) and subtracts a portion of the z vector, sin(a). The z vector is also diminished by the cos(a) factor but a portion, sin(a) of the y vector is added to the z vector. The x vector is unaffected by the rotation. Look at the transformation matrix as being the identity matrix, with 1's down the diagonal and zeros elsewhere. The rotation splits the 1 into separate sine and cosines.

Observe that multiplication of the wind vector by the transformation matrix has the desired result.

```
>  with(linalg):
>  UVW := matrix([[U],[V],[W]]);# the wind speed vector
```

$$UVW := \begin{bmatrix} U \\ V \\ W \end{bmatrix}$$

```
>  TA := matrix([[1,0,0],[0,cos(a),-sin(a)],[0,sin(a),cos(a)]]);
```

$$TA := \begin{bmatrix} 1 & 0 & 0 \\ 0 & \cos(a) & -\sin(a) \\ 0 & \sin(a) & \cos(a) \end{bmatrix}$$

```
>  evalm(TA&*UVW);
```

$$\begin{bmatrix} U \\ \cos(a)\,V - \sin(a)\,W \\ \sin(a)\,V + \cos(a)\,W \end{bmatrix}$$

Similarly, right handed rotation about the y-axis by an angle b diminishes the x vector by the factor cos(b) and subtracts a portion, sin(b), of the z vector from it. The z vector is also diminished by the factor cos(b) but sin(b) of the x vector is added to it. The y vector is untouched.

```
>  TB := matrix([[cos(b),0,-sin(b)],[0,1,0],[sin(b),0,cos(b)]]);
```

$$TB := \begin{bmatrix} \cos(b) & 0 & -\sin(b) \\ 0 & 1 & 0 \\ \sin(b) & 0 & \cos(b) \end{bmatrix}$$

Lastly, rotation about the z-axis by the angle c diminishes both the x vector and the y vector by cos(c). The x vector component, sin(c) further diminishes the y vector and the y vector contributes portion sin(c) to the x vector.

```
>  TC := matrix([[cos(c),sin(c),0],[-sin(c),cos(c),0],[0,0,1]]);
```

$$TC := \begin{bmatrix} \cos(c) & \sin(c) & 0 \\ -\sin(c) & \cos(c) & 0 \\ 0 & 0 & 1 \end{bmatrix}$$

A Combined Transformation Matrix

Importantly, all these rotations have additive effects. That is, a rotation around one axis followed by a rotation about another axis gives the same effect as both rotations occurring together. Application to the general case of all three rotations can be done by a combined transform.

```
>  (TA &* TB) &* TC;
```

$$(TA \&* TB) \&* TC$$

```
>   T := evalm(%);
```

$T :=$
$[\cos(b)\cos(c),\ \cos(b)\sin(c),\ -\sin(b)]$
$[-\sin(a)\sin(b)\cos(c) - \cos(a)\sin(c),\ -\sin(a)\sin(b)\sin(c) + \cos(a)\cos(c),$
$-\sin(a)\cos(b)]$
$[\cos(a)\sin(b)\cos(c) - \sin(a)\sin(c),\ \cos(a)\sin(b)\sin(c) + \sin(a)\cos(c),$
$\cos(a)\cos(b)]$

The final redoing of the statistics also requires the transpose of this matrix, to be applied to the right side of the statistics diad.

```
>   Tt := transpose(T);
```

$$Tt := \begin{bmatrix} \cos(b)\cos(c),\ -\sin(a)\sin(b)\cos(c) - \cos(a)\sin(c),\ \cos(a)\sin(b)\cos(c) - \sin(a)\sin(c) \\ \cos(b)\sin(c),\ -\sin(a)\sin(b)\sin(c) + \cos(a)\cos(c),\ \cos(a)\sin(b)\sin(c) + \sin(a)\cos(c) \\ -\sin(b),\ -\sin(a)\cos(b),\ \cos(a)\cos(b) \end{bmatrix}$$

These transformation matrices will be needed later, so they are saved in an appropriate directory.

```
>   currentdir();
```
$$\text{"c:\\\\mplbook\\\ scottVOL1\\\\turbulence"}$$
```
>   save TA,TB,TC,T,Tt, "trans.m";
```

5.6.2   Analysis of Turbulence Data

The turbulence in fluid flows is somewhat statistical with some systematic properties. The usual approach is to take a sample over a time period known to include most of the fluctuating statistics and presume that the statistical properties in that time period are relative stationary, at least in time, that the statistics are 'frozen'. The statistics are direction dependent or 'anisotropic'; they vary in space or are 'nonhomogeneous'. Whatever they are, they have some three dimensionality and randomness, with a little regularity.

The scene is an oval playing field, at about 3PM and a 10 metre tower holds three high-resolution Gil propeller anemometers, set so that one tends to line up with the main wind and the others are in coordinate directions. Each has a good cosine response to wide angles which means that it only responds to wind in the given coordinate direction. Generators in the heads produce analogue voltages that are recorded every 0.1 sec.; they can detect wind speed variations of less than 0.1 m/sec. The sky is only slightly overcast; the wind is generally from the west or southwest and the Murdoch oval is experiencing a sea breeze.

Approximately 9000 data points were collected over about a 15 minute interval, a sufficient time for statistical validity and within the statistical 'window'. The problem is that the average wind direction is not known until after the data have been collected. It is not possible to orient the 3-headed anemometer before the experiment; it must be corrected for direction 'a posteriori'. The data were collected at two levels, about 5 m height, the 'lower' level; and about 10 m, the 'upper' level; here we will work only with the 'upper' data. First we locate the data.

```
> currentdir();
```
> “C:\\Program Files\ \Maple V Release 5.1\\BIN.WNT”
```
> currentdir("c:/mplbook/scottVOL1/turbulence"):
> currentdir();
```
> “c:\\mplbook\\ scottVOL1\\turbulence”

Have a look at the file in a convenient editor.

```
> ssystem("edit");
```

$$[0, \text{“”}]$$

This is a bit of a trial and error and an experience. The length of the file is such that one can not look at it all very easily and most editors will either refuse to read it or truncate it. You can read the file with readline, a do loop and a semicolon; it scrolls fast and you can get a brief look as the data flash past. The stop button is effective with input or output; it quickly stops the scrolling of data. Next, after fclose, organise another read with readline to count and summarise the data.

```
> filenm := "161454.dat";
```

$$filenm := \text{“161454.dat”}$$

```
> fl := fopen(filenm,READ,TEXT);
```

$$fl := 0$$

```
> s := readline(fl);
```

Read several lines to get a feel for the data.

$$s := \text{“Turbulence data 16/4/96 14:54”}$$

```
> for i from 1 to infinity
> while s<>0 do;
> s:= readline(fl);
> if (i < 10 or
> i mod 1000 = 0 or
> i > 9000) then;
> print('row '.i,s); fi;
> od:
```

$$row\ 1,\ \text{“”}$$

$$row\ 2,\ \text{“}\quad\quad upper\quad\ —\quad\ lower\text{”}$$

$$row\ 3,\ \text{“}\quad x\quad\ y\quad\ z\ —\ x\quad\ y\quad\ z\text{”}$$

row 4, " 589 -212   -98   394 -212   17"
row 5, " 616 -217 -101   386 -217   22"
row 6, " 645 -220 -102   405 -221   28"
row 7, " 671 -226 -104   401 -222   29"
row 8, " 694 -222 -117   373 -231   31"
row 9, " 739 -219 -119   376 -234   35"
row 1000, " 300 -142   -18   483 -132   11"
row 2000, " 873 -239   -40   862 -339   121"
row 3000, " 387   -21 -124   550 -201   20"
row 4000, " 102 -230   -17   90 -276   -54"
row 5000, " 238 -396   -8   341 -257   -53"
row 6000, " 524   -59   -74   380 -341   -28"
row 7000, " 343 -204   8   285 -125   151"
row 8000, "   29 -521   -48   336 -702   25"
row 9000, " 217 -376   -19   256 -459   60"
row 9001, " "
row 9002, 0

```
>  nlines = i;
```

$$nlines = 9003$$

It seems the counter went one beyond the reading of the end-of-file and there were 3 remaining lines in the header, making a total of 8997 lines. Next we rescan the data, using sscanf( ) to pick off the lower and upper data set.

Go watch telly

This is much more time consuming.

```
>  fclose(fl);

>  fl := fopen(filenm,READ,TEXT);
```

$$fn := 0$$

```
>  upper:=[]:  lower:=[]:  i := 0:

>  us := NULL: ls := NULL:
>  while(not feof(fl)) do
>  ll := readline(fl);
>  lst := sscanf(ll,"%d%d%d%d%d%d");
>  if nops(lst)=6 then
>  i := i+1;
>  us:=us,[op(1..3,lst)];
>  ls:=ls,[op(4..6,lst)];
>  fi;
>  od:
```

```
Error, (in sscanf) input and format strings expected
```

The data collection is complete. The error message is because sscanf( ) tried to operate on the end-of-file marker.

```
> upper := [us]:  lower := [ls]:
> print('there are '.i.' lines');
```

there are 8997 lines

```
> fclose(fn);
> upper[8997],lower[8997]; upper[8997][1];
```

$$[217, -376, -19], [256, -459, 60]$$

217

Next, have a look. Plot the data and save it for further use.

Conversion to minutes of time

```
> w1:=NULL: w2:=NULL: w3:=NULL:
> for i from 4000to 4500 do;
> w1:=w1,evalf([i/600,upper[i][1]],5);
> w2:=w2,evalf([i/600,upper[i][2]],5);
> w3:=w3,evalf([i/600,upper[i][3]],5);
> if (i mod 1000 = 0) then print(i); fi;
> od:
```

4000

More than an hour is required to plot all the data. Best to sit and cycle through it, looking for extraordinary variability and trends.

```
> plot([[w1],[w2],[w3]],colour=[red,blue,green],
> linestyle=[2,3,4]); PP := %:
```

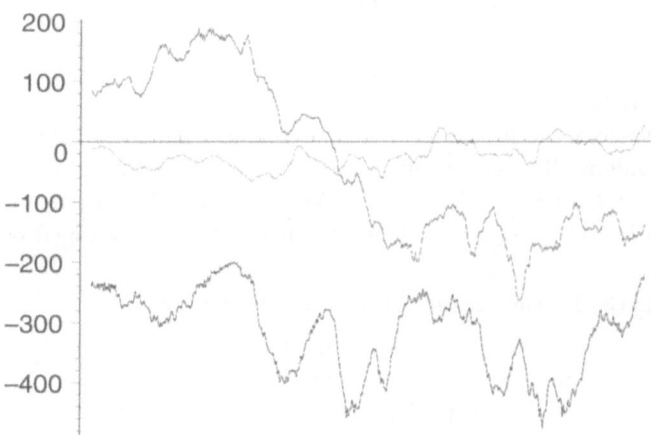

The data goes in the current directory

```
> currentdir();
```

"c:\\mplbook\\ scottVOL1\\turbulence"

```
> save upper,lower,"turb.m";
> read "turb.m";
```

Averages and Deviations

Next we run through the data to get the averages and deviations from the averages. Normally we should first convert the data to sensible units, m/s. Here we don't do this because integers can be manipulated without error in Maple; the raw data from the data logger arrived in an integer format. Scaling the data into recognisable units does allow some understanding during data manipulation; errors are less likely to be made and propagated.

Unit conversions are at the end of this section.

```
> with(stats[describe]);
```

[*coefficientofvariation, count, countmissing, covariance, decile, geometricmean, harmonicmean, kurtosis, linearcorrelation, mean, meandeviation, median, mode, moment, percentile, quadraticmean, quantile, quartile, range, skewness, standarddeviation, sumdata, variance*]

```
> Ubar := mean([seq(op(1,upper[i]),i=1..8997)]);
```
$$Ubar := \frac{2579903}{8997}$$

```
> Vbar := mean([seq(op(2,upper[i]),i=1..8997)]);
```
$$Vbar := \frac{-755382}{2999}$$

```
> Wbar := mean([seq(op(3,upper[i]),i=1..8997)]);
```
$$Wbar := \frac{-76228}{2999}$$

```
> #with(linalg);
```

The statistics we seek are in the form of a covariance matrix, 9 values, of which 6 are independent. The covariance matrix or 'statistical diad' lists all the covariances or correlations between the deviations of the 3 components of velocity. The form below presumes that all the quantities are averaged over the whole data set.

```
> uvw:=matrix(3,3,[[uu,uv,uw],[vu,vv,vw],[wu,wv,ww]]);
```
$$uvw := \begin{bmatrix} uu & uv & uw \\ vu & vv & vw \\ wu & wv & ww \end{bmatrix}$$

Large symbols stand for the full velocity components in the 3 directions; small symbols stand for the deviations.

```
> u = U - 'Ubar'; v = V - 'Vbar'; w = W - 'Wbar';
```
$$u = U - Ubar$$
$$v = V - Vbar$$
$$w = W - Wbar$$

```
> i:='i';
```

$$i := i$$

```
>   data := matrix(8997,3,[seq(
>   [op(1,upper[i])-Ubar,
>   op(2,upper[i])-Vbar,
>   op(3,upper[i])-Wbar],
>   i=1..8997)]):

>   data[8888,1];
```

$$\frac{1090873}{8997}$$

```
>   tdata := transpose(data):
```

The covariance matrix or statistical diad is formed by the product of the above transpose with the data matrix. Division by the number of points normalises the value. The whole procedure may be time consuming.

```
>   rdiad := evalm(tdata&*data);
```

$$rdiad := \begin{bmatrix} \dfrac{2874985918448}{8997} & \dfrac{322124464024}{2999} & \dfrac{-16370794891}{2999} \\[2mm] \dfrac{322124464024}{2999} & \dfrac{836773098626}{2999} & \dfrac{28263819838}{2999} \\[2mm] \dfrac{-16370794891}{2999} & \dfrac{28263819838}{2999} & \dfrac{27229649826}{2999} \end{bmatrix}$$

```
>   diad := map(x->evalf(x/8997,5),rdiad);
```

$$diad := \begin{bmatrix} 35517. & 11938. & -606.73 \\ 11938. & 31012. & 1047.5 \\ -606.73 & 1047.5 & 1009.2 \end{bmatrix}$$

```
>   currentdir();

>   save Ubar,Vbar,Wbar,diad, "diad.m";
```

5.6.3 The Required Rotations

The transform mathematically rotates the data so that the statistics are correct. We use the average winds over the period.

```
>   UVW := matrix(3,1,[[Ubar],[Vbar],[Wbar]]);
```

$$UVW := \begin{bmatrix} \dfrac{2579903}{8997} \\[2mm] \dfrac{-755382}{2999} \\[2mm] \dfrac{-76228}{2999} \end{bmatrix}$$

```
>   sol:=evalm(T&*UVW);
```

sol :=

$$\left[\frac{2579903}{8997} \cos(b) \cos(c) - \frac{755382}{2999} \cos(b) \sin(c) + \frac{76228}{2999} \sin(b) \right]$$

$$\left[-\frac{2579903}{8997} \sin(a) \sin(b) \cos(c) - \frac{2579903}{8997} \cos(a) \sin(c) \right.$$

$$\left. + \frac{755382}{2999} \sin(a) \sin(b) \sin(c) - \frac{755382}{2999} \cos(a) \cos(c) + \frac{76228}{2999} \sin(a) \cos(b) \right]$$

$$\left[\frac{2579903}{8997} \cos(a) \sin(b) \cos(c) - \frac{2579903}{8997} \sin(a) \sin(c) \right.$$

$$\left. - \frac{755382}{2999} \cos(a) \sin(b) \sin(c) - \frac{755382}{2999} \sin(a) \cos(c) - \frac{76228}{2999} \cos(a) \cos(b) \right]$$

If we are to be aligned with the mean wind on rotation, the first term must equal the overall, mean wind strength; the second and third terms must be zero.

```
>  sol[1,1]=sqrt(Ubar^2+Vbar^2+Wbar^2);
```

$$\frac{2579903}{8997} \cos(b) \cos(c) - \frac{755382}{2999} \cos(b) \sin(c) + \frac{76228}{2999} \sin(b) =$$

$$\frac{1}{26982003} \sqrt{106521472153515068581}$$

```
>  sol[2,1]=0;
```

$$-\frac{2579903}{8997} \sin(a) \sin(b) \cos(c) - \frac{2579903}{8997} \cos(a) \sin(c) + \frac{755382}{2999} \sin(a) \sin(b) \sin(c)$$

$$-\frac{755382}{2999} \cos(a) \cos(c) + \frac{76228}{2999} \sin(a) \cos(b) = 0$$

```
>  sol[3,1]=0;
```

$$\frac{2579903}{8997} \cos(a) \sin(b) \cos(c) - \frac{2579903}{8997} \sin(a) \sin(c) - \frac{755382}{2999} \cos(a) \sin(b) \sin(c)$$

$$-\frac{755382}{2999} \sin(a) \cos(c) - \frac{76228}{2999} \cos(a) \cos(b) = 0$$

```
>  eqns := [%,%%,%%%];
```

$$eqns := [\frac{2579903}{8997}\cos(a)\sin(b)\cos(c) - \frac{2579903}{8997}\sin(a)\sin(c)$$

$$-\frac{755382}{2999}\cos(a)\sin(b)\sin(c) - \frac{755382}{2999}\sin(a)\cos(c) - \frac{76228}{2999}\cos(a)\cos(b) = 0,$$

$$-\frac{2579903}{8997}\sin(a)\sin(b)\cos(c) - \frac{2579903}{8997}\cos(a)\sin(c)$$

$$+\frac{755382}{2999}\sin(a)\sin(b)\sin(c) - \frac{755382}{2999}\cos(a)\cos(c) + \frac{76228}{2999}\sin(a)\cos(b) = 0,$$

$$\frac{2579903}{8997}\cos(b)\cos(c) - \frac{755382}{2999}\cos(b)\sin(c) + \frac{76228}{2999}\sin(b) =$$

$$\frac{1}{26982003}\sqrt{106521472153515068581}]$$

```
> solve(convert(eqns,set),{a,b,c}):  convert(%,list);
```

$$[c = \arctan(-\frac{2266146}{44635}\%1, \frac{289}{5}\%1), b = \arctan($$

$$-\frac{11791317182725}{7251068517136354964139049 2}\%1\sqrt{471652687309} + \frac{1}{685823316},$$

$$\frac{2358263436545}{317078086667032016413}\%1), a = a]$$

$$\%1 := \text{RootOf}(471652687309\_Z^2 - 79691329)$$

```
> map(allvalues,%):  evalf(%);
```

$$[c = -.7207434960, c = 2.420849158, b = .06649879789, b = 1.604008928, a = a]$$

The larger angles are unacceptable, considering that we know the instruments were responding mostly to the main effects of the wind. Also, it appears that the angle a, the roll, is undetermined by the set of equations.

```
> subs([b=0.06649879789,c=-.7207434960], eqns):
```

```
> evalf(%);
```

This is a solution

$$[0 = 0, 0 = 0, 382.5114094 = 382.5114096]$$

Exercise* 5.9 Check the Statistics

Repeat the calculation using one or two procedures that reread in the data and get the statistics from the wind data, without using matrix operations. The combined operation can be done using no real computer memory. The data is read once to get the averages, and read again with each line being interpreted as it is read.

Exercise* 5.10 Simplified Angle Calculation

It seems the angles calculated are given by equations:

```
> arccos(Ubar/sqrt(Ubar^2+Vbar^2)): evalf(%); # Jaw angle c
```

```
> arccos(sqrt(Ubar^2+Vbar^2)/sqrt(Ubar^2+Vbar^2+Wbar^2));
```

Consider these angles. Are they algebraic solutions of the equations? Also, 'the answers' above are solutions and the angles are small, but are there other solutions; do the angles have the right signs? The arccos may be ambiguous.

5.6.4   Rotating the Statistics

Choosing the simplest solution, the transform matrix becomes floating point numbers. The transformation allows a restructuring of both sides of the statistical diad. The diad, conceptually, is made up of a vertically aligned vector of velocities multiplied on the right by a horizontally aligned vector of velocities; with the row-across, column-down method of multiplication, this makes a 3x3 matrix, the diad or cross correlation matrix. Multiplying the diad on the left by the transform can be visualised as converting the vertical vector; multiplying the diad on the right by the transpose of the transform, the horizontal vector. The full conversion of the statistical matrix requires a left multiplication by the transform and a right multiplication by the transpose of the transform.

```
> ss:= {a=0,b=0.06649879789,c=-.7207434960};
```
$$ss := \{a = 0,\ b = .06649879789,\ c = -.7207434960\}$$

```
> Ts := evalf(subs(ss,evalm(T)));
```
$$Ts := \begin{bmatrix} .7496546916 & -.6584848271 & -.06644979811 \\ .6599434542 & .7513152715 & 0 \\ .04992474811 & -.04385310930 & .9977897696 \end{bmatrix}$$

```
> Tts := evalf(transpose(Ts));
```
$$Tts := \begin{bmatrix} .7496546916 & .6599434542 & .04992474811 \\ -.6584848271 & .7513152715 & -.04385310930 \\ -.06644979811 & 0 & .9977897696 \end{bmatrix}$$

```
> evalm(diad);
```
$$\begin{bmatrix} 35517. & 11938. & -606.73 \\ 11938. & 31012. & 1047.5 \\ -606.73 & 1047.5 & 1009.2 \end{bmatrix}$$

```
> evalm(Ts&*diad&*Tts): corrected_diad := evalf(%,5);
```
$$corrected\_diad := \begin{bmatrix} 21777. & 3739.1 & 235.96 \\ 3739.1 & 44812. & 636.46 \\ 235.96 & 636.46 & 948.52 \end{bmatrix}$$

There is considerable change in most of the statistics. It is well to check out the effect of this transformation, first by back calculation of the effects on the average wind, second by recalculating the data, and the statistics

```
> evalm(UVW);
```

$$\begin{bmatrix} \dfrac{2579903}{8997} \\ \dfrac{-755382}{2999} \\ \dfrac{-76228}{2999} \end{bmatrix}$$

```
> evalm(Ts&*UVW);
```

$$\begin{bmatrix} 382.5114094 \\ 0 \\ 0 \end{bmatrix}$$

```
> evalf(sqrt(Ubar^2+Vbar^2+Wbar^2));
```

$$382.5114096$$

The generating of a new set of transformed data recognises that data is naturally in rows and collects each velocity vector as a horizontal row vector. Then to keep from double transposing, the vector is transformed by multiplying by Tts on the right.

```
> row(data,5000); evalm(%&*Tts);
```

$$\left[\dfrac{272146}{8997}, \dfrac{-357247}{2999}, \dfrac{55235}{2999}\right]$$

$$[99.89214731, -69.53589076, 25.11112032]$$

```
> matrix(5,3,[seq(evalm(row(data,i)&*Tts),i=1..5)]);
```

$$\begin{bmatrix} 205.1460675 & 229.4278569 & -59.08086161 \\ 228.8785177 & 243.4897539 & -60.50699717 \\ 252.6604080 & 260.3741683 & -59.92540991 \\ 276.2352385 & 273.0248064 & -60.35982735 \\ 291.7072045 & 291.2087669 & -72.35823760 \end{bmatrix}$$

```
> matrix(5,3,[seq(evalm(row(data,i)&*Tts),i=3001..3005)]);
```

Adjust the range to see any 5 lines of the calculation

$$\begin{bmatrix} -47.83360149 & 235.2446577 & -82.94405269 \\ -45.86914351 & 236.5645446 & -89.82873159 \\ -49.33291087 & 233.9247708 & -83.04390219 \\ -55.33014840 & 228.6452232 & -83.44330017 \\ -59.59629669 & 230.3305973 & -81.72298257 \end{bmatrix}$$

```
> matrix(8997,3,[seq(evalm(row(data,i)&*Tts),i=1..8997)]):
```

```
> newdata := evalm(%):
```

```
> tnewdata:=evalm(transpose(%)):
```

```
> corrected_diad := evalm(corrected_diad);
```

$$corrected\_diad := \begin{bmatrix} 21777. & 3739.1 & 235.96 \\ 3739.1 & 44812. & 636.46 \\ 235.96 & 636.46 & 948.52 \end{bmatrix}$$

```
> evalm(tnewdata&*newdata):
```

```
> another_diad := map(x->evalf(x/8997,5),%);
```

$$another\_diad := \begin{bmatrix} 21777. & 3739.1 & 235.94 \\ 3739.1 & 44813. & 636.47 \\ 235.94 & 636.47 & 948.49 \end{bmatrix}$$

The properly
aligned data has
exactly the same
statistical diad

```
> #currentdir("c:/mplbook/scottVOL1/turbulence");
```

```
> #read "trans.m";
```

5.6.5   Wind Units

This whole calculation was done with only a set of integers, to keep the errors
minimal and satisfy Maple. In reality, the numbers have units that we can
relate to.

```
> convunits := x->x/2048*18.36*m/s;
```

Maple can deal
with units

$$convunits := x \rightarrow .008964843750 \frac{x\,m}{s}$$

```
> map(convunits,UVW):
```

```
> matrix(3,1,[['Ubar'],['Vbar'],['Wbar']])=evalf(%,3);
```

$$\begin{bmatrix} Ubar \\ Vbar \\ Wbar \end{bmatrix} = \begin{bmatrix} 2.57\,\dfrac{m}{s} \\ -2.26\,\dfrac{m}{s} \\ -.228\,\dfrac{m}{s} \end{bmatrix}$$

```
> map(convunits@@2,diad):  evalm(uvw) = evalf(%,3);
```

$$uvw = \begin{bmatrix} 2.85\,\dfrac{m^2}{s^2} & .959\,\dfrac{m^2}{s^2} & -.0488\,\dfrac{m^2}{s^2} \\ .959\,\dfrac{m^2}{s^2} & 2.49\,\dfrac{m^2}{s^2} & .0842\,\dfrac{m^2}{s^2} \\ -.0488\,\dfrac{m^2}{s^2} & .0842\,\dfrac{m^2}{s^2} & .0811\,\dfrac{m^2}{s^2} \end{bmatrix}$$

```
> map(convunits@@2,corrected_diad):  evalm(uvw)=evalf(%,3);
```

$$uvw = \begin{bmatrix} 1.75\,\dfrac{m^2}{s^2} & .301\,\dfrac{m^2}{s^2} & .0190\,\dfrac{m^2}{s^2} \\[2ex] .301\,\dfrac{m^2}{s^2} & 3.60\,\dfrac{m^2}{s^2} & .0512\,\dfrac{m^2}{s^2} \\[2ex] .0190\,\dfrac{m^2}{s^2} & .0512\,\dfrac{m^2}{s^2} & .0762\,\dfrac{m^2}{s^2} \end{bmatrix}$$

The data show rather large turbulence intensities (diagonal terms) with the lateral (y-direction) component comparable to the longitudinal component (x-direction); the magnitudes of the winds are also similar in the two directions. The vertical turbulence level ww is small as is the vertical wind; this is expected from the presence of the ground. These statistics are not significantly altered by rotation of the statistics into the direction of the mean wind. However, the other statistics are dramatically changed. The statistic uw has changed sign and is about half the magnitude; it is $\overline{u\prime w\prime}$, the covariance of the longitudional and the vertical wind, and represents the shear force effected by the vertical shearing of the horizontal wind. The expected result is a negative one which represents an extraction of momentum or wind strength by the ground. A positive value is possible well above the ground and probably represents the presence of a low-level jet, below the level of measurement. Windspeeds would tend to be diminishing as one goes vertically, perhaps an effect of the sea-breeze. This is all speculation, of course, but shows the importance of getting the statistical correlation matrix correct.

Exercise 5.11 Matrix Multiplication

The above 'redoing' of the wind data, to correct for misalignment is applied to individual rows of the data matrix, the separate wind speed vectors. The procedure allows peeking at some part of the calculation before getting into the full 8997 line calculation. Perform the operation using the full matrix calculation.

```
>  evalm(data&*Tts):
```

Exercise 5.12 Properties of the Transforms

Evaluate T &* Tt and Ts &* Tts as well as their determinants. They are unity matrices and some are identities.

Exercise* 5.13 'lower' Wind Data

Repeat the calculations with the lower wind data. Perhaps there is a significant difference in the shear stress, as predicted by the covariance of u and w.

5.7   Struvite: Fertilizer and Kidney Stone

Maple is not alone as a symbolic manipulator and language in this world. In fact, it can work quite effectively as a complimentary program with other languages. Volume 2 considers conversion and companion use of Maple in detail. Here we show the benefit of having Maple do the symbolism, which is straightforwardly coded into Fortran. The scene is the solution and precipitation of struvite. The whole process is based on simplicity, calculations of the expected liquid and solid equilibria, coupled with the necessary mass balance and electroneutrality expressions.

The lumps found in wastewater, the soil and kidneys, owe their being to the multiple species that interact and settle into solid forms. Multiple substitutions can portray the chemical equilibria between the many species, producing large polynomials with coefficients that are complex combinations of the equilibrium constants. It is usual for the pH to be a driving argument and common for there to be more than one argument, following Gibb's Phase rule.

Waste systems and kidney stones often contain significant amounts of the sparingly soluble rock phosphate, struvite($MgNH_4PO_4 \cdot 6H_2O$). Since it contains nitrogen as well as phosphorus, it is a valuable slow-release fertilizer. It is a gritty contaminant of canned seafoods. As a phosphate rock, it naturally derives from guano and some islands in the Pacific are primary sources of it. With recent experimental data on the system at 25C and 30C, its physical chemistry is becoming clear. The concept that it only forms from 1:1:1 ratios of the constituents is flawed; it forms in a wide range of conditions. With a full consideration of the ramifications of electroneutrality, mass balance and common ion effects, understanding emerges[26][28]. With Maple, a fully flexible algorithm can be generated. The equations are presented below; the following Maple worksheet proceeds with symbolic and graphical design calculations.

The equilibrium calculation assumes that:

(1) Species are limited to those that can be derived from water, magnesium, ammonia, and phosphate. In the liquid, species included are Mg^{2+}, $Mg(OH)^+$, NH_3, NH_4^+, PO_4^{3-}, HPO_4^{2-}, $H_2PO_4^-$, H_3PO_4, $MgNH_4PO_4$, $MgHPO_4$, $MgPO_4^-$, $MgH_2PO_4^+$, $NH_4PO_4^{2-}$, $(NH_4)_2PO_4^-$, $NH_4HPO_4^-$, as well as OH^-, H^+, and H_2O itself. There is an allowance for the presence of excess non-reactive positive or negative ions, *e. g.*, Na^+ or Cl^-.

(2) The effect of ionic strength I on activity coefficients is cared for by an *ad hoc* form that allows the calculation of the activities of individual ions:

$$\gamma_i = 0.5 z_i z_e \log \sqrt{I}$$

where $z_e = \prod z_j^{y_j}$ is the equivalent charge of the environmental field of ion

i, $y_j = \frac{m_j}{\sum m_j}$ is the fraction of charge j to the 'pool' of ions of charge opposite in sign to ion i. $\prod$ is the overall product from ions that have the opposite charge to i. $\sum m_j$ is the sum over all ions of charge j [28].

(3) Only one solid phase is formed with all solid species mutually soluble in one another. The mole fractions of these species are identical to their activities.

The nature of this algorithm evolves from the fact that two independent input variables are necessary to fully describe the system. This is apparent from Gibb's Phase Rule for the case of a constant temperature and pressure, with four independently variable components and two phases. It is found that the solutions are polynomial functions to high powers in H and Mg (at least 11th power in H); hence these two ion concentrations are appropriate independent variables, input to the algorithm.

Maple and Struvite Chemistry

Maple is useful for solving the large polynomial equations that are common in these chemical equilibria. The formation of struvite is a particularly good example where the algebraic form is well generated. However, the numerical calculations necessary to reveal the multivariate aspects of the problem are too consuming in exact computer algebra. Then, Maple can produce computer codes for other languages. In this case Fortran can deal with the large expressions relatively quickly, and produce a solution.

5.7.1 The Chemical Equilibria

First we write down the equations for the equilibria, with the intent of assignment. That is, the variables we would rather not know about are presented on the left side, assigned values, and then ignored. Note that we have intentionally left out some; they are easily calculated from the primary arguments, H and Mg. These confuse the presentation and decrease the efficiency of calculation, so they are added later.

Efficiency is important here and, after the symbolic calculations are complete, it is recommended that further calculations be done with an efficient, floating point calculator; that is, a language that does efficient calculations in floating point, like Fortran. Statements are included here to prepare the Fortran statements that can be used in a program to process the rather gross with a little amounts of data that can be produced. careful editing

The primary calculating variable will be NH3, the 'dissolved ammonia' concentration. This nitrogen-containing species is non-ionic and is often called 'dissolved ammonium hydroxide', or simply 'ammonium hydroxide'; either interpretation makes no real difference to the calculations.

A listing of
appropriate K
values is below

```
>  restart:
>  MNP := K1*Mg*NH4*PO4;       #Struvite
>  NH4 := K2*NH3*H;            #Ammonium
```

It is equally possible to use ammonium ion as a primary calculating variable. Then the system would be altered to leave out this statement and put in a similar one for NH3.

```
>  H3P := K3*H2P*H;            #Phosphoric acid
>  H2P := K4*HP*H;             #Dihydrogen phosphate
>  HP  := K5*PO4*H;            #Hydrogen phosphate
>  MP8 := K6*Mg^3*PO4^2;       #Magnesium phosphate with 8 H2O
>  MP22 := K7*Mg^3*PO4^2;      #Magnesium phosphate with 22 H2O
>  MHP   := K10*Mg*HP;         #Magnesium hydrogen phosphate
>  MHPd  := K12*MHP;
>  #dissolved Magnesium hydrogen phosphate
```

$$MNP := K1\ Mg\ NH4\ PO4$$
$$NH4 := K2\ NH3\ H$$
$$H3P := K3\ H2P\ H$$
$$H2P := K4\ HP\ H$$
$$HP := K5\ PO4\ H$$
$$MP8 := K6\ Mg^3\ PO4^2$$
$$MP22 := K7\ Mg^3\ PO4^2$$
$$MHP := K10\ Mg\ K5\ PO4\ H$$
$$MHPd := K12\ K10\ Mg\ K5\ PO4\ H$$

Most of the literature doesn't include this last species. Wrigley et al[28] found this to be an important adjunct, together with two other ignored species:

```
>  MP    := K13*Mg*PO4;        #Magnesium phosphate ion
>  MH2P  := K14*Mg*H2P;        #Magnesium dihydrogen phosphate ion
```

$$MP := K13\ Mg\ PO4$$
$$MH2P := K14\ Mg\ K4\ K5\ PO4\ H^2$$

5.7.2 The Electroneutrality Equation

The coupling equation that drives the whole requires that the solution be electrically neutral. Note that the concentrations of ions with multiple charges have, accordingly, been given an appropriate multiplier. This is set up as should most of the equations in Maple, as an expression that would equal zero. The term Exions cares for added positive ions or, alternately, the excess of positive ions *Cations* over negative ions *Anions*, as single charged species. It can give quite a bias to the charges and cause a shift in the equilibria, with more or less struvite precipitating.

```
>   ffff:=Exions+H+NH4+2*Mg+MgOH+MH2P-OH-3*PO4-2*HP-H2P-MP;
```

$$\textit{ffff} := Exions + H + K2\,NH3\,H + 2\,Mg + MgOH + K14\,Mg\,K4\,K5\,PO4\,H^2$$
$$- OH - 3\,PO4 - 2\,K5\,PO4\,H - K4\,K5\,PO4\,H^2 - K13\,Mg\,PO4$$

Next, by experiment it was found that PO4 is an easy Anion to remove here so we use this condition to remove it from the unknown list.

This is the result of Trial and Error

```
>   PO4 := solve(ffff,PO4);
```

$$PO4 := -\frac{Exions + H + K2\,NH3\,H + 2\,Mg + MgOH - OH}{K14\,Mg\,K4\,K5\,H^2 - 3 - 2\,K5\,H - K4\,K5\,H^2 - K13\,Mg}$$

Converting to Fortran

Here we pause to collect codes for our Fortran program. This has no real bearing on the Maple codes other than to supply an alternative and efficient calculator, later.

```
>   #readlib(fortran):
>   #Activates the Fortran conversion library

>   #fortran(PO4,optimized,filename='PO4.for');
>   #can produce fortran codes in the file
```

5.7.3   Solid Phase Mass Balance

It is usual in such calculations to ignore the solid phase. Of course, this is a nonsense considering that we intend to establish how much struvite will precipitate. The balance simply adds up the mole fractions of constituents, and the total should come to one. Note that, by convention, mole fractions are used for solid phase constituents and molar units for liquid phase constituents.

Again, an algebraic expression

```
>   ggggg := MNP+MP8+MP22+MHP+MgOH2-1;
>   #Overall balance in solid phase
```

$$ggggg := -\frac{K1\,Mg\,K2\,NH3\,H\,\%2}{\%1} + \frac{K6\,Mg^3\,\%2^2}{\%1^2} + \frac{K7\,Mg^3\,\%2^2}{\%1^2} - \frac{K10\,Mg\,K5\,\%2\,H}{\%1}$$
$$+ MgOH2 - 1$$
$$\%1 := K14\,Mg\,K4\,K5\,H^2 - 3 - 2\,K5\,H - K4\,K5\,H^2 - K13\,Mg$$
$$\%2 := Exions + H + K2\,NH3\,H + 2\,Mg + MgOH - OH$$

5.7.4   The Solution

A little cleanup and the above expression gives the solution we seek.

```
> gggg := normal(ggggg):
```

This is the *secret bit*

```
> ggg := op(1,gggg);
> #Need to check that this is the right term
```

Because the whole is equal to zero, we have the same solution if we remove any non-zero terms that are multipliers. There is a need to check this carefully, because Maple may well put the terms in a difficult order.

The output is the large expression on the opposite page

```
> indets(ggg) minus {K.(1..14),Exions,OH,MgOH,MgOH2};
```

$$\{H, NH3, Mg\}$$

Exions, OH, MgOH and MgOH2 and the known constants

We notice that, other than the terms we have purposely left out, only the arguments H and Mg remain. The NH3 is to be removed using this expression. Remember, again, that two arguments drive this system; we have chosen H and Mg. Examining the expression, we find that it is a simple quadratic expression in the dissolved NH3, of the form

```
> a*x^2 + b*x + c = 0;
```

where a, b and c are a little tedious

$$a\,x^2 + b\,x + c = 0$$

```
> gg:=collect(ggg,NH3):
```

If a is zero, pick a different operand, above

```
> a:=coeff(gg,NH3,2);
```

$$a := K6\,Mg^3\,K2^2\,H^2 + 3\,K1\,Mg\,K2^2\,H^2 + K7\,Mg^3\,K2^2\,H^2 - K1\,Mg^2\,K2^2\,H^4\,K14\,K4\,K5$$
$$+ K1\,Mg^2\,K2^2\,H^2\,K13 + K1\,Mg\,K2^2\,H^4\,K4\,K5 + 2\,K1\,Mg\,K2^2\,H^3\,K5$$

```
> b:=coeff(gg,NH3,1);
```

$$ggg := -9 + 12\,MgOH2\,K5\,H + 4\,K6\,Mg^4\,Exions + K6\,Mg^3\,Exions^2 + 6\,MgOH2\,K13\,Mg$$
$$- 12\,K5\,H - 6\,K13\,Mg + MgOH2\,K13^2\,Mg^2 + 4\,MgOH2\,K5^2\,H^2$$
$$+ 2\,K6\,Mg^3\,Exions\,H + 2\,K6\,Mg^3\,Exions\,MgOH - 2\,K6\,Mg^3\,Exions\,OH$$
$$+ 2\,K6\,Mg^3\,H\,MgOH - 2\,K6\,Mg^3\,H\,OH - 2\,K6\,Mg^3\,MgOH\,OH$$
$$+ 2\,K7\,Mg^3\,Exions\,H + 2\,K7\,Mg^3\,Exions\,MgOH - 2\,K7\,Mg^3\,Exions\,OH$$
$$+ 2\,K7\,Mg^3\,H\,MgOH - 2\,K7\,Mg^3\,H\,OH - 2\,K7\,Mg^3\,MgOH\,OH$$
$$+ 2\,K10\,Mg\,K5^2\,H^3 + 6\,K10\,Mg^2\,K5\,H + 3\,K10\,Mg\,K5\,H^2 + 4\,K10\,Mg^2\,K5^2\,H^2$$
$$- 4\,K5\,H\,K13\,Mg + 6\,MgOH2\,K4\,K5\,H^2 + 4\,MgOH2\,K5^2\,H^3\,K4$$
$$+ MgOH2\,K4^2\,K5^2\,H^4 + 9\,MgOH2 + 6\,K14\,Mg\,K4\,K5\,H^2 - 6\,K4\,K5\,H^2$$
$$+ K6\,Mg^3\,H^2 + 4\,K6\,Mg^4\,H + 4\,K6\,Mg^4\,MgOH - 4\,K6\,Mg^4\,OH$$
$$+ K6\,Mg^3\,MgOH^2 + K6\,Mg^3\,OH^2 + K7\,Mg^3\,Exions^2 + 4\,K7\,Mg^4\,Exions$$
$$+ K7\,Mg^3\,H^2 + 4\,K7\,Mg^4\,H + 4\,K7\,Mg^4\,MgOH - 4\,K7\,Mg^4\,OH$$
$$+ K7\,Mg^3\,MgOH^2 + K7\,Mg^3\,OH^2 - 4\,K5^2\,H^3\,K4 - K4^2\,K5^2\,H^4 + 4\,K6\,Mg^5$$
$$+ K1\,Mg\,K2\,NH3\,H^4\,K4\,K5 - 2\,K1\,Mg\,K2\,NH3\,H^2\,OH\,K5$$
$$+ 2\,K1\,Mg\,K2\,NH3\,H^3\,K5 + K1\,Mg^2\,K2\,NH3\,H^2\,K13$$
$$+ 2\,K1\,Mg\,K2\,NH3\,H^2\,Exions\,K5 + 6\,K1\,Mg^2\,K2\,NH3\,H + 3\,K1\,Mg\,K2\,NH3\,H^2$$
$$+ 3\,K1\,Mg\,K2^2\,NH3^2\,H^2 + 2\,K1\,Mg\,K2^2\,NH3^2\,H^3\,K5$$
$$+ K1\,Mg\,K2\,NH3\,H^3\,Exions\,K4\,K5 + 2\,K1\,Mg^2\,K2\,NH3\,H^3\,K4\,K5$$
$$+ K1\,Mg\,K2\,NH3\,H^3\,MgOH\,K4\,K5 - K1\,Mg\,K2\,NH3\,H^3\,OH\,K4\,K5$$
$$+ 2\,K1\,Mg^3\,K2\,NH3\,H\,K13 + K1\,Mg^2\,K2\,NH3\,H\,Exions\,K13$$
$$+ 4\,K1\,Mg^2\,K2\,NH3\,H^2\,K5 + 2\,K1\,Mg\,K2\,NH3\,H^2\,MgOH\,K5$$
$$+ K1\,Mg^2\,K2\,NH3\,H\,MgOH\,K13 - K1\,Mg^2\,K2\,NH3\,H\,OH\,K13$$
$$- K1\,Mg^2\,K2\,NH3\,H^4\,K14\,K4\,K5 - K1\,Mg^2\,K2^2\,NH3^2\,H^4\,K14\,K4\,K5$$
$$+ K1\,Mg\,K2^2\,NH3^2\,H^4\,K4\,K5 + K1\,Mg^2\,K2^2\,NH3^2\,H^2\,K13$$
$$- 2\,K1\,Mg^3\,K2\,NH3\,H^3\,K14\,K4\,K5 - K1\,Mg^2\,K2\,NH3\,H^3\,Exions\,K14\,K4\,K5$$
$$- K1\,Mg^2\,K2\,NH3\,H^3\,MgOH\,K14\,K4\,K5 + K1\,Mg^2\,K2\,NH3\,H^3\,OH\,K14\,K4\,K5$$
$$+ 3\,K1\,Mg\,K2\,NH3\,H\,Exions + 3\,K1\,Mg\,K2\,NH3\,H\,MgOH$$
$$- 3\,K1\,Mg\,K2\,NH3\,H\,OH + 2\,K6\,Mg^3\,Exions\,K2\,NH3\,H + 4\,K6\,Mg^4\,K2\,NH3\,H$$
$$+ 2\,K6\,Mg^3\,K2\,NH3\,H\,MgOH - 2\,K6\,Mg^3\,K2\,NH3\,H\,OH$$
$$+ 2\,K6\,Mg^3\,K2\,NH3\,H^2 + K6\,Mg^3\,K2^2\,NH3^2\,H^2 + 2\,K7\,Mg^3\,Exions\,K2\,NH3\,H$$
$$+ 4\,K7\,Mg^4\,K2\,NH3\,H + 4\,K7\,Mg^5 - 4\,K5^2\,H^2 - K13^2\,Mg^2$$
$$+ 2\,K7\,Mg^3\,K2\,NH3\,H\,MgOH - 2\,K7\,Mg^3\,K2\,NH3\,H\,OH$$
$$+ 2\,K7\,Mg^3\,K2\,NH3\,H^2 + K7\,Mg^3\,K2^2\,NH3^2\,H^2 + K10\,Mg\,K5^2\,H^4\,K4$$
$$- 2\,K10\,Mg\,K5^2\,H^2\,OH + K10\,Mg^2\,K5\,H^2\,K13 + 2\,K10\,Mg\,K5^2\,H^2\,Exions$$
$$+ 3\,K10\,Mg\,K5\,H^2\,K2\,NH3 + 2\,K10\,Mg\,K5^2\,H^3\,K2\,NH3$$
$$+ K10\,Mg\,K5^2\,H^3\,Exions\,K4 + 2\,K10\,Mg^2\,K5^2\,H^3\,K4$$
$$+ K10\,Mg\,K5^2\,H^3\,MgOH\,K4 - K10\,Mg\,K5^2\,H^3\,OH\,K4 + 2\,K10\,Mg^3\,K5\,H\,K13$$
$$+ K10\,Mg^2\,K5\,H\,Exions\,K13 + 2\,K10\,Mg\,K5^2\,H^2\,MgOH$$
$$+ K10\,Mg^2\,K5\,H\,MgOH\,K13 - K10\,Mg^2\,K5\,H\,OH\,K13$$
$$- K10\,Mg^2\,K5^2\,H^4\,K14\,K4 - K10\,Mg^2\,K5^2\,H^4\,K2\,NH3\,K14\,K4$$
$$+ K10\,Mg\,K5^2\,H^4\,K2\,NH3\,K4 + K10\,Mg^2\,K5\,H^2\,K2\,NH3\,K13$$
$$- 2\,K10\,Mg^3\,K5^2\,H^3\,K14\,K4 - K10\,Mg^2\,K5^2\,H^3\,Exions\,K14\,K4$$
$$- K10\,Mg^2\,K5^2\,H^3\,MgOH\,K14\,K4 + K10\,Mg^2\,K5^2\,H^3\,OH\,K14\,K4$$
$$+ 3\,K10\,Mg\,K5\,H\,Exions + 3\,K10\,Mg\,K5\,H\,MgOH - 3\,K10\,Mg\,K5\,H\,OH$$
$$- K14^2\,Mg^2\,K4^2\,K5^2\,H^4 + 4\,K14\,Mg\,K4\,K5^2\,H^3 + 2\,K14\,Mg\,K4^2\,K5^2\,H^4$$
$$+ 2\,K14\,Mg^2\,K4\,K5\,H^2\,K13 - 2\,K4\,K5\,H^2\,K13\,Mg + 4\,MgOH2\,K5\,H\,K13\,Mg$$
$$- 6\,MgOH2\,K14\,Mg\,K4\,K5\,H^2 + MgOH2\,K14^2\,Mg^2\,K4^2\,K5^2\,H^4$$
$$- 4\,MgOH2\,K14\,Mg\,K4\,K5^2\,H^3 - 2\,MgOH2\,K14\,Mg\,K4^2\,K5^2\,H^4$$
$$- 2\,MgOH2\,K14\,Mg^2\,K4\,K5\,H^2\,K13 + 2\,MgOH2\,K4\,K5\,H^2\,K13\,Mg$$

$b := 2\,K7\,Mg^3\,K2\,H^2 + 2\,K6\,Mg^3\,K2\,H^2 + 4\,K7\,Mg^4\,K2\,H + 3\,K1\,Mg\,K2\,H^2$

$\qquad + 6\,K1\,Mg^2\,K2\,H + 4\,K6\,Mg^4\,K2\,H + 2\,K7\,Mg^3\,K2\,H\,MgOH$

$\qquad - 2\,K7\,Mg^3\,K2\,H\,OH + 3\,K10\,Mg\,K5\,H^2\,K2 + 2\,K10\,Mg\,K5^2\,H^3\,K2$

$\qquad - K10\,Mg^2\,K5^2\,H^4\,K2\,K14\,K4 + K10\,Mg^2\,K5\,H^2\,K2\,K13$

$\qquad + K10\,Mg\,K5^2\,H^4\,K2\,K4 - 2\,K6\,Mg^3\,K2\,H\,OH + 2\,K7\,Mg^3\,Exions\,K2\,H$

$\qquad + K1\,Mg\,K2\,H^3\,Exions\,K4\,K5 + K1\,Mg\,K2\,H^3\,MgOH\,K4\,K5$

$\qquad - K1\,Mg\,K2\,H^3\,OH\,K4\,K5 + 2\,K1\,Mg^2\,K2\,H^3\,K4\,K5 + K1\,Mg\,K2\,H^4\,K4\,K5$

$\qquad + 4\,K1\,Mg^2\,K2\,H^2\,K5 + 2\,K1\,Mg\,K2\,H^2\,Exions\,K5 + K1\,Mg^2\,K2\,H\,Exions\,K13$

$\qquad + 2\,K1\,Mg\,K2\,H^2\,MgOH\,K5 + K1\,Mg^2\,K2\,H\,MgOH\,K13 + K1\,Mg^2\,K2\,H^2\,K13$

$\qquad - 2\,K1\,Mg\,K2\,H^2\,OH\,K5 - K1\,Mg^2\,K2\,H\,OH\,K13 + 2\,K6\,Mg^3\,Exions\,K2\,H$

$\qquad + 2\,K6\,Mg^3\,K2\,H\,MgOH + 2\,K1\,Mg\,K2\,H^3\,K5 - 2\,K1\,Mg^3\,K2\,H^3\,K14\,K4\,K5$

$\qquad - K1\,Mg^2\,K2\,H^4\,K14\,K4\,K5 - K1\,Mg^2\,K2\,H^3\,Exions\,K14\,K4\,K5$

$\qquad - K1\,Mg^2\,K2\,H^3\,MgOH\,K14\,K4\,K5 + K1\,Mg^2\,K2\,H^3\,OH\,K14\,K4\,K5$

$\qquad + 2\,K1\,Mg^3\,K2\,H\,K13 + 3\,K1\,Mg\,K2\,H\,Exions + 3\,K1\,Mg\,K2\,H\,MgOH$

$\qquad - 3\,K1\,Mg\,K2\,H\,OH$

> `c:=coeff(gg,NH3,0);`

$c := 2\,K6\,Mg^3\,Exions\,MgOH + 2\,K6\,Mg^3\,Exions\,H - 2\,K6\,Mg^3\,Exions\,OH$

$\qquad + 2\,K6\,Mg^3\,MgOH\,H - 2\,K6\,Mg^3\,MgOH\,OH - 2\,K6\,Mg^3\,H\,OH$

$\qquad + 2\,K7\,Mg^3\,Exions\,MgOH + 2\,K7\,Mg^3\,Exions\,H - 2\,K7\,Mg^3\,Exions\,OH$

$\qquad + 2\,K7\,Mg^3\,MgOH\,H - 2\,K7\,Mg^3\,MgOH\,OH - 2\,K7\,Mg^3\,H\,OH$

$\qquad + 2\,K10\,Mg\,K5^2\,H^3 + 4\,K10\,Mg^2\,K5^2\,H^2 + 3\,K10\,Mg\,K5\,H^2 + 6\,K10\,Mg^2\,K5\,H$

$\qquad - 4\,K5\,H\,K13\,Mg + MgOH2\,K4^2\,K5^2\,H^4 + 6\,MgOH2\,K4\,K5\,H^2$

$\qquad - 4\,MgOH2\,K5^2\,H^3\,K14\,Mg\,K4 + MgOH2\,K14^2\,Mg^2\,K4^2\,K5^2\,H^4$

$\qquad - 2\,MgOH2\,K14\,Mg^2\,K4\,K5\,H^2\,K13 - 2\,MgOH2\,K14\,Mg\,K4^2\,K5^2\,H^4$

$\qquad + 2\,MgOH2\,K13\,Mg\,K4\,K5\,H^2 - 6\,MgOH2\,K14\,Mg\,K4\,K5\,H^2$

$\qquad + K10\,Mg\,K5^2\,H^3\,Exions\,K4 + K10\,Mg\,K5^2\,H^3\,MgOH\,K4$

$\qquad - K10\,Mg\,K5^2\,H^3\,OH\,K4 + 2\,K10\,Mg^2\,K5^2\,H^3\,K4 + K10\,Mg\,K5^2\,H^4\,K4$

$\qquad + 2\,K10\,Mg\,K5^2\,H^2\,Exions + K10\,Mg^2\,K5\,H\,Exions\,K13$

$\qquad + 2\,K10\,Mg\,K5^2\,H^2\,MgOH + K10\,Mg^2\,K5\,H\,MgOH\,K13 + K10\,Mg^2\,K5\,H^2\,K13$

$\qquad - 2\,K10\,Mg\,K5^2\,H^2\,OH - K10\,Mg^2\,K5\,H\,OH\,K13 - 2\,K10\,Mg^3\,K5^2\,H^3\,K14\,K4$

$\qquad - K10\,Mg^2\,K5^2\,H^4\,K14\,K4 - K10\,Mg^2\,K5^2\,H^3\,Exions\,K14\,K4$

$\qquad - K10\,Mg^2\,K5^2\,H^3\,MgOH\,K14\,K4 + K10\,Mg^2\,K5^2\,H^3\,OH\,K14\,K4$

$\qquad + 2\,K10\,Mg^3\,K5\,H\,K13 + 3\,K10\,Mg\,K5\,H\,Exions + 3\,K10\,Mg\,K5\,H\,MgOH$

$\qquad - 3\,K10\,Mg\,K5\,H\,OH + 4\,MgOH2\,K5\,H\,K13\,Mg + 4\,MgOH2\,K5^2\,H^3\,K4$

$\qquad + 4\,K5^2\,H^3\,K14\,Mg\,K4 - K14^2\,Mg^2\,K4^2\,K5^2\,H^4 + 2\,K14\,Mg^2\,K4\,K5\,H^2\,K13$

$\qquad + 2\,K14\,Mg\,K4^2\,K5^2\,H^4 - 2\,K13\,Mg\,K4\,K5\,H^2 + 4\,MgOH2\,K5^2\,H^2$

$\qquad + MgOH2\,K13^2\,Mg^2 + 6\,MgOH2\,K13\,Mg + 12\,MgOH2\,K5\,H + 4\,K7\,Mg^5$

$\qquad - 4\,K5^2\,H^2 - K13^2\,Mg^2 + K6\,Mg^3\,H^2 + K6\,Mg^3\,Exions^2 + 4\,K6\,Mg^4\,Exions$

$\qquad + 4\,K6\,Mg^4\,MgOH + 4\,K6\,Mg^4\,H - 4\,K6\,Mg^4\,OH + K6\,Mg^3\,MgOH^2$

$\qquad + K6\,Mg^3\,OH^2 + K7\,Mg^3\,H^2 + K7\,Mg^3\,Exions^2 + 4\,K7\,Mg^4\,Exions$

$\qquad + 4\,K7\,Mg^4\,MgOH + 4\,K7\,Mg^4\,H - 4\,K7\,Mg^4\,OH + K7\,Mg^3\,MgOH^2$

$\qquad + K7\,Mg^3\,OH^2 - 4\,K5^2\,H^3\,K4 - K4^2\,K5^2\,H^4 + 6\,K14\,Mg\,K4\,K5\,H^2 - 6\,K4\,K5\,H^2$

$\qquad + 9\,MgOH2 + 4\,K6\,Mg^5 - 9 - 6\,K13\,Mg - 12\,K5\,H$

```
>  #fortran(a,optimized,filename='a.for'):

>  #fortran(b,optimized,filename='b.for'):

>  #fortran(c,optimized,filename='c.for'):

>  indets(a) minus {K.(1..14)};
```
$$\{H, Mg\}$$
```
>  indets(b) minus {K.(1..14)};
```
$$\{Exions, MgOH, H, Mg, OH\}$$
```
>  indets(c) minus {K.(1..14)};
```
$$\{Exions, MgOH, MgOH2, H, Mg, OH\}$$
```
>  vvv := % union %% union %%%;
```
$$vvv := \{Exions, MgOH, MgOH2, H, Mg, OH\}$$

It is considered that we know the concentrations H and Mg along with extra ions Exions; the additional conditions, below, supply the OH, MgOH and MgOH2. We have a quadratic solution for the concentration of NH3.

The Fortran program is organised so that the auxillary variable assignments occur directly after specifying values for H and Mg.

5.7.5 Assign the Right Variables

Assign NH3

Next solve the equation for NH3 and assign NH3. Then assign the auxillary variables; this minimises the calculations and memory use.

```
>  sss := solve(gg,NH3):nops([%]);
```
$$2$$

```
>  NH3 := op(1,[sss]):
```
Do not back up here! It means a complete restart at the first.

The correct operand is found by Trial and Error

Assign the auxillary variables

These assignments need to be done in the order given, as one 'feeds' into another.

```
>  OH := K11/H;
```
$$OH := \frac{K11}{H}$$

```
>  MgOH := K8*Mg*OH;
```
$$MgOH := \frac{K8 \; Mg \; K11}{H}$$

```
>  MgOH2 := K9*Mg*OH^2;
```

$$MgOH2 := \frac{K9 \, Mg \, K11^2}{H^2}$$

The Equilibrium Constants

Next we collect and substitute in the equilibrium constants. Note that the ones given are more appropriate for 30C but knowledge of many of the constants is meager. Below is a listing of appropriate K values. Note the comments after the values. Some data are from standard texts but many are from Wrigley et al[28].

Ks1 = 5.9D14 ! Struvite formation (good fit)
Ks2 = 1.0D09 ! Ammonium (literature)
Ks3 = 1.5D02 ! Phosphoric acid (literature)
Ks4 = 1.6D07 ! Dihydrogen phosphate(literature)
Ks5 = 1.8D12 ! Hydrogen phosphate (fit)
Ks6 = 1.0D25 ! Magnesium phosphate with 8 H2O (poor fit)
Ks7 = 7.9D22 ! Magnesium phosphate with 22 H2O (ratio, literature)
Ks8 = 5.0D04 ! Magnesium hydroxide ion (fit)
Ks9 = 5.0D10 ! Magnesium hydroxide (literature)
Ks10 = 5.0D05 ! Magnesium hydrogen phosphate (fit)
Ks11 = 1.4D-14 ! Water dissociation
Ks12 = 7.0D-1 ! Dissolved MHP (poor fit)
Ks13 = 1.3D+3 !Magnesum phosphate ion formation(L&L)
Ks14 = 3.2D+1 !Magnesium dihydrogen phosphate ion formation(L&L)
Ks15 = 2.7D-5 ! Dissolved Struvite(fit)
Ks16 = 8.0D+3 ! Ammonium phosphate ion formation NP(poor fit)
Ks17 = 7.0D+6 ! Diammonium phosphate ion formation N2P(guess)
Ks18 = 2.0D+1 ! NHP ion formation(guess)

We proceed to substitute in the set of constants. Remember that we have not cared for Exions and the values of the constants are generally unknowns in the expressions for the assigned variables.

```
>    constants :=
>    [K1=6*10^14,K2=1*10^9,K3=150,
>    K4=16*10^6,K5=18*10^11,K6=10^25,
>    K7=79*10^21,K8=5*10^4,K9=5*10^10,
>    K10=5*10^5,K11=14/10^15,K12=7/10,
>    K13=13000,K14=32] :
```

5.7.6   Save your Results

At this point we have a series of algorithms with which to calculate all the given species in terms of the H and Mg ion concentrations, the equilibrium constants,

```
> species :=['MNP','NH3','NH4','PO4','OH',
> 'H3P','H2P','HP','MP','MP8','MP22','MHP',
> 'MHPd','MH2P','MgOH','MgOH2'];
```

$$species := [MNP, NH3, NH4, PO4, OH, H3P, H2P, HP, MP, MP8, MP22, MHP, MHPd, MH2P, MgOH, MgOH2]$$

```
> evalf(constants,4);
```

$$[K1 = .6000\,10^{15}, K2 = .1000\,10^{10}, K3 = 150., K4 = .1600\,10^{8}, K5 = .1800\,10^{13},$$
$$K6 = .1000\,10^{26}, K7 = .7900\,10^{23}, K8 = 50000., K9 = .5000\,10^{11}, K10 = 500000.,$$
$$K11 = .1400\,10^{-13}, K12 = .7000, K13 = 13000., K14 = 32.]$$

For further reference, later and in this section, we save all the values in binary Maple format, for easy recall.

```
> save MNP,NH3,NH4,PO4,OH,H3P,H2P,HP,MP,MP8,MP22,MHP,
> MHPd,MH2P,MgOH,MgOH2, "species.m";
```

```
> save constants, "consts.m";
```

Note that they will be recalled as assignments, eg, Maple will read OH:=K11/H; After typing read "species.m"; you should find

```
> restart:
```

```
> read "species.m"; read "consts.m";
```

```
> OH;
```

$$\frac{K11}{H}$$

The other assignments are more complex.

Look in the Maple directory under bin or bin.win for binary files

5.7.7 Plotting the Effects of H and Mg

To plot the data, the information must be mostly numeric. First we substitute in the values for the constants.

```
> ss := subs(constants,NH3):
```

Next, put in a few values and see if the selections are correct. Specifically, we consider a plot of NH3 as a function of H and Mg, with H along the abscissa and Mg along the ordinate, all for the case when Exions=0. For convenience and scaling, we use 'p' values.

```
> conditions := {Exions=0,H=10^(-pH),Mg=10^(-pMg)};
```

Importantly, the semicolon and output display acknowledge that Maple knows what is wanted, and lets us know that the primary arguments H and Mg have not been corrupted.

$$conditions := \{Exions = 0,\ H = 10^{(-pH)},\ Mg = 10^{(-pMg)}\}$$

We really don't
want to look at
this now

```
> s := subs(conditions,ss):
> subs(pH=7,pMg=4,s):evalf(%);
```

If this is negative,
go back to where
NH3 was defined

.00003108343203

If this number is .00003108343203, it is correct; otherwise, use the alternative root for the assignment of NH3. We can go ahead with a plot. It takes about 10 minutes with a 250MHz Intel chip on a Toshiba laptop, operating Windows 95, with 64 Megs of RAM.

```
> with(plots):
> rngs := pH=2..12,pMg=0..10:
> opts := grid=[100,100],contours=[$ 1..10]:
> titl := title="Dissolved NH3 Concentrations",
> titlefont=[TIMES,BOLD,16],thickness=2:
> contourplot(-log10(s),rngs,opts,titl);
```

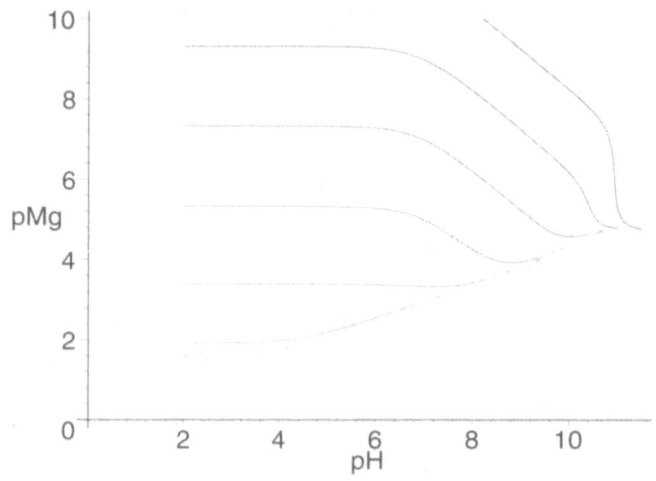

Dissolved NH3 Concentrations

Remember that this plot is presented as pNH3 values; read upward, each line is a factor of 10 lower in the concentration of NH3.

Next we repeat this calculation with a rather large excess of positive ions. This is about 1/10 the concentration of sea water. The ions considered tend to alter the portion of anions and cations through effects on the electroneutrality equation. Note that the salt added must be an acid or base or otherwise have ions common with ions already present.

```
>   conditions := {Exions=.05,H=10^(-pH),Mg=10^(-pMg)};
```

$$conditions := \{Exions = .05,\ H = 10^{(-pH)},\ Mg = 10^{(-pMg)}\}$$

```
>   s := subs(conditions,ss):
>   subs(pH=7,pMg=4,s):evalf(-log10(%));
```
6.287636024

```
>   with(plots):
>   titl := title="pNH3 with Exions = .05",
>   titlefont=[TIMES,BOLD,16],thickness=2:
>   contourplot(-log10(s),rngs,opts,titl);
```

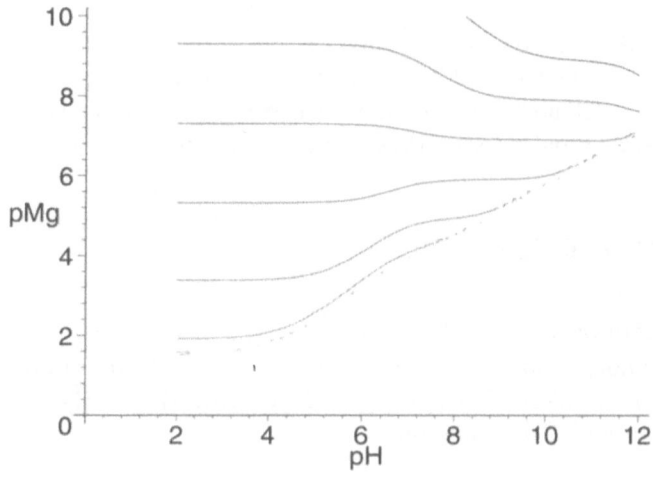

pNH3 with Exions = .05

The area of the graph where the lines converge and there are 'limiting' lines is an area of 'impossible' combinations. That means that some or all of the conditions set in the equilibria are not present. Probably there is no solid, and the solution requires the presence of solid. The above shows a significant increase in that area and, probably, a more soluble situation. If precipitation were desired (waste treatment), we would avoid that area and would not like an excess of positive ions. Of course, if we have a problem with kidney stones, a few more positive ions would be beneficial. Let us hope the body knows this already!

Exercise* 5.14 Plot Other Species

Consider the amount of phosphorus in the liquid Plqd; in waste water, this is an indicator of possible pollution to our waterways. The same with Nitrogen, Nlqd; with the possible exception of blue-green algae, algal blooms arise often from excess levels of nitrogen. What of the solid constitution; is it high in the valuable slow-release fertilizer, Struvite?

Exercise* 5.15 Remove the solids

Redo the above calculations without the formation of any solid. That is, remove any reference to the solid mass balance. This also means, with the Gibb's Phase Rule, that one only needs to know the H concentration to specify all the other species concentrations. Plot values for the concentration of dissolved NH_3 in the solution; compare with the above calculation.

Exercise* 5.16 Effect of Exions

Follow through the calculation with known values for H and Mg and several values of Exions, with H and Mg in the same ranges as given above. What is the effect of Exions, as a parameter? Present the results in a lucid graphical form.

5.7.8   Design Calculation

```
>  restart:
>  read "species.m":  read "consts.m":
```

What we want to know is how much struvite-laden solid will precipitate, in the ground or sewage or kidneys, given a certain concentration in the input. Simply, this is a solution of the equations

```
>  eeq := '{P=Plqd+f*Psolid,M=Mlqd+f*Msolid,N=Nlqd+f*Nsolid}';
```

where P,M and N are known. The unknown quantity f gram-atoms of solid/litre is expected to precipitate. To get into this, we first assign all the variables we know, as a function of the two inputs, H and Mg. This, now, means substitution for the constants and Exions.

$$eeq := \{P = Plqd + f\,Psolid,\ M = Mlqd + f\,Msolid,\ N = Nlqd + f\,Nsolid\}$$

Liquid Mass Balance

Have a peep at each of these with a ; look for stray variables

```
>  Nlqd := NH4 + NH3:
>  #     Nitrogen balance in the liquid
>  Plqd := PO4  + HP + H3P + H2P + MHPd + MP + MH2P:
>  #     Phosphorus balance in the liquid
>  Mlqd := Mg + MgOH + MHPd + MP + MH2P:
>  #     Magnesium balance in the liquid
```

Solid Mass Balance

Again, have a peep

```
>  Nsolid := MNP+MHP: #gram atoms Nitrogen/mole solid
>  Psolid := MNP+2*MP8+2*MP22+MHP: #gram atoms P/mole solid
>  Msolid := MNP+3*MP8+3*MP22+MHP: #gram atoms Mg/mole solid
```

All the quantities should only be functions of the constants, Exions, H and Mg.

```
>  eeq := {P=Plqd+f*Psolid,M=Mlqd+f*Msolid,N=Nlqd+f*Nsolid}:
```

This set of three equations should give us the three unknowns, H, Mg and f, when we know Nsolid, Psolid and Msolid. Let us try fsolve with numerical values for P, N and M.

```
>  subs({P = 1/10000, N = 1/10000, M = 1/10000},eeq):
>  ssol := subs(constants,%):
>  sol := subs(Exions=0,ssol):
>  #solve(sol,{H,Mg,f}):
>  #fsolve(sol,{H=1e-5,Mg=1e-5,f=.5},1..infinity):
```

Neither solve nor fsolve can cope with the expression. There are many different powers, intermixed. For some idea as to why this occurred, play around with the statements below and have a good look; change colons for semicolons. In each operation, the operands are probably in a different order.

```
>  indets(sol):  convert(%,list):  nops(%); op(1..3,%%);
```
$$4$$
$$Mg, H, f$$
```
>  op(4,%%%):  op(1,%):  # a sqrt is removed
>  collect(%,[H,Mg]):  expand(%):
>  coeffs(%,[H,Mg],'t'):  t;
```

Mg, Mg^3, Mg^2, Mg^5, Mg^4, Mg^6, $H^2 Mg$, $H^2 Mg^2$, $Mg\, H$, $Mg^3 H$, $Mg\, H^3$, $H^2 Mg^4$, $\dfrac{Mg^2}{H^2}$,

$Mg^3 H^2$, $H^4 Mg^2$, $Mg^4 H$, $H^3 Mg^3$, $Mg^5 H$, $H^4 Mg^5$, $H^2 Mg^5$, $H^4 Mg^4$, $H^6 Mg^4$,

$H^6 Mg^3$, $H^4 Mg^3$, $H^6 Mg^2$, $H^4 Mg$, $H^5 Mg^2$, $Mg^2 H$, $H^5 Mg$, $H^6 Mg$, $H^5 Mg^3$, $H^3 Mg^4$,

$\dfrac{Mg^3}{H}$, $H^3 Mg^2$, $Mg^6 H^2$, $\dfrac{Mg^4}{H^2}$, $\dfrac{Mg^4}{H}$, $Mg^6 H$, $H^3 Mg^5$, $\dfrac{Mg^5}{H^2}$, $\dfrac{Mg^3}{H^2}$, $\dfrac{Mg^6}{H^2}$, $\dfrac{Mg^5}{H}$, $\dfrac{Mg^6}{H}$,

$H^3 Mg^6$, $\dfrac{Mg^2}{H}$, $H^5 Mg^4$, $H^5 Mg^5$, $H^4 Mg^6$

To make this calculation possible, and easier for Maple, we remove the variable f. Then there are only two equations remaining, {eq1, eq2}, that have the variables {H, Mg}.

```
>   solve(P=Pl+f*Ps,f);
```
$$\frac{P - Pl}{Ps}$$
```
>   eq1 := subs(f=%,Ml+f*Ms-M);
```
$$eq1 := Ml + \frac{(P - Pl)\, Ms}{Ps} - M$$
```
>   eq2 := subs(f=%%,Nl+f*Ns-N);
```
$$eq2 := Nl + \frac{(P - Pl)\, Ns}{Ps} - N$$

The expressions get more complex when we substitute for some of the running variables

```
>{Pl=Plqd,Ps=Psolid,Ml=Mlqd,Ms=Msolid,Nl=Nlqd,Ns=Nsolid}:
>   ggg :=subs(%,eq1):     fff := subs(%%,eq2):
```

Try a selected input condition, with equal numbers of gram-atoms of all the nutrients.

Test values to check the algorithm

```
>   subs({P = 1/10000, N = 1/10000, M = 1/10000},ggg):
>   subs(constants,%):
>   subs(Exions=0,%):
>   ggg := normal(%):
>   subs({P = 1/10000, N = 1/10000, M = 1/10000},fff):
>   subs(constants,%):
>   subs(Exions=0,%):
>   fff := normal(%):
>   #fsolve({ggg,fff},{H=1e-5,Mg=1e-5},1..infinity):
```

The use of solve or fsolve still does not work. This required five minutes to get nowhere. A little more exploration looks for needless multipliers; both expressions are a series of added expressions.   indexsave

```
>   numer(fff):  whattype(%):
>   denom(%%), whattype(%%);# have a look
```
$$1, +$$
```
>   numer(ggg):  whattype(%):
>   denom(%%), whattype(%%);# have a look
```
$$1, +$$

Save the equations so we can use them later.

```
>   currentdir("c:/mplbook/mylib");
```
$$\text{“c:} \backslash\backslash \text{Program Files} \backslash \ \backslash \text{Maple V Release 5.1} \backslash\backslash \text{BIN.WNT”}$$

```
>   currentdir();
```
$$\text{``c:}\backslash\backslash\text{mplbook}\backslash\backslash\text{ mylib''}$$
```
>   save fff,ggg, "design.m";
```

Scan for a solution

The only requirement, of course, is to find the H and Mg values that cause the expressions fff and ggg to be zero. The approach will be to scan for possible solutions. In reality, we are 'trying the water', before we jump into the 'mud'. To minimise time and gain maximal accuracy, double precision, hardware floating point evaluations are used.

```
>   subs([H=1/10^5,Mg=1/10^5],ggg):  evalhf(%);
```

Except for special combined forms, evalhf has to have a numerical argument and it doesn't take a second argument.

$$-.6865539273150367$$

```
>   seq(evalf(subs([H=1/10^i,Mg=1/10^i],fff),6),i=3..12);
```

$-1.49109, -.0906678, -.0120641, .00854915, .0288747, .0451874, .116407, .817912,$
$8.17519, 112.184$

```
>   i := 'i':  j:='j':
>   fvalues:=array(4..11,4..11):
>   gvalues:=array(4..11,4..11):
```

Be prepared for a wait here. With only evalf( ) cals, this takes about 15 minutes on an IBM RISK Unix machine operating at about 100MHz. On a 1998 DOS system, a 250MHz laptop and with evalhf( ) cals, it takes less than 5 minutes.

Stand up and stretch during this calculation

```
>   for i from 4 to 11 do;
>   for j from 4 to 11 do;
>   subs([H=1/10^i,Mg=1/10^j],fff);
>   evalhf(%); fvalues[i,j]:=evalf(%,5);
>   subs([H=1/10^i,Mg=1/10^j],ggg);
>   evalhf(%); gvalues[i,j]:=evalf(%,5);
>   od:  od:
>   convert(fvalues,matrix):  evalf(%,4);
```

$$\begin{bmatrix} -.09067 & -.1179 & -.3197 & -.9956 & -3.145 & -9.945 & -31.45 & -99.46 \\ -.04667 & -.01206 & .005629 & .03181 & .1051 & .3338 & 1.056 & 3.340 \\ -.04305 & -.01011 & .008549 & .03999 & .1306 & .4144 & 1.311 & 4.146 \\ -.03146 & -.007092 & .006245 & .02888 & .09423 & .2989 & .9457 & 2.991 \\ -.01420 & -.002715 & .003224 & .01392 & .04519 & .1433 & .4532 & 1.433 \\ -.005269 & -.0002113 & .003268 & .01152 & .03678 & .1164 & .3681 & 1.164 \\ .005855 & .002487 & .008825 & .02661 & .08250 & .2592 & .8179 & 2.585 \\ -.4546 & .06870 & .1384 & .1775 & .3366 & .8844 & 2.635 & 8.175 \end{bmatrix}$$

Read down for pH and across for pMg

```
>   convert(gvalues,matrix):  evalf(%,3);
```

$$\begin{bmatrix} -2.15 & -6.96 & -22.2 & -70.3 & -222. & -703. & -2220. & -7030. \\ -.212 & -.687 & -2.19 & -6.93 & -21.9 & -69.4 & -219. & -694. \\ -.0213 & -.0687 & -.219 & -.693 & -2.19 & -6.94 & -22.0 & -69.4 \\ -.00230 & -.00745 & -.0237 & -.0750 & -.237 & -.751 & -2.37 & -7.51 \\ -.000460 & -.00136 & -.00432 & -.0137 & -.0432 & -.137 & -.432 & -1.37 \\ -.000747 & -.000379 & -.00124 & -.00392 & -.0124 & -.0392 & -.124 & -.392 \\ -.00836 & -.0000315 & -.000349 & -.00119 & -.00385 & -.0122 & -.0388 & -.123 \\ -.0114 & .000710 & .0000502 & -.000171 & -.000946 & -.00360 & -.0121 & -.0389 \end{bmatrix}$$

It is quite clear for the given input;

> $\{P = 1/10000, N = 1/10000, M = 1/10000\}$:

The pH is between 10 and 11 and pMg between 4 and 5.

> fvalues[10,4],fvalues[11,4];
> gvalues[10,5],gvalues[11,5];

.0058547, −.45462

−.000031536, .00070991

> first_go := [H=1/10^10,Mg=1/10^5];

$$first\_go := [H = \frac{1}{10000000000}, Mg = \frac{1}{100000}]$$

5.7.9 A Double Newton-Raphson Technique

or fff and ggg   Above are two separate polynomial equations f=0 and g=0 in two variables, the H and Mg concentrations. We use surrogates x and y for the solutions and a technique that is little known; a 'derivative-chasing' technique that solves both equations iteratively. The idea is encapsulated in the equations of differentials below ; use finite difference estimates and solve them to get ever-diminishing values of f and g. Enter the desired changes in df and dg and calculate the required changes in dx and dy.

> 'df = Diff(f(x,y),x)*dx + Diff(f(x,y),y)*dy';
> 'dg = Diff(g(x,y),x)*dx + Diff(g(x,y),y)*dy';

$$df = (\frac{\partial}{\partial x} f(x, y)) \, dx + (\frac{\partial}{\partial y} f(x, y)) \, dy$$

$$dg = (\frac{\partial}{\partial x} g(x, y)) \, dx + (\frac{\partial}{\partial y} g(x, y)) \, dy$$

Specifically, the values of x and y are unknown; they are the solution to the set of equations with the expression f=0 and the expression g=0. The derivatives fx, fy, gx, and gy are calculated as symbolic expressions. A trial and error begins; rough, guess values of x and y are used to obtain numerical values of f and g and the derivatives. To gain a solution, we substitute df=-f and dg=-g into the equations below. Solving these equations for dx and dy gives the changes to x and y that should lead to a better estimate of a solution. The changes are applied, and the calculation repeated. Provided the changes are small and the equations are smooth enough, a solution should be found.

The partial derivatives are defined by symbols

```
>   'fx=Diff(f(x,y),x),fy=Diff(f(x,y),y)';
>   'gx=Diff(g(x,y),x),gy=Diff(g(x,y),y)';
```

$$fx = \tfrac{\partial}{\partial x}\, f(x,\, y),\; fy = \tfrac{\partial}{\partial y}\, f(x,\, y)$$

$$gx = \tfrac{\partial}{\partial x}\, g(x,\, y),\; gy = \tfrac{\partial}{\partial y}\, g(x,\, y)$$

In this complex worksheet some variables need to be unassigned. Care should be taken to unassign names that are no longer needed and protect names that are important. The following checks that the variables above are assigned.

Checking the worksheet

```
>   map2(has,[anames()],['f','g','x','y']);
```
$$[false,\ false,\ false,\ false]$$
```
>   map2(has,[anames()],['fx','gx','fy','gy']);
```
$$[false,\ false,\ false,\ false]$$
```
>   #unassign('f','g','x','y','fx','gx','fy','gy');
>   with(linalg):  A := matrix(2,2,[[fx,fy],[gx,gy]]);
```

$$A := \begin{bmatrix} fx & fy \\ gx & gy \end{bmatrix}$$

The purpose is to solve the equation AX=B where all are matrices.

```
>   X := matrix(2,1,[[dx],[dy]]); B := matrix(2,1,[[-f],[-g]]);
```

$$X := \begin{bmatrix} dx \\ dy \end{bmatrix}$$

$$B := \begin{bmatrix} -f \\ -g \end{bmatrix}$$

```
>   inverse(A);
```

$$\begin{bmatrix} -\dfrac{gy}{-fx\,gy + fy\,gx} & \dfrac{fy}{-fx\,gy + fy\,gx} \\[3mm] \dfrac{gx}{-fx\,gy + fy\,gx} & -\dfrac{fx}{-fx\,gy + fy\,gx} \end{bmatrix}$$

```
>   multiply(%,B);
```

$$\begin{bmatrix} \dfrac{gy\,f}{-fx\,gy + fy\,gx} - \dfrac{fy\,g}{-fx\,gy + fy\,gx} \\[4mm] -\dfrac{gx\,f}{-fx\,gy + fy\,gx} + \dfrac{fx\,g}{-fx\,gy + fy\,gx} \end{bmatrix}$$

```
> change := map(factor,%); # combines denominators
```

$$change := \begin{bmatrix} -\dfrac{-gy\,f + fy\,g}{-fx\,gy + fy\,gx} \\[2ex] \dfrac{-gx\,f + fx\,g}{-fx\,gy + fy\,gx} \end{bmatrix}$$

indexinterupt

```
> numer(change[1,1])/Denominator:
```

The large letter versions F and G will become the numerical, evaluated versions of f and g.

```
> dx := subs({f=F,g=G,fy=Fy,gy=Gy},%);
```

$$dx := \frac{Gy\,F - Fy\,G}{Denominator}$$

```
> numer(change[2,1])/Denominator:
```

```
> dy := subs({f=F,g=G,fx=Fx,gx=Gx},%);
```

$$dy := \frac{-Gx\,F + Fx\,G}{Denominator}$$

The x and y are surrogates for H and Mg, so

```
> f  := subs({H=x,Mg=y},fff):
```

```
> g  := subs({H=x,Mg=y},ggg):
```

```
> Fx := evalf(diff(f,x),20):
```

```
> Fy := evalf(diff(f,y),20):
```

```
> Gx := evalf(diff(g,x),20):
```

```
> Gy := evalf(diff(g,y),20):
```

For efficiency, the denominator is separated out and evaluated as a precision floating point number. Convergence problems probably occur as zeros in the denominator. These terms form the determinant of the cross derivative matrix or the 'Wronskian'.

```
> denof := evalf(Fy*Gx-Fx*Gy,20):
```

Also consider
command line
Maple

Remember the struggle is to solve for the amount of struvite solid formed from just a single solution with known amounts of Magnesium, Nitrogen and Phosphorus. The calculation proved too slow to work effectively in a loop; with a small machine, it is more than possible that the calculation will go adrift. Expect to hit the break/interrupt key or the stop button with two taps in rapid succession.

Run the loop line-by-line until you reach the end; then recycle at the top. This gives a running output and one can see where the calculation is faltering. The data are displayed at the end in the sequences datax and datay. Hopefully, your calculation converges as well as this one. Note that we have carefully used the guess from the last section as a start to minimise spurious calculations.

```
>   datax:=NULL: datay:=NULL:
>   #unassign('x','y'):
>   first_go;
```

$$[H = \frac{1}{10000000000}, \; Mg = \frac{1}{100000}]$$

```
>   xys := map(rhs,first_go):

>   i:=0:
>   #********begin cycle********
>   #for i from 1 to 10 do;
>   x:=xys[1]; y:=xys[2];
```

$$x := .1533422326 \, 10^{-9}$$
$$y := .00002647000725$$

```
>   F:=evalhf(f); G:=evalhf(g);
```

$$F := -.1495204802419464 \, 10^{-7}$$
$$G := .1868495680476990 \, 10^{-8}$$

```
>   Denominator:= evalhf(denof);
```

$$Denominator := .6781608674747173 \, 10^{8}$$

```
>   evalf([dx,dy],20); # expected changes in x and y
```

$$[.20623430155221794352 \, 10^{-14}, \, -.29535849010842304623 \, 10^{-9}]$$

Lists can be added

```
>   [x,y] + %;
```

$$[.1533442949 \, 10^{-9}, .00002646971189]$$

```
>   xys := %;
```

$$xys := [.1533442949 \, 10^{-9}, .00002646971189]$$

```
>   i:   i:= %+1;
```

$$i := 5$$

```
>   datax:=datax,[i,xys[1]]:   datay:=datay,[i,xys[2]]:
>   evalf([datax],4); evalf([datay],4);
```

$$[[1., .1405 \, 10^{-9}], [2., .1620 \, 10^{-9}], [3., .1549 \, 10^{-9}], [4., .1533 \, 10^{-9}], [5., .1533 \, 10^{-9}]]$$
$$[[1., .00001777], [2., .00002440], [3., .00002638], [4., .00002647], [5., .00002647]]$$

```
>   #od;
```

There is convergence and a possible solution for the H and Mg concentrations in a solution with equal numbers of gram-atoms per litre, {M=1/100000 ,N=1/100000, P=1/100000}. Still, the procedures were cumbersome; we seek an alternative way. The solution seems more straightforward using the logarithmic form, the 'p' form of the concentrations.

To the next chapter. A technique of 'Clicking on a Solution' gets the solution in a flash, using active plotting (section 6.1, page 313).

Exercise 5.17 Remaining Error

Check the remaining error in f (or fff) and g (or ggg) when

>{H=0.1533*10^9,Mg=0.0002647};

with the above values of M, N and P.

Exercise* 5.18 Newton-Raphson Loop

Calculate the value of f (or fff) or g (or ggg) in each loop and, using a while construction of the form

```
>    #while abs(f)>1/10^8 or abs(g)>1/10^8 or i<10 do;
```

Close the above loop. Be patient, the calculation takes some time.

Exercise* 5.19 Using pH and pMg as Arguments

Replace the arguments in the entire calculation with -log10( ) variations. This should give a better range and better convergence.

Exercise* 5.20 Inverse Interpolation

Fit some of the data in the vicinity of the solution to a multivariate quadratic function, using the technique given in section 4.3.5, page 218. Reverse the the dependence and independence with H and Mg becoming dependent variables as functions of fff and ggg. Once the fit is complete, the solution is when fff and ggg are zero.

Exercise* 5.21 Constitution of the Solids

What do the results mean? The concentrations of H and Mg are known in a solution with $\frac{1}{10^4}$ gram atom/litre each of M, N and P in the initial solution. How much solid precipitate is formed in gram atoms per litre? What is the constitution of the solid precipitate? List all the solid species and calculate their concentrations, as gram-atoms per gram-atom of solid. Look up the molecular weights and turn these numbers into grams of solid per litre. Note that f, the gram-atoms of solid per litre is the joining quantity. With kidney stones, it seems critical that you know this, along with the pH and pMg of the solution. This result is quite alkaline, not a pH you would find in your body.

```
>    evalf(-log10(1.533*10^(-10)),4);
```
                        9.814

Exercise* 5.22 N-R without Derivatives

In the last section, the first equation in df and dg ultimately produce equations for the required changes in dx and dy that are proportional to f and g. If, then, f and g are zero, no change is required and we have a solution. That means that it is not necessary to have or calculate the derivatives fx, fy, gx and gy. Any function of x or y or, indeed, any function or even constants 'written in' for fx, fy, gx and gy may lead to a solution. It is difficult to calculate derivatives and this alternate approach can be made to work. You might try only calculating the derivatives occasionally.

Exercise* 5.23 Fortran or C Program

Use some of your language skills to produce a program that uses the Maple algorithms. Calculate various species concentrations based on the quadratic,

```
>   a*x^2 + b*x + c = 0;
```

where x is the dissolved ammonia concentration. It will be necessary to save the output of Maple appropriately, perhaps with optimized algorithms.

```
>   fortran(a,optimized);
```

```
>   save %, "a.for";
```

The file a.for is edited before pasting into your program. Remember to include the species OH, $MgOH$ and $MgOH_2$. With the program you can output lists to be read into Maple and plotted. Alternatively, you can plot the data within Fortran or C. An ambitious attack of the problem might complete the design calculation over a range of possible scenarios.

Many variables and many equations are no problem to Maple. The linear problems presented, matrix and statistical, from breakfast foods to river flows to turbulence are completed easily with little constraint on the variable numbers or the size of expressions. The other side is a highly complicated non-linear problem; the struvite problem is more difficult. In fact, Maple is a superb front-end solver to generates a symbolism in which floating point numbers may be crunched. An alternative, using active plotting, is presented in the next chapter.

Acknowledgements for Chapter 5

The concatenation derives from Char et al, *Maple V Langauge Reference Manual* [8], pg 184; the use of anames and unames from Greg Fee of Simon Fraser University. Stoichiometric calculations and isolve come from Tony Scott's article in MapleTech, Issue 7 [27]. Stewart Greenhill used the transformation for correcting turbulence statistics in developing the Murdoch Turbulence Probe, as part of his PhD at Murdoch University, School of Environmental Science. The turbulence data were retrieved in the field by Adrian Blockley as part of Murdoch's Air Pollution Meteorology Unit. The struvite material was part of work done by Kelvin Webb and Tim Wrigley for their PhD theses, School of Environmental Sciences, Murdoch University. Peter Kloeden of the Fachbereich Mathematik, Johann Wolfgang Goethe University, suggested the use of arbitrary functions or constants in a Newton-Raphson scheme, as a way of dealing with difficult derivatives.

Chapter 6

Active Plotting and Spreadsheets

Solutions from Pictures

Extracting Information, Digitizing

Throwing Curves About

In and out of Spreadsheets

Active Plotting and Spreadsheets

Übersicht

Pick a curve or two; look to see where they cross; you have a solution. Look at an image anywhere on the screen; capture it in lines and points and redraw it, any size or with any sort of additional information. Grab a curve; put it onto a graph, directly; no numbers. With your environmental accounting forms, Maple can offer as good a sheet as any, with infinite precision. This chapter finishes off with some of the recent developments in Maple, and some new companion software.

Guidewords:

Curve fitting, Matching Solutions; digitisation, posting, capture, images, pictures, fscanf( ), sscanf( ), feof( ), pipe, tracing, sketching; .eps, .gif; smartplots, solar radiation, radian day, Heaviside( ); spreadsheet, Context Menu CM, relative referencing, absolute referencing, ~A1 to ~G10; 'not' symbol, arrow, arrow and box, activate, drag and drop; parse( ), insert( ), Sparse( ), Eparse( ); traperror( ), SearchText( )

This is a closing look at what Maple can offer in an active sheet, principally plotting. The scientist views most of the world as a series of lines, points or surfaces; it happens this way because we must put numerical values on what we see or how it moves, even though it may spoil the beauty within. First, implicit drawing of curves is used to solve the burdensome problem of the last chapter. Then, a program is introduced to capture the simplest features of an image, inside or outside of Maple. Lastly, some of the data are played collectively using Maple's spreadsheet.

At this point it is a good idea to review some of the recent terms that relate to worksheets, menus and spreadsheets

```
>   ?glossary
>   ?update
```

6.1   Clicking on a Solution

```
>   currentdir("c:/mplbook/mylib");
```
> "C:\\Program Files\ \Maple V Release 5.1\\BIN.WNT"

```
> currentdir();
```
$$\text{``c:}\backslash\backslash\text{mplbook}\backslash\backslash \text{ mylib''}$$

Start where we left off, last chapter, page 307. We were trying to solve for the amount of struvite that might precipitate from a given mixture, that is, form kidney stones.

Within the right directory we read the data; the read statement 'puts back' or assigns the objects specified during the save. Specifically this is fff and ggg, the expressions that define the H and Mg concentrations. They are the two equations that must be solved to know H and Mg. We review plots by simply attaching it to our worksheet, using with( ). That means we can use the shortened names to call on the commands.

```
> read "design.m";
```

```
> with(plots);
```

[*animate, animate3d, animatecurve, changecoords, complexplot, complexplot3d, conformal, contourplot, contourplot3d, coordplot, coordplot3d, cylinderplot, densityplot, display, display3d, fieldplot, fieldplot3d, gradplot, gradplot3d, implicitplot, implicitplot3d, inequal, listcontplot, listcontplot3d, listdensityplot, listplot, listplot3d, loglogplot, logplot, matrixplot, odeplot, pareto, pointplot, pointplot3d, polarplot, polygonplot, polygonplot3d, polyhedra_supported, polyhedraplot, replot, rootlocus, semilogplot, setoptions, setoptions3d, spacecurve, sparsematrixplot, sphereplot, surfdata, textplot, textplot3d, tubeplot*]

Next, we substitute p-surrogates for Mg and H; they tend to vary more evenly. From here, search for values of pH and pMg which satisfy the two equations. The display of each of the two implicit expressions in different colours allows a graphical solution.

```
> e1 := subs({H=10^(-pH),Mg=10^(-pMg)},fff):
```

```
> e2 := subs({H=10^(-pH),Mg=10^(-pMg)},ggg):
```

```
> rng := pH=6..12,pMg=3..7:
```

```
> P1 := implicitplot(e1,rng,colour=blue):
```

```
> P2 := implicitplot(e2,rng,colour=green):
```

```
> display(P1,P2);
```

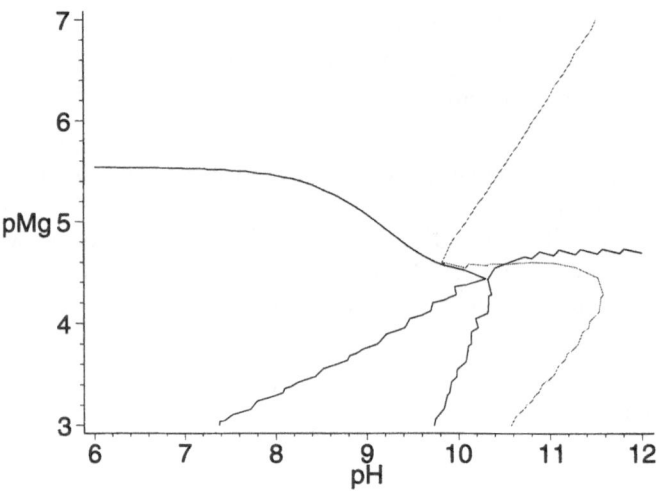

It appears there are 2 possible solutions or crossing points. The plot is active; click on it. An active tracker appears in the little window just above the Maple window on the left; clicking with the mouse shows the position of the tip of the arrow in plotting coordinates. Find the points where the curves seem to cross. One curve clearly crosses; the other may cross but appears to be tangent. With the arrow precisely on the crossing point, click the mouse; the position of the arrow tip appears in the little tracker window. Now, highlight the numbers in the window; use copy Ctrl-c. You will have collected the position of the two crossing curves in proper plotting coordinate values. Paste the values into the next execution group with Ctrl-v. You have a solution to the problem; the values of pH and pMg satisfy the two equations. The only ambiguity seems to be the other possible crossings. Replot with a different range.

```
> [10.49, 4.58]: sol1 := [pH=%[1],pMg=%[2]];
```
$$sol1 := [pH = 10.49,\ pMg = 4.58]$$

```
> rng := pH=9.7..10.5,pMg=4.2..4.8:

> P3 := implicitplot(e1,rng,colour=blue):

> P4 := implicitplot(e2,rng,colour=green):

> display(P3,P4);
```

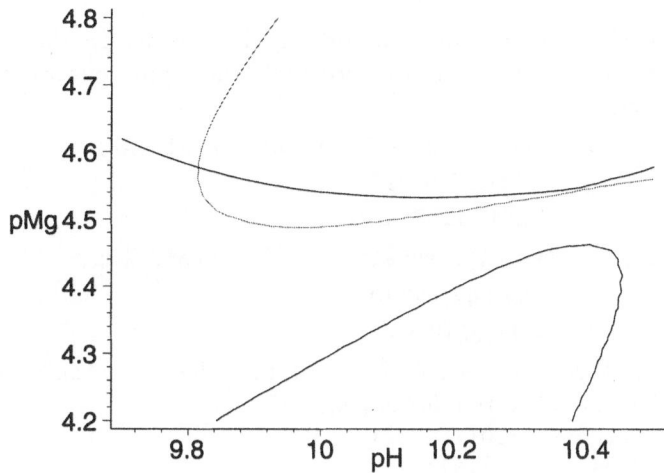

```
>    [9.81, 4.58]:  sol2 := [pH=%[1],pMg=%[2]];
```

$$sol2 := [pH = 9.81, pMg = 4.58]$$

This shows another possible solution; there still could be a third, where the curves nearly meet.

```
>    rng := pH=10.2..10.5,pMg=4.5..4.6:
>    implicitplot(e1,rng,colour=blue):
>    implicitplot(e2,rng,colour=green):
>    display(%,%%);
```

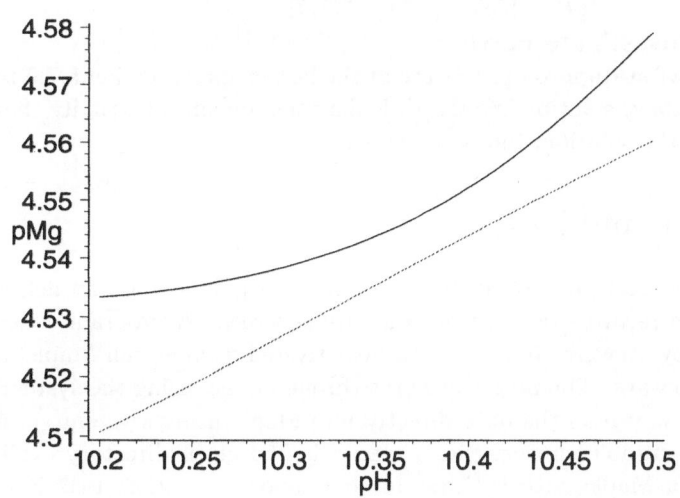

No connection. We have two possible solutions; check them out. Do the solutions satisfy the original equations? Substituting into the expressions shows that they come very close.

> `subs(sol1,e1):  evalf(%);  subs(sol1,e2):  evalf(%);`

$$.00002770642360$$
$$.0001422727275$$

> `subs(sol2,e1):  evalf(%);  subs(sol2,e2):  evalf(%);`

$$.6948421422\,10^{-5}$$
$$-.1378319480\,10^{-5}$$

Lastly, we compare these possible solutions with the values from the struvite design calculation, end of section 5.7.9, page 307.

> `sols :=evalf([pH=-log10(1.533/10^10),`
> `pMg=-log10(2.647/10^4)],4);`

$$sols := [pH = 9.814, pMg = 3.577]$$

> `subs(sols,e1):  evalf(%);  subs(sols,e2):  evalf(%);`

$$.01577709437$$
$$-.02317399158$$

An overview of all the values suggests that the 'solution' from the last chapter shows the largest 'error'. The hand-picking of the crossing points gives one some confidence. It seems that the pMg value is probably not around 3.5; it is likely that the pH is 9.81 with the pMg , 4.58.

> `sol1;sol2;solns;`

$$[pH = 10.49,\ pMg = 4.58]$$
$$[pH = 9.81,\ pMg = 4.58]$$
$$[pH = 9.814,\ pMg = 3.577]$$

Exercise* 6.1 Investigate Further

Apply the Newton-Raphson procedure of the last chapter, section 5.7.9 to confirm that the above solution is best. Calculate the amount of struvite that precipitates from the solution, the parameter f.

6.2   Capture and Post

Data from pictures and graphs can become lines and points when an active graphical program recovers points from the visible screen. A program called capture, written by Stewart Greenhill, can be activated from within Maple; it is launched in two ways. The first way works through a file, using the system( ) call; the second way puts the data directly into Maple using ssystem( ). A variant of the executable capture.exe is placed in the given directory. It is run from inside of Maple with the system commands.   The task is to extract sufficient information from the image of Perry Lakes (as presented in section 4.1, page 181) to replot WestLake. The picture is in file perrylakes.eps or perrylakes.gif that can be viewed with many different packages. In fact, any file, or even an image of the screen itself can be used as a source of information. The data are digitised and made accessible to numerical analysis and representation. In this case we have opened it with Photoshop but Microsoft Word or Gimp from Linux would work equally well. The image is on the Windows 95 screen with the window in the left side of the screen to avoid the banners, scrollbars and menus. The program is located in c:/mplbook/image-capture, so we first set our current directory.

```
>  currentdir("c:/mplbook/image-capture"):
```

```
>  currentdir();
```

$$\text{``c:}\backslash\backslash\text{mplbook}\backslash\backslash \text{ image-capture''}$$

The system command and operating system processes redirections and the data is directed into file tt. This doesn't need Maple to work and can run directly from the DOS prompt. The data are viewed in the editor; later, the data can be read into Maple with readline or readdata. When the first command is activated, a transparent screen is placed over the whole Windows 95 screen. Pointing and clicking on 7 points leads to the following results. Typing q quits the program and typing u updates the transparent screen.

```
>  system("capture > tt");
```

$$0$$

```
>  system("edit tt");
```

$$0$$

```
>  #ssystem("capture");
```

[0, "505, 171\n485, 225\n468, 276 \n469, 276\n493, 352\n549, 343 \n550, 343\n"]

The third statement uses the ssystem command instead; the data are directly fed into the Maple output. This is a list with 2 elements. If we look at only the second element, the 0 response is omitted but the string nature of the output with the carriage returns \n remain.

```
>  #ssystem("capture")[2];
```

"91, 337\n135, 341\n171, 339 \n206, 342\n246, 343\n278, 344 \n323, 346\n"

An alternative program capturef, reads the data directly into the file "pipefile"

```
>  ssystem("capturef");
```

$$[0, \text{`` ''}]$$

```
>  system("edit pipefile");
```

$$0$$

capture and capturef use u for update and q for quit

The sequence of read and decode with sscanf( ) or fscanf( ) allows a formatting of the data.

The loop presents all the data from the file

```
> fl := fopen("pipefile",READ);
```
$$fl := 0$$

```
> readline(fl);
```
$$\text{``126,311''}$$

```
> sscanf(%, "%d,%d"); # parses the string
```
$$[126, 311]$$

```
> fscanf(fl, "%d,%d"); # parses a line from the file
```
$$[127, 311]$$

```
> feof(fl);
```

We have not yet reached the End of File

$$false$$

```
> while not feof(fl) do print(fscanf(fl, "%d,%d")) od;
```
$$[153, 305]$$
$$[192, 305]$$
$$[223, 313]$$
$$[254, 323]$$
$$[\,]$$
$$[\,]$$
$$[\,]$$
$$[\,]$$

```
> fclose(fl);
```

The whole idea seems to work. However, the above shows some rubbish that was read after the data points were collected. Organising a loop that reads forever allows that the data in the file can be continuously displayed while they are being read.

Continuous Reading of Data

The following routine launches the program capturew that collects [x,y] points and transfers them into Maple. First, organise your image so you can see the important pieces on the screen, including some scales or axes for calibration. If you want to see what Maple is doing, it will be necessary to leave some room for Maple, perhaps to the right of the screen. Next, execute the routine. It will produce at least one DOS window that may obscure your image. If so, wait a moment and then push the tab. This updates the screen and the window should disappear. If all looks right, you are set up to collect data from the image. This consists of moving the arrow cursor so that the tip is exactly on a point. Then you click with the left mouse button and a little red x will appear. Repeating this operation allows the capture of an entire picture. Each time you click with the mouse, another red x should appear. These are your data and, if you don't want problems resorting or whatever, take the data in the way they will be used. For instance, if you want an image of two touching circles, do each individually with single, consistent and consecutive points around the circumference. If you do half of one, half of the other, the other half of the first and the other half of the second, or collect random points, the line drawing will be a mess and take a lot of work to clean up.

You can 'paint' with rapid movements

Note that you should be a little careful not to put too many identical points on the screen by being jittery or moving the mouse too much with the left button held down. It makes a mess and it is difficult to organise the data; though they can be edited out after collection. Also, if you want resolution, blow up the image so that it is a decent size and pixel resolution is not a problem. Some images are of very high resolution.

This routine launches a capture window, a transparent window that covers the whole screen; an MS/DOS window opens and then closes. The use of the tab key updates the capture window and removes any remaining windows started with the program. The file fl, called "pipefile" in the current directory is opened for capturing the data. The continuous loop only allows the printing (output) of usable data and uses the empty list to terminate the loop. Closing the file finishes the routine.

With capturew tab updates the screen and esc quits

```
>   ssystem("capturew -f pipefile"):
>   fl := fopen("pipefile", READ):
>   result := 0:
>   while not (result = []) do
>   result := fscanf(fl, "%d,%d,%*s");
>   if (not (result = 0)) and (not (result = ⌊⌋)) then
>   print(result);
>   fi;
>   od:
>   fclose(fl);
```

$$[280, 259]$$
$$[283, 249]$$
$$[273, 245]$$
$$[266, 240]$$
$$[251, 233]$$
$$[238, 234]$$
$$[206, 249]$$

The tracing process works from a transparent window generated by capture that initially fills the whole screen. That window captures the image in back, whatever is there when the update is evoked; pushing the tab key refreshes or updates the window. To see what Maple is doing while you are tracing, normalise the window by pushing the middle button in the upper left corner of the screen. You can then resize and move the 'transparent' window wherever you like. If you leave the window full size or partially covering something you want to see, you can update the image in the window by pressing the tab key. This also allows 'capture' of images that are changing; pushing the tab simply takes a snapshot and holds it in the transparent window.

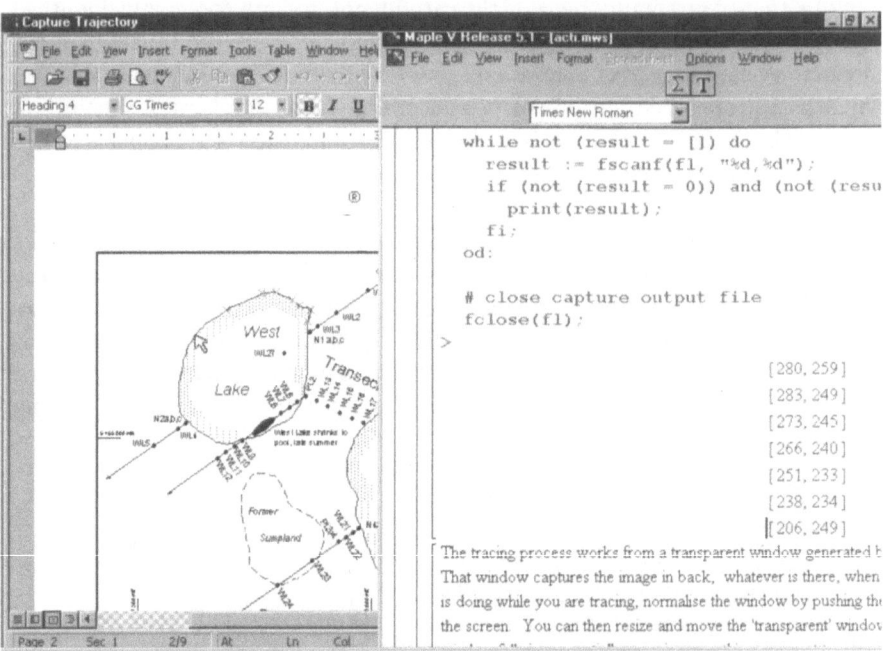

For further use, highlight the output and copy below. Rehighlight and use the right mouse button to convert to a list. The result shown is a decent representation of the upper side of WestLake, except for the fact that the pixelated output measures vertically downward and we have no scale. No matter. Back to the scale on the lower right of the figure. Evoke capture , as above; there are 53 pixels in the 100 m scale.

This can be done outside of Maple

```
> R0 := [[280, 259], [283, 249], [273, 245],
> [266, 240], [251, 233], [238, 234], [206, 249]];
```

$$R0 := [[280, 259], [283, 249], [273, 245], [266, 240], [251, 233], [238, 234], [206, 249]]$$

```
> plot(R0);
```

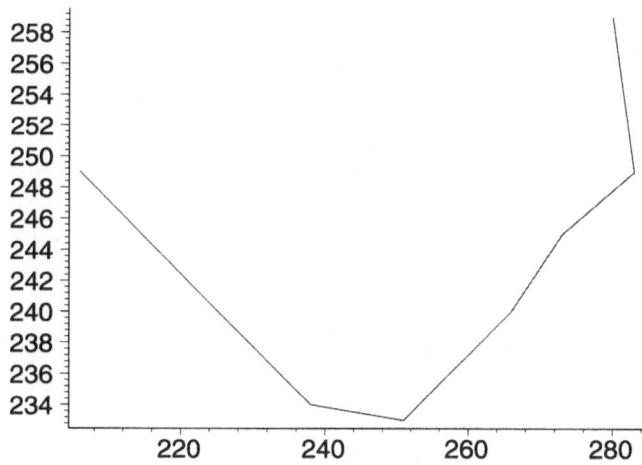

```
>   'map(x->(300-x)/53*100,op(i,R0))';
```

$$\mathrm{map}\!\left(x \to \frac{30000}{53} - \frac{100}{53}\,x,\ \mathrm{op}(i,\,R0)\right)$$

```
>   seq(evalf(%,4),i=1..nops(R0));
```

indexop indexmap indexseq—ii indexplot!titlefont—ff

[37.74, 77.36], [32.08, 96.23], [50.94, 103.8], [64.15, 113.2], [92.45, 126.4], [117.0, 124.5], [177.4, 96.23]

```
>   R1 := [%]:
```

```
>   titlefont=[TIMES,BOLD,16]:
>   labels=["W-E metres","S-N metres"]:
>   plot(R1,title="Top of WestLake",%,%%);
```

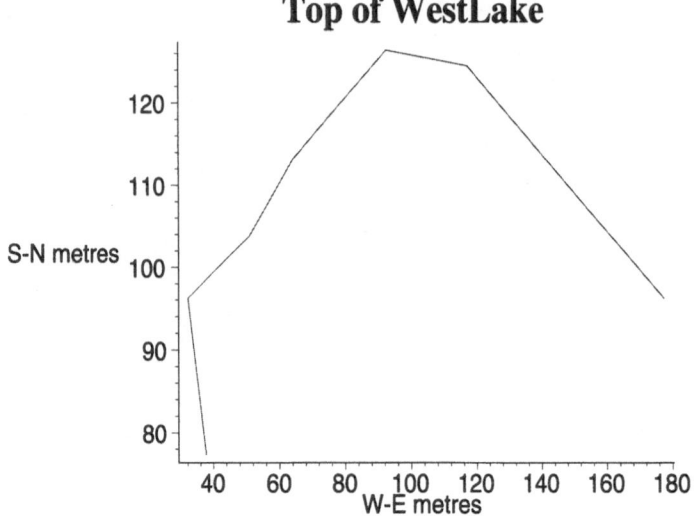

Capture with NT

Named pipes are a means for communicating information between processes. A named pipe can be opened like a regular file, but its contents are determined dynamically by a controlling process. The following command launches capture and outputs the resulting points to a pipe called "\\.\pipe\test"

```
> ssystem(capturew -p test)
```

In Maple, the named pipe can be opened using:

```
> fl := fopen("\\.\pipe\test");
```

A program can read from the pipe using fscanf, and will receive an end of file condition when the capture program has closed the pipe, when the user exits the capture program with Esc. indexcapture!exit!Esc

```
> while not feof(fl) do
> result := fscanf(fl, "%d,%d,%*s");
> print(result);
> od:
> fclose(fl);
```

Posting

The same approach can be used to post positions of bore holes or simply for drawing the boundaries of a paddock, house, shed or whatever. Then, importantly, we need to know something about the point. A variation of the same program, by Stewart Greenhill, is set up with a procedure called post for this purpose. It puts whatever is typed on the keyboard alongside the posting, it is input with the clicking of the mouse. Note that the transparent window that is produced is updated by the use of tab and the program is quit with an esc. One should normalise the transparent window, clicking on the centre square button in the upper right corner; then shift it around and size it so that it takes in the image of concern. This should leave the Maple window visible enough to see the output as it comes in, the data you are collecting and the posting information.

Set up your image on the upper left side of the screen and activate the procedure before starting. Then remove the # from the next statement and click the execute button or use the CR. Wait a moment when the black window appears; then push the tab; it will disappear and the screen image will be updated.

```
>   post := proc() local fl,result,dd;
>   currentdir("c:/mplbook/image-capture");
>   ssystem("capturew -f pipefile");
>   fl := fopen("pipefile", READ);
>   result := 0:  dd := NULL;
>   while not (result = []) do
>   result := fscanf(fl, "%d,%d,%s\n");
>   if (not (result = 0)) and (not (result = [])) then
>   print(result); dd := dd,%;
>   fi;
>   od:
>   fclose(fl);
>   RETURN([dd]);
>   end:

>   #post(); data:=%:  data;
```

$$[97, 207, \text{``WL12''}]$$
$$[112, 197, \text{``WL11''}]$$
$$[128, 186, \text{``WL10''}]$$
$$[141, 175, \text{``WL9''}]$$
$$[211, 127, \text{``WL8''}]$$
$$[227, 117, \text{``WL7''}]$$
$$[244, 105, \text{``WL6''}]$$
$$[258, 96, \text{``PL2''}]$$
$$[229, 303, \text{``0''}]$$
$$[318, 303, \text{``100''}]$$

[[97, 207, "WL12"], [112, 197, "WL11"], [128, 186, "WL10"], [141, 175, "WL9"], [211, 127, "WL8"], [227, 117, "WL7"], [244, 105, "WL6"], [258, 96, "PL2"], [229, 303, "0"], [318, 303, "100"]]

```
>   nops(data);
```

10

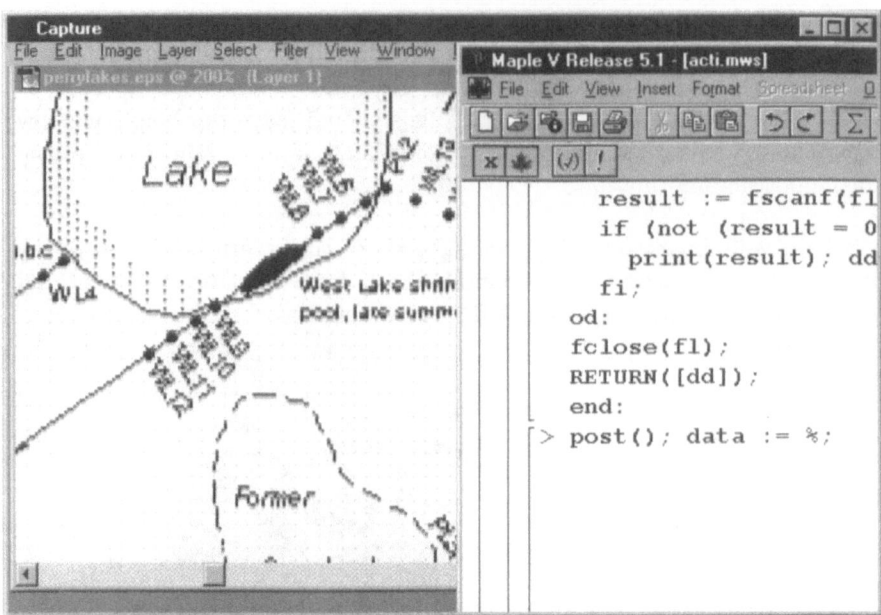

As this is an active sheet, an extra copy of the posted data is copied below. This allows replacement in case it is destroyed.

```
>   data :=[[97, 207, "WL12"], [112, 197, "WL11"],
>   [128, 186, "WL10"], [141, 175, "WL9"], [211, 127, "WL8"],
>   [227, 117, "WL7"], [244, 105, "WL6"], [258, 96, "PL2"],
>   [229, 303, "0"], [318, 303, "100"]]:
```

Referring back to section 4.1, these particular data from the picture of Perry Lakes (from perrylakes.eps) are displaced but have a built-in calibration, with 318-229 or 89 pixels for 100 horizontal metres. The last posting is taken from the scale at the bottom of the map.

The post( ) procedure accepts no errors in what is typed in; it leaves a 'tag' on the data unless it is overwritten by a blank; and the massage by fscanf( ) accepts commas and spaces as delimiters. This means that, if you don't want a comment attached to a point, you should enter a blank; also you should normally limit your comments or postings to single words, without commas. However, all the data do go to the file called "pipefile"; the errors or other variations can be edited in an editor.

Exercise* 6.2 Capture Procedure
Enclose the code above and make a smooth procedure called capture.

Exercise 6.3 Draw a Picture in Lines
Using the image of WestLake (or another lake) in Microsoft Word, Corel Draw, Photoshop or another package, digitise the full surrounds of the lake, put in the titles and boundaries so that the picture resembles the scanned picture. Plot the results. Check the size of the file exported to .gif or .eps by Maple. To export the file see section 1.5, page 34 and sections 1.6 and 3.6; you might want to use

```
> plotsetup(ps,plotoutput="c:plot.ps");
```

6.3   Smartplots

A drag-and-drop environment that is a plot; a plot that lives and can take on the Maple defaults of points or line, range, style, colour or orientation. Look at a simple plot, in smartplot( ). Note the little 'Live' in the upper right corner.

```
> smartplot(cos(x));
```

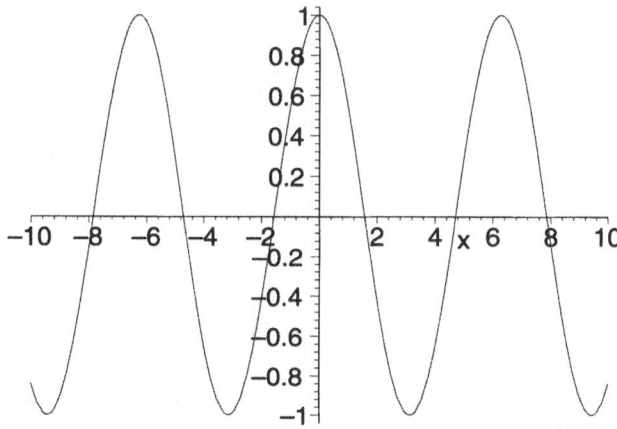

[>

This creates a place for Maple input

   Use the [> button to make a blank execution group, just below the plot; click on the [> and add a few blank spaces. Maximise the window and organise your working space so that you see both the smartplot and the execution group. Click on the graph and move the mouse pointer over the red curve. The pointer should cause the curve to change color; at the bottom the screen, the left side of the lower bar should acknowledge that cos(x) has been selected. Keep the left mouse button held down and 'drag' the curve away; with the left mouse button down and the arrow moving off the curve, the arrow changes into a 'not' symbol, a 'slashed circle'. As with ordinary 'drag and drop' this means you have 'grabbed' the curve. Now, move the 'slashed circle' off the plot and position it properly within the blank execution group; the pointer should now have changed shape again; an arrow with an attached box should make it clear what you are doing. Release the mouse. Voila! The curve disappears from the graph and reappears as a Maple input.

Copying and Removing

Reexecute the statement smartplot(cos(x)); as usual, the default menu setting Options/Replace Output remakes the smartplot and replaces the vacant plot. Repeat the 'drag and drop' operation; if the 'arrow and box' is over inert text, the cos(x) appears as inert text; if the 'arrow and box' is placed over a spreadsheet cell (see section 6.4), the cos(x) appears there. Holding down the Ctrl key while 'dragging' allows you to copy the cos(x) without destroying the curve; the 'arrow and box' shows + to acknowledge that the cos(x) is to be copied instead of moved. If you want to remove the curve, select it (point the tip of the arrow at it and see the colour change) and press the Del key.

Replacing

Notice that the cos(x) below the smartplot doesn't have the necessary semicolon. Add the semicolon and use the CR; the Maple output is echoed in blue. Now, highlight the blue Maple output with the mouse. The output is 'put' back into the vacant graph with an ordinary 'drag and drop'; that is, collect it with the left mouse button held down and place it in the graph window. Just for practice, collect the cos(x) curve again and put it back, as a Maple output. It works, but the output may no longer be blue. The smartplot 'plots' the Maple output and curves on the graph can be 'reborn' as Maple input.

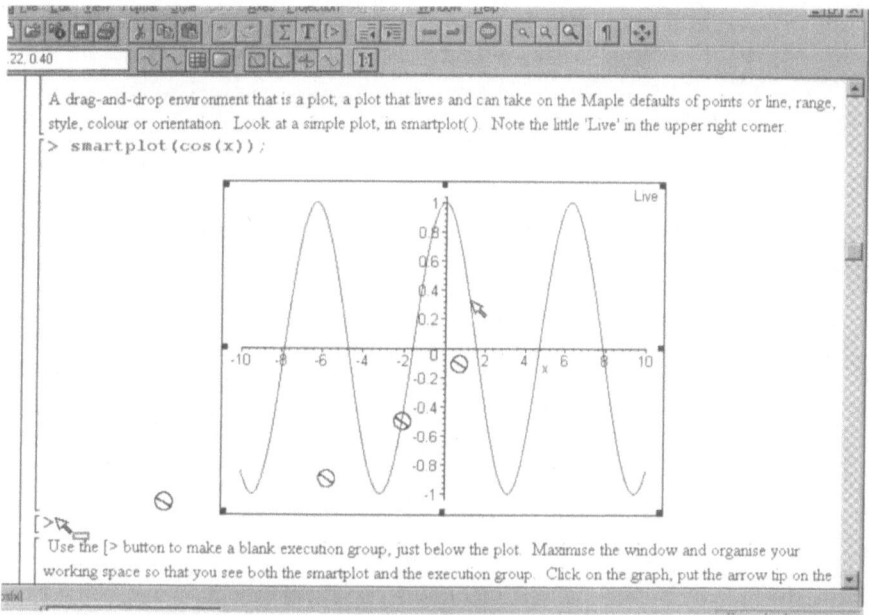

Changing and Matching

Smartplot( ), plot( ) as well as PLOT( ) structures can be rescaled and shifted on the worksheet. Clicking on the plot forms a box around the plot with eight little black squares; moving the mouse arrow to one of the little black squares produces a double-headed arrow that allows you to 'drag' the little square; the side or corner increases or decreases to change the size or aspect ratio of the plot. Most plots can also be shifted right or left using the main menu Format/Left Justify or Format/Right Justify. A smartplot( ) also allows you to change the features of selected curves so that it appears different from the others. This is done by simply selecting the curve and using the right mouse button. Choose the colour or style you want.

Curves can just be 'thrown onto' a smartplot and the effects of different structure or arguments can then be compared or used to fit data. This 'tweaking', with active intervention and replotting, gives another route for interpretaion and presentation of data. Most of the features of smartplot( ) are also available in 3D. However, a plot of 2D curves cannot have a surface, but a 3D surface can contain curves.

Select with the arrow over curve; look for a colour change

```
>   smartplot(cos(x),sin(y));
```

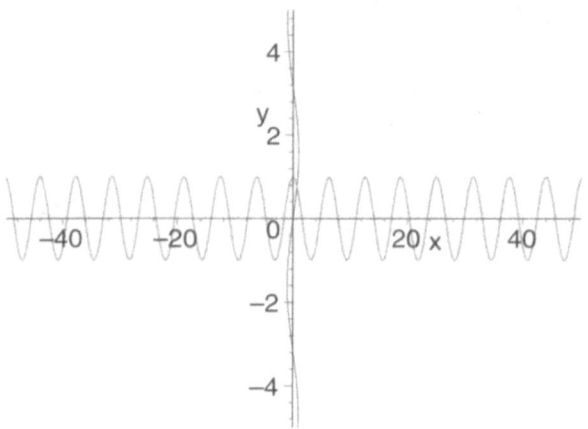

Solar Radiation Maximum

The following solar radiation fluxes were collected through the day; nighttime values are not shown and were effectively zero. The data were recorded manually and are roughly average values for the time intervals; they seem to be a truncated sinesoidal function. What is the expected maximum value of solar radiation flux that day? The sequence contains a listing of the time interval in hours and the average values measured in $\frac{watts}{m^2}$.

```
>   [6,7+11/60,107],[7+11/60,9+19/60,294],[9+19/60,11+28/60,577],
>   [11+28/60,14+31/60,608],[14+31/60,16+30/60,382],[16+30/60,18,132];
```

$$[6, \frac{431}{60}, 107], [\frac{431}{60}, \frac{559}{60}, 294], [\frac{559}{60}, \frac{172}{15}, 577], [\frac{172}{15}, \frac{871}{60}, 608], [\frac{871}{60}, \frac{33}{2}, 382],$$

$$[\frac{33}{2}, 18, 132]$$

```
>   rdat := [%]:
```

The data are best plotted as a histogram or bar graph. We won't use the histogram plotting capabilities of Maple here but make a smartplot compatible function that effectively does the same thing. It uses the classical step function of Heaviside; this is zero for all arguments less than zero and 1 for all arguments greater than zero. Heaviside allows us to have disjointed or 'pasted together' functions. First we make a procedure cut to take a 'piece' out of any function. We immediately see that the negative of the cosine, taken over a day of 2*Pi radians, gives the right sort of function.

Use the right mouse button to change the Axes/Ranges

```
>   cut := proc(x,x1,x2)
```

```
>   Heaviside(x-x1)*Heaviside(x2-x);

>   end;
```
$cut := \mathbf{proc}(x,\ x1,\ x2)\,\text{Heaviside}(x - x1) \times \text{Heaviside}(x2 - x)\,\mathbf{end}$

```
>   smartplot(cut(x,Pi/2,3*Pi/2)*500*(-cos(x)));
```

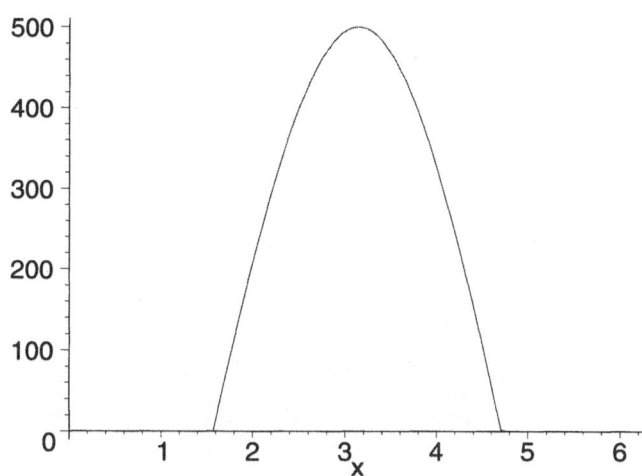

Next we alter the data and make a histogram, for a 2*Pi radian day.

```
>   change2rad := proc(ls)
>   [2*Pi/24*op(1,ls),2*Pi/24*op(2,ls),
>   op(3,ls)]; end;
```
$change2rad := \mathbf{proc}(ls)\,[1/12 \times \pi \times \text{op}(1,\ ls),\ 1/12 \times \pi \times \text{op}(2,\ ls),\ \text{op}(3,\ ls)]\,\mathbf{end}$

```
>   raddat := map(change2rad,rdat);
```

$$raddat := [[\frac{1}{2}\,\pi,\ \frac{431}{720}\,\pi,\ 107],\ [\frac{431}{720}\,\pi,\ \frac{559}{720}\,\pi,\ 294],\ [\frac{559}{720}\,\pi,\ \frac{43}{45}\,\pi,\ 577],\ [\frac{43}{45}\,\pi,\ \frac{871}{720}\,\pi,\ 608],$$
$$[\frac{871}{720}\,\pi,\ \frac{11}{8}\,\pi,\ 382],\ [\frac{11}{8}\,\pi,\ \frac{3}{2}\,\pi,\ 132]]$$

```
>   histo := proc(x,dat::listlist)
>   local n,ss,i,y,rng;
>   ss := [];
>   for i to nops(dat) do;
>   op(i,dat); y:=op(3,%); rng:=op(1..2,%%);
>   ss:=[op(ss),y*cut(x,rng)];
>   od;
>   convert(ss,'+');
>   end;
```

$histo := \mathbf{proc}(x, \, dat :: listlist)$

$\mathbf{local}\, n, \, ss, \, i, \, y, \, rng;$

$\quad ss := [\,] \, ;$

$\quad \mathbf{for}\, i \,\mathbf{to}\, \text{nops}(dat)\, \mathbf{do}$

$\qquad \text{op}(i, \, dat)\, ; \, y := \text{op}(3, \, \%)\, ; \, rng := \text{op}(1..2, \, \%\%)\, ; \, ss := [\text{op}(ss), \, y \times \text{cut}(x, \, rng)]$

$\quad \mathbf{od};$

$\quad \text{convert}(ss, \, ` + `)$

$\mathbf{end}$

> `histo(x,raddat);`

$$107\, \text{Heaviside}(x - \frac{1}{2}\,\pi)\, \text{Heaviside}(\frac{431}{720}\,\pi - x)$$

$$+\, 294\, \text{Heaviside}(-\frac{431}{720}\,\pi + x)\, \text{Heaviside}(\frac{559}{720}\,\pi - x)$$

$$+\, 577\, \text{Heaviside}(-\frac{559}{720}\,\pi + x)\, \text{Heaviside}(\frac{43}{45}\,\pi - x)$$

$$+\, 608\, \text{Heaviside}(-\frac{43}{45}\,\pi + x)\, \text{Heaviside}(\frac{871}{720}\,\pi - x)$$

$$+\, 382\, \text{Heaviside}(-\frac{871}{720}\,\pi + x)\, \text{Heaviside}(\frac{11}{8}\,\pi - x)$$

$$+\, 132\, \text{Heaviside}(-\frac{11}{8}\,\pi + x)\, \text{Heaviside}(-x + \frac{3}{2}\,\pi)$$

> `smartplot(%);`

The Axes/Ranges
and Style/Line
Width are altered

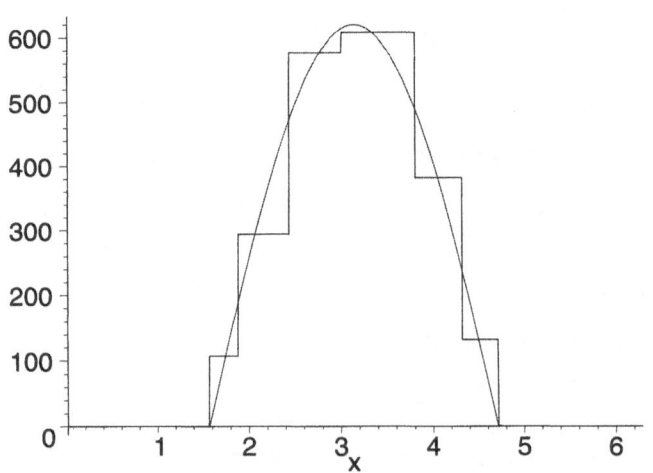

> `smartplot(cut(x,Pi/2,3*Pi/2)*620*(-cos(x)));`

The solution uses a 'drag and drop' or 'cut and paste' technique to get a fit, by eye. The last statement is left unexecuted. Clean up the smartplot by altering Axes/Ranges and the Style/Line Width/Medium with the right mouse button. Adjust the multiplier to achieve a 'best' fit. A trial and error 'eyeball' fit suggests that the maximum solar radiation on that day was around $620 \frac{watts}{m^2}$

The phase may be adjusted by appending to the cos argument

Exercise 6.4 Powers of the Sine

Use smartplot( ) to get plots of ever-greater powers of the sine using the Ditto operation, by using 'drag and drop' with ever-higher powers into the plot. This operation could be done with any function and can be fast because smartplot doesn't replot each time.

```
>  s := sin(x):
>  smartplot(%):
>  s; s:=%^2:
```

Exercise 6.5 Crossed Lines for Data Points

Redo the histogram plot, above, with crossed line X's for data points. First, collect the data and use the average time as the abscissa and the given radiation flux values for the ordinate. From there make two lines of slope Pi/4 that go through a given point; one of positive slope and one of negative slope. The procedure Xp( ) below should do this and 'cut' an appropriate piece out of each line to leave an X shaped symbol at the given point; del is half the symbol size. The given smartplot is only written for one point; xi and yi are the coordinates. When all the data are plotted together, the graph will need a revised Range, as well as different Line Styles and Colours.

```
>  Xp := proc(x,xi,yi,del)
>  cut(x-xi,-del,del);
>  yi*((x-xi)+1)*%,
>  yi*(-(x-xi)+1)*%;
>  end;
```

$Xp := \mathbf{proc}(x, \xi, yi, del) \, cut(x - \xi, -del, del); \, yi \times (x - \xi + 1) \times \%, \, yi \times (-x + \xi + 1) \times \% \, \mathbf{end}$

```
>  smartplot(Xp(x,xi,yi,0.1));
```

6.4 Spreadsheets

A tabular presentation form to give easy visualisation. In Maple the syntax is very similar to that of Excel, Quatro Pro or Lotus 1-2-3. Typical features of spreadsheets are included, such as referring to the contents of other cells and copying the contents of a cell to new cells. The real difference lies in the capability of essentially infinite precision, algebraic or symbolic calculations, and the use of the multiple types available within Maple. In addition, the effects of simultaneous, symbolic calculations in a line or two dimensional grid is possible.

The spreadsheet resides within a Maple worksheet and has access to all the parameters that have been assigned. There are special menu functions, including a new menu line at the bottom, the Context Menu. Exchange between a spreadsheet and the worksheet can use the usual cut and paste, and drag and drop; interchanges of other than single cell contents are in the form of a matrix.

Setting up the Spreadsheet

or push Alt-i
or push Alt-r

Go to the Insert menu and drag the mouse down to Spreadsheet. A Spreadsheet should appear. The spreadsheet is probably too small; click inside the spreadsheet and move the mouse cursor near an edge; you have found the edge when an arrow appears; move just outside the spreadsheet and click again; a border should appear with 8 small, active black squares; moving the mouse cursor to one of these brings forth a double-headed arrow; dragging the edge, or better, the lower right corner, allows resizing of the spreadsheet.

Spreadsheet Menu

The double-headed, diagonal arrow resizes the spreadsheet

Click inside the Spreadsheet

If you are not happy with the size of the cells, want the spreadsheet size to match the grid, or otherwise don't like the look of the spreadsheet, go to the spreadsheet menu. Enter text and equations by dropping, pasting, or typing information into cells. Properties allows you to change the colour, alignment inside cells, precision of the calculation and decimal points displayed, as well as the evaluation, floating or symbolic. You can change the entire spreadsheet by selecting the upper left corner; a column by selecting A, B, C or whatever; a row by selecting 1,2, 3 or whatever; a cell by simply clicking on it.

Context Sensitive Menu

Activating or 'clicking on' the spreadsheet alters the lower menu bar. This menu is set up to be sensitive to context and is called the Context Menu CM; from the left, it has a fill, evaluate all stale cells, accept and evaluate the input, an undo when you are unhappy with the change you have made, and importantly, a long edit line on the right. The edit line is the place to put your input, once you have settled or clicked on a given cell. The line is accepted when you use CR but, be cautious, Maple will tend to 'smooth' your input and won't accept bad syntax. The captured image, below, shows the two entries for each cell: The number 13000/53 appears in cell E4; it is really an output as all cells in the spreadsheet are fully evaluated. The statement that produced the number 13000/53 appears in the Context Menu edit line, just above the window. The long bar to the right is the input; it shows the syntax of the input and all editing should be done in that line. The buttons just to the left of the edit line are fill, evaluate all, evaluate input and undo.

Single Entries

Individual pieces of data may be entered by 'clicking on' or 'activating' the cell and typing in the data. However, 'out-of-date' hatching does appear and one must be aware that a return executes the statement in the cell. To avoid this it is easier to work in the Context Menu CM edit line, the long white bar on the right; it appears when you click within the spreadsheet. If you want to place an entry directly into a cell, it should be in the form of Maple Output; then it can be 'cut and pasted' or 'drag and dropped' directly into a cell. Remember that the visible entries in the spreadsheet are most like Maple Output. After entry and activation; click on evaluate input in the Context Menu or press the Enter key. During an edit, if you click on undo before you press CR, it will undo the latest change; the edit field in the Context bar returns to the original cell content.

Ditto

The spreadsheet does acknowledge the Ditto or % operation, the aforesaid, both in the spreadsheet and in the worksheet. This is one way of transferring rather awkward objects such as matrices, lists or paragraphs to the spreadsheet. These items do need to be Maple Output, however, and 'what went before' or 'previously' or 'above' is now at least 3D, including up and down and the worksheet.

Drag and Drop

The ordinary 'drag and drop' works by clicking on a cell or highlighting a group of cells, moving the mouse cursor just outside the highlighted area and back in; at the edge of the area the mouse cursor should turn into our familiar arrow; then we can drag the whole lot to a different point on the spreadsheet; the new location should be at the tip of the arrow. Drag and drop seems quite open to error so be aware that the button on the toolbar, Edit/undo or the left button on the Context Menu are there; they should be able to undo any accidental changes–but don't count on it.

Delete uses highlight and the Del key

Fill

Entering data in a column or row requires some sort of indexing, usually starting with a sequence of numbers. On a spreadsheet, this is done with Fill. Fill starts by entering the first number, in this case 1, in the upper left corner of the spreadsheet, row 1, column A. Thinking of column or vertical data entry, we select column A or highlight a sufficient number of rows to suit our purposes. Fill is activated either with the Spreadsheet/Fill, the right mouse button or the icon at the leftmost end of the Context Menu. We Fill in the first column with the numbers 1 to 4, with Fill/Detailed/, a Step Size of 1 and a Stop Value of 4, to correspond to our range of interest.

Automatic Data Entry

With indices placed within the spreadsheet, variables assigned outside the spreadsheet can be brought inside by simply executing the cell. In this case, we type data[~A1] into cell B1. The ~ means that this is a reference to a cell. Fill down, activate by dragging the left mouse button over the first cell and the remaining cells you want filled; complete the fill by activating the left most CM button or, with the mouse cursor on the highlighted cells, press the right mouse button. Click on Fill/Down to complete the entry of the names for the data points. Look at an individual cell entry by clicking on it to see the unexecuted entry in the Context Menu edit bar. Executing the cells, in this case one at a time, completes the data entry. In the figure below, cells B3 and B4 have not been executed. Of course the worksheet statement, below, must be executed to bring the silly names through into the spreadsheet.

Click on, with the left mouse button

Execution occurs with CR

```
>  data := [Arron,Bilke,Clave,Dilta];
```
$$data := [Arron,\ Bilke,\ Clave,\ Dilta]$$

	A	B	C	D	E
1	1	*Arron*			
2	2	*Bilke*			
3	3	$data_3$			
4	4	$data_4$			
5					

Retrieving Data from Cells

Highlight the information you want, use Ctrl-c or the right mouse button or Edit/Copy from the menu. Move to the position in the worksheet that you want the information. Ctrl-v or right mouse Paste or Edit/Paste delivers the information, but again, most easily in the edit bar of the Context Menu. Note the format; we have highlighted cells B2 through B4 from the spreadsheet above and paste it into the execution group below.

```
>   MATRIX([[Bilke], [Clave], [Dilta]]);
```
$$\left[\begin{array}{c} Bilke \\ Clave \\ Dilta \end{array} \right]$$

From Digitizer to Spreadsheet

Revision of our capture routine allows that we collect data from an image and place it into the spreadsheet.

Again, when you run the procedure, consider adjusting the size of the transparent screen to normal, moving it about, and updating it with the tab key. This allows us to see the data coming in while we run the arrow cursor around the image of the lake. Any image will do; the image is in the file perrylakes.eps or perrylakes.gif, which can be viewed with most painting or photographic packages, including Microsoft Word. In this instance, we have collected a rough set of data just to demonstrate the spreadsheet.

```
> capture := proc() local fl,result,dd;
> currentdir("c:/mplbook/image-capture");
> ssystem("capturew -f pipefile");
> fl := fopen("pipefile", READ);
> result := 0:  dd := NULL;
> while not (result = []) do
> result := fscanf(fl, "%d,%d,%*s");
> if (not (result = 0)) and (not (result = [])) then
> print(result); dd := dd,%;
> fi;
> od:
> fclose(fl);
> RETURN([dd]);
> end;
```

$capture := \mathbf{proc}()$
$\mathbf{local}\, fl,\, result,\, dd;$
  currentdir("c:/mplbook/image-capture") ;
  ssystem("capturew -f pipefile") ;
  $fl := \text{fopen}(\text{"pipefile"},\, READ)\,;$
  $result := 0\,;$
  $dd := NULL\,;$
  $\mathbf{while\ not}\ (result = [])\,\mathbf{do}$
    $result := \text{fscanf}(fl,\ \text{"%d,%d,%*s"})\,;$
    $\mathbf{if\ not}\ (result = 0\ \mathbf{or}\ result = [])\,\mathbf{then}\,\text{print}(result)\,;\ dd := dd,\,\%\ \mathbf{fi}$
  $\mathbf{od};$
  $\text{fclose}(fl)\,;$
  $\text{RETURN}([dd])$
$\mathbf{end}$

```
> #capture():  data := %;
```

[241, 182]   Deactivated to keep from destroying the information

[296, 188]

[365, 203]

[404, 215]

[467, 235]

[520, 249]

[573, 276]

$data := [[241, 182], [296, 188], [365, 203], [404, 215], [467, 235], [520, 249], [573, 276]]$

```
> data:=[[241,182],[296,188],[365,203],
> [404,215],[467,235],[520,249],[573,276]]; # for data recovery
```

To enter these data in the spreadsheet, put sequential numbers in the first column either by highlighting about 10 squares (to A10) with the mouse or using the down arrow key along with the shift key; then select Spreadsheet/Fill/Detailed, use the right mouse menu or click in the left-most part of the Context Menu; fill in the requester, using a Step Size of 1 and a Stop Value of 10. Clicking the right mouse within the spreadsheet or using the lowest item on the spreadsheet menu also allows the 'resize to grid' function, which makes room for the cells of interest.

The arrow key with the shift key has a smooth action

Now, click on cell B1 and enter data(~A1) into the Context Menu edit window, on the right. The first data point numbers data[1] should appear in the spreadsheet. Click on the head of the column or highlight cells in the column, including B1 to B7. Again, fill down; the relative referencing of the spreadsheet will automatically adjust the index so that the data are collected sequentially into the spreadsheet. Undefined cells are then deleted by highlighting and using the del key.

You may need to reactivate the data statement

WestLake	[x, y values]		xpos		ypos	Assignments
	pixels		metres		metres	
1	[241, 182]	$\frac{24100}{53}$	454.7169811	$\frac{13600}{53}$	256.6037736	
2	[296, 188]	$\frac{29600}{53}$	558.4905660	$\frac{13000}{53}$	245.2830189	
3	[365, 203]	$\frac{36500}{53}$	688.6792453	$\frac{11500}{53}$	216.9811321	
4	[404, 215]	$\frac{40400}{53}$	762.2641509	$\frac{10300}{53}$	194.3396226	
5	[467, 235]	$\frac{46700}{53}$	881.1320755	$\frac{8300}{53}$	156.6037736	
6	[520, 249]	$\frac{52000}{53}$	981.1320755	$\frac{6900}{53}$	130.1886792	
7	[573, 276]	$\frac{57300}{53}$	1081.132075	$\frac{4200}{53}$	79.24528302	
Factor	$\frac{100}{53}$					

Resizing cells

Now that we have an idea as to how much data were collected, the cells can be resized using the right mouse button or spreadsheet/row or spreadsheet/column, to a size of about 80 pts. push the upper left square button on the spreadsheet to select the whole spreadsheet and use spreadsheet/column to change the width of all cells. Also, we like titles on the columns; highlight the first row, the square 1 button, go to spreadsheet/row/insert ; insert two rows and add titles. the calibration factor to convert to metres is placed in the bottom; it is a little crowded; place the cursor just after the 10 for row ten, when it is located just right, a double horizontal line icon appears; then the line can be grabbed and moved to format the row. Again, resize to grid.

Calibration and Referencing

Go to cell C3, enter ~B10*op(1,~B3); the op(1,~B3) refers to the x value; the application of the $ to the horizontal and vertical references make the reference absolute; so as to always refer to the same cell. Highlight cell C3 and the cells below; fill down; the x data points are complete, except we would like to see them in decimals. The spreadsheet menu and the right mouse menu have a properties submenu to set up floating point arithmetic specifics. Massage the data using evalf( ); enter into cell D3 and fill downward. With the column still highlighted, in the Spreadsheet/properties menu, select floating point, 10 Digits for the calculation and 1 decimal for display. Repeat for the y data, using op(2, B3) and adjust the data so it reads in the appropriate direction with an offset of 600 metres.

Assignment

The data may need to be accessible outside of the spreadsheet, in the worksheet. There are many ways of doing this. If only a few values are required, single assignments of selected values may suffice, using assign( ). But usually access to more data is required, at least some of the output columns or rows. Two ways are presented, one rather dirty and quick; the other requires a judicious assignment of variables.

Assignment as a Matrix

The first way is clear, collect the spreadsheet itself and make it an array; it already is a type of array so this isn't very difficult. The statement below allows flexibility, including collection as a list of lists. Click on the upper left corner of the spreadsheet. Use Ctrl-c to copy it; paste it in the blank space in the statement. Activate the execution group with CR. A mess will appear. Type Ctrl-z or use the Edit menu to Undo Typing. All the mess disappears; activating the next statement selects the rows and columns of interest and makes an orderly matrix.

```
> with(linalg):

> array(op(evaln(        ))):
> SM := submatrix(%,1..10,1..6):
> seq(SM[1,i],i=1..6);
```

submatrix is part
of linalg

$$\text{“WestLake”}, [x, y\ values], 0, xpos, 0, ypos$$

```
> eval(SM);
```

“WestLake”	$[x, y\ values]$	0	$xpos$	0	$ypos$
0	$pixels$	0	$metres$	0	$metres$
1	[241, 182]	$\frac{24100}{53}$	454.7169811	$\frac{13600}{53}$	256.6037736
2	[296, 188]	$\frac{29600}{53}$	558.4905660	$\frac{13000}{53}$	245.2830189
3	[365, 203]	$\frac{36500}{53}$	688.6792453	$\frac{11500}{53}$	216.9811321
4	[404, 215]	$\frac{40400}{53}$	762.2641509	$\frac{10300}{53}$	194.3396226
5	[467, 235]	$\frac{46700}{53}$	881.1320755	$\frac{8300}{53}$	156.6037736
6	[520, 249]	$\frac{52000}{53}$	981.1320755	$\frac{6900}{53}$	130.1886792
7	[573, 276]	$\frac{57300}{53}$	1081.132075	$\frac{4200}{53}$	79.24528302
Factor	$\frac{100}{53}$	0	0	0	0

Notice that the array has defaulted to 10 digit floating point numbers. If desired, single decimal figures can be regenerated in the fourth and sixth columns with searchtext( ) and parse( ).

```
> convert(1081.13075,symbol); searchtext('.',%);
> parse(substring(%%,1..%+1));
```

$$1081.13075$$
$$5$$
$$1081.1$$

The second way involves the use of assignment within the spreadsheet. For individual variables required in the worksheet, this is always possible with an assignment using, say, the last row in the spreadsheet. Following this statement by a CR results in the necessary execution and assignment.

```
> assign(Factor=~$B$10)
```

Assigning a Whole Row

For the assignment of an entire row, a few simple choices must be made. It is not possible to retrieve the name of a given cell in the spreadsheet; the level of evaluation is always complete. Hence we recover the name from the initial row. In both of the cases below, we copy the statement under 'Assignments' in the spreadsheet, into cell G3. The first statement gives an extended name of the column; the second statement gives a direct association in terms of positioning in the spreadsheet, using lower case letters. In this case both versions were copied into the spreadsheet and filled down through row 10. When this is followed with Spreadsheet/Evaluate Selection the sequential values are assigned. After the assignments, the column can be erased and used again. The assignments are one way so they remain available to the worksheet, compatible with values in the spreadsheet at the time of assignment. Unassignments can be handled in a similar way, using statements like unassign(evaln(d.˜A3)). Also, be aware that any cell hatched or marked 'out of date' should be activated or executed before usage.

```
>   assign(evaln(xpos.˜A3)=˜D3)

>   assign(evaln(d.˜A3)=˜D3)

>   xpos1;
                        454.7169811

>   d1;
                        454.7169811
```

Placing Lists and Arrays into Spreadsheets

Below we enter a list of data, a title; then, our matrix SM. The title is entered as data in a row, lining the indices horizontally. This means putting a 1 in A1 and Fill/Detailed /Right Step Size 1 Stop Value 10; then, ttl[˜A1] is entered into cell A2 and filled right through F2; the data appear in row 2. Putting the matrix SM back into the spreadsheet requires vertical indices in column A; highlight the column and Fill/Detailed/Downward; enter eval(SM[˜$A1,˜B$1]) into cell B3; highlight exactly from B3 to B12 with shift-down arrow; fill down; highlight all cells in the rectangle from B3 to F12; fill right. Lastly, we can make it look nice by altering the number of digits using Spreadsheet/Properties/Precision and remove the border with Spreadsheet/Show Border.

click on A
Step Size 1 Stop
Value 10

The spreadsheet
doesn't look nice

```
>   ttl := [''','','Data','converted','to','Metres'];
            ttl := [ , , Data, converted, to, Metres]

>   SM[1,1];
                        "WestLake"
```

1	2	3	4	5	6
2		Data	converted	to	Metres
3	$[x, y\ values]$	0	xpos	0	ypos
4	pixels	0	metres	0	metres
5	[241, 182]	$\dfrac{24100}{53}$	454.7169811	$\dfrac{13600}{53}$	256.6037736
6	[296, 188]	$\dfrac{29600}{53}$	558.4905660	$\dfrac{13000}{53}$	245.2830189
7	[365, 203]	$\dfrac{36500}{53}$	688.6792453	$\dfrac{11500}{53}$	216.9811321
8	[404, 215]	$\dfrac{40400}{53}$	762.2641509	$\dfrac{10300}{53}$	194.3396226
9	[467, 235]	$\dfrac{46700}{53}$	881.1320755	$\dfrac{8300}{53}$	156.6037736
10	[520, 249]	$\dfrac{52000}{53}$	981.1320755	$\dfrac{6900}{53}$	130.1886792
	[573, 276]	$\dfrac{57300}{53}$	1081.132075	$\dfrac{4200}{53}$	79.24528302
	$\dfrac{100}{53}$	0	0	0	0

Printing a Spreadsheet

The filled-in entries of a spreadsheet are printed as a fully rectangular spreadsheet with entry A1 in the upper left-hand corner. All entries that contain 'spaces' are considered as entries in a spreadsheet so a 'border' will be printed if entries on the outer edges are filled with 'spaces'. If it is desired to hide the headers, go to the Spreadsheet/Show Border and remove the tick.

Fonts in Spreadsheets

The 2D output font is used in the spreadsheet; if the style for 2D output is changed under 'Format/Styles' from the menu bar, the entries in a spreadsheet will change. Provided the spreadsheet is active when the font change is made, the font is specific to the spreadsheet, not the whole Maple worksheet.

Plotting from the Spreadsheet

Maple is a powerful tool in dealing with data and displaying data; there are many algorithms for data handling, interpretation and statistics. Here we take the simplest view of the spreadsheet above and plot column D as xpos and column F as ypos. Rather than do assignments, as shown above, we use a collection technique to create a list of x and y positions. In the next two input lines we collect cells D5 to D11 and F5 to F11, respectively. Highlight, copy with Ctrl-c, paste with Ctrl-v. With a large number of points, the 'dirty'

Highlight with a shift key and an appropriate arrow

technique above collects the whole spreadsheet and delivers the data quickly; here it is quick and clear with a simple copy and paste. After copy, edit the input line before execution or, as here, convert the matrices to listlists.

Add a colon, and use CR

```
>  MATRIX([[454.7169811], [558.4905660], [688.6792453],
>  [762.2641509], [881.1320755], [981.1320755], [1081.132075]]):
>  MATRIX([[256.6037736], [245.2830189], [216.9811321],
>  [194.3396226], [156.6037736], [130.1886792], [79.24528302]]):
>  xll := convert(%%,listlist);
```

$xll := [[454.7169811], [558.4905660], [688.6792453], [762.2641509], [881.1320755],$
$[981.1320755], [1081.132075]]$

```
>  yll := convert(%%,listlist);
```

$yll := [[256.6037736], [245.2830189], [216.9811321], [194.3396226], [156.6037736],$
$[130.1886792], [79.24528302]]$

```
>  xypts := zip((x,y)->[op(x),op(y)],xll,yll);
```

$xypts := [[454.7169811, 256.6037736], [558.4905660, 245.2830189],$
$[688.6792453, 216.9811321], [762.2641509, 194.3396226],$
$[881.1320755, 156.6037736], [981.1320755, 130.1886792],$
$[1081.132075, 79.24528302]]$

```
>  plots[pointplot](xypts);
```

plot works just as well

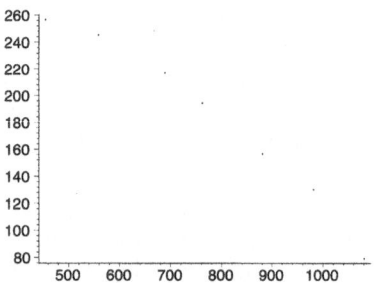

6.5   I/O from Spreadsheets

Maple 6 does this automatically

The exchange of data from one set of software to another is full of pitfalls. Much of the material in Maple can be moved from one window to another with a simple edit, copy and paste, either with the menu or HotKeys but, with Spreadsheets, it is important to get the matrix of information right, retaining the missing information, string information as well as proper precision in the numbers. With Maple 6, much of the difficulty has been removed and one can, more or less, highlight either spreadsheet and copy and paste the data matrix into the other. In the following, robust and semi-transparent ways of completing the exchange are presented, for use with earlier versions of Maple, to present the underlying structures, and allow exchanges that may still be difficult.

Importing from Excel

This may be done with an ASCII file having spaces or commas as delimiters (see sections 4.1, 4.2 and 4.3.4). Here we consider a simple cut and paste. The data are simply cut from a column of an Excel spreadsheet. As a demonstration, we edit the data to make an assignment statement and place the data between double quotes to make a string. The string sv is a vertical column; the string sh is a horizontal row.

```
>  sv := "   3.722
>  3.645
>  3.602
>  3.546
>  3.506
>  3.515
>  3.510
>  3.119
>  3.144
>  ";
```

$$sv := \text{`` } 3.722\backslash n3.645\backslash\ n3.602\backslash n3.546\backslash n3.506\backslash n3.515\ \backslash n3.510\backslash n3.119\backslash n3.144\ \backslash n\ \text{''}$$

```
>  sh := "   WL15.7222.8083.645
>  ";
```

$$sh := \text{`` } WL1\backslash t5.722\backslash t2.808\ \backslash t3.645\backslash n\ \text{''}$$

The procedure Sparse edits the string and puts it in the form of a listlist with extra list brackets around the elements to make sure that commas are not misinterpreted. Each character is checked in turn and substitutions make an appropriate Maple listlist. The start and the end require special consideration and the case of multiple rows. Sparse( ) parses the \t and \n string delimiters to make a listlist construction inside Maple. The companion program put( ) goes inside the spreadsheet to fetch the data.

Note that the data set collected must end with a \n or CR. Also, if the data contain mixed numbers and letters (as "152.1 m2"), the parse(%) statement will produce an error; if so, it is possible to massage the central block of statements to convert a space into a "*" Lastly, note the inappropriate use of 'is' as a local variable; 'is' being a function that is part of the assume facility.

```
> Sparse := proc(x::string)
> local n,xx,y,yy,i,ii,is,ie,ss,flag;
> n := length(x); y:="";
> for is from 1 to n while x[is]=" " do; od;
> # senses leading blanks
> for ii from n to 0 by -1 while x[ii]=" " do; od;
> # senses trailing blanks
> if x[ii]<>"\n" then RETURN("Error--not Excel copy") fi;
> # must have \n or CR at end of string
> ie := ii-1; flag := false;
> # check for multiple rows
> for i from is to ie do; xx := x[i]; yy:=xx;
> if  xx = "\t" then  yy:="],[" fi;
> if  xx = "\n" then  flag := true; yy:="]],[[" fi;
> y:=cat(y,yy); od;
> cat("[[",y[1..length(y)],"]]");
> if flag then "[".%."]" else % fi;
> parse(%);
> end;
```

$$Sparse := \mathbf{proc}(x::string)$$

$$\mathbf{local}\, n,\, xx,\, y,\, yy,\, i,\, ii,\, is,\, ie,\, ss,\, flag;$$

$$n := \text{length}(x)\,;$$

$$y := \text{""}\,;$$

$$\mathbf{for}\, is\, \mathbf{to}\, n\, \mathbf{while}\, x_{is} = \text{" "}\, \mathbf{do}\, \mathbf{od}\,;$$

$$\mathbf{for}\, ii\, \mathbf{from}\, n\, \mathbf{by}\, -1\, \mathbf{to}\, 0\, \mathbf{while}\, x_{ii} = \text{" "}\, \mathbf{do}\, \mathbf{od}\,;$$

$$\mathbf{if}\, x_{ii} \neq \text{" \textbackslash n"}\, \mathbf{then}\, \text{RETURN}(\text{"Error–not Excel copy"})\, \mathbf{fi}\,;$$

$$ie := ii - 1\,;$$

$$flag := false\,;$$

$$\mathbf{for}\, i\, \mathbf{from}\, is\, \mathbf{to}\, ie\, \mathbf{do}$$

$$xx := x_i\,;$$

$$yy := xx\,;$$

$$\mathbf{if}\, xx = \text{"\textbackslash t"}\, \mathbf{then}\, yy := \text{"],["}\, \mathbf{fi}\,;$$

$$\mathbf{if}\, xx = \text{"\textbackslash n"}\, \mathbf{then}\, flag := true\,;\, yy := \text{"]],[["}\, \mathbf{fi}\,;$$

$$y := \text{cat}(y,\, yy)$$

$$\mathbf{od}\,;$$

$$\text{cat}(\text{"[["},\, y_{1..\text{length}(y)},\, \text{"]]"})\,;$$

$$\mathbf{if}\, flag\, \mathbf{then}\, \text{"[".%."["}\, \mathbf{else}\, \%\, \mathbf{fi}\,;$$

$$\text{parse}(\%)$$

$$\mathbf{end}$$

```
> Sparse(sh); MH := %:
```

$$[[WL1], [5.722], [2.808], [3.645]]$$

```
> Sparse(sv); MV := %:
```

$$[[[3.722]], [[3.645]], [[3.602]], [[3.546]], [[3.506]], [[3.515]], [[3.510]], [[3.119]],$$
$$[[3.144]]]$$

The procedure put( ) enters the elements of M into the spreadsheet. A
string is collected from Excel and converted (parsed) by Sparse( ) to a listlist,

so that it is ready for insertion into the Maple spreadsheet; this is copied directly into the spaces in the last statement below; a CR produces the fully parsed listlist M. With row 1 and column A as indices, put(~$A2,~B$1,M) is typed into the edit bar into the spreadsheet location 2,B. Filling down and right completes the transfer from Excel.

```
>  put := proc(i::posint,j::posint,x::listlist) local err,y,r;
>  err :='improper op or subscript selector'; y := NULL;
>  r := traperror(op([1,1],x));
>  if r=err then RETURN fi;
>  if type(r,list) then
>  r := traperror(op(op([i,j],x)));
>  if r=err then RETURN else y:=r fi;
>  elif i=1 then
>  r := traperror(op(op(j,x)));
>  if r=err then RETURN else y:=r fi;
>  else RETURN;
>  fi;
>  y;
>  end;
```

$put := \mathbf{proc}(i::posint, \ j::posint, \ x::listlist)$
$\mathbf{local} \ err, \ y, \ r;$

$\quad err := \text{'}improper \ op \ or \ subscript \ selector\text{'};$

$\quad y := NULL;$

$\quad r := \text{traperror}(\text{op}([1, \ 1], \ x));$

$\quad \mathbf{if} \ r = err \ \mathbf{then} \ RETURN \ \mathbf{fi};$

$\quad \mathbf{if} \ \text{type}(r, \ list) \ \mathbf{then}$

$\qquad r := \text{traperror}(\text{op}(\text{op}([i, \ j], \ x))); \ \mathbf{if} \ r = err \ \mathbf{then} \ RETURN \ \mathbf{else} \ y := r \ \mathbf{fi}$

$\quad \mathbf{elif} \ i = 1 \ \mathbf{then} \ r := \text{traperror}(\text{op}(\text{op}(j, \ x))); \ \mathbf{if} \ r = err \ \mathbf{then} \ RETURN \ \mathbf{else} \ y := r \ \mathbf{fi}$

$\quad \mathbf{else} \ RETURN$

$\quad \mathbf{fi};$

$\quad y$

$\mathbf{end}$

```
>  seq(put(1,i,MH),i=1..4);
```
$$WL1, \ 5.722, \ 2.808, \ 3.645$$

```
>  seq(put(j,1,MV),j=1..9);
```
$$3.722, \ 3.645, \ 3.602, \ 3.546, \ 3.506, \ 3.515, \ 3.510, \ 3.119, \ 3.144$$

```
>  M := Sparse("    "):
```

	1	2	3	4
1	PL1	7.079	2.880	3.722
2	WL1	5.722	2.808	3.645
3	WL2	5.480	2.776	3.602
4	WL3	5.308	2.735	3.546
5	N1a	4.857	2.710	3.506
6	N1b	4.945	2.715	3.515
7	N1c	5.013	2.711	3.510
8	N2a	5.083	2.381	3.119
9	N2b	5.138	2.399	3.144
10	N2c	5.214	2.405	3.142
11	WL4	5.806	2.341	3.013
12	WL5	7.426	2.260	2.817
13	WL6	3.427	2.682	3.447

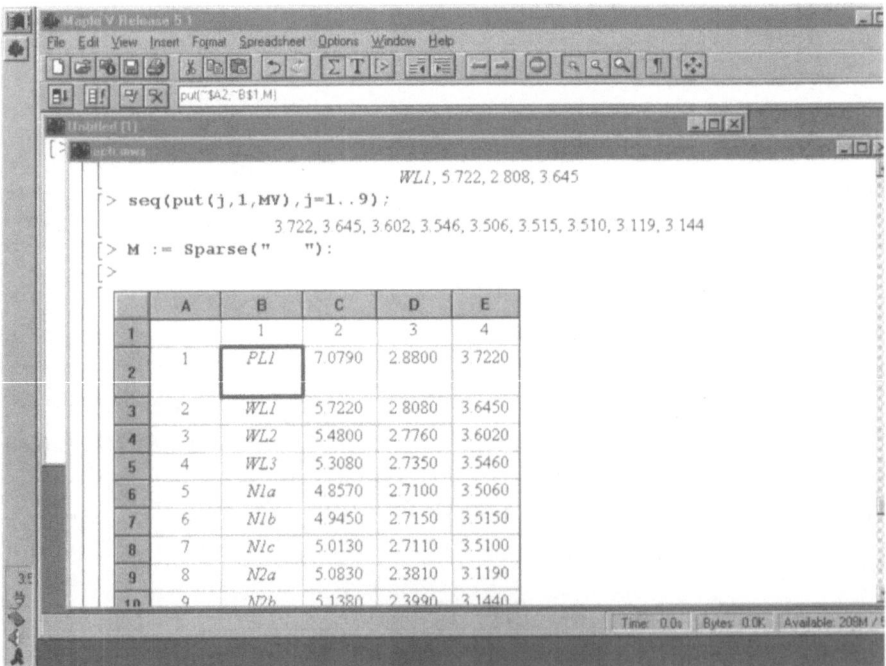

The procedure put( ) is purposely made robust so as not to corrupt the spreadsheet when the op commands are out of range. The special use of traperror( ) prevents awkward error remarks in the output. It will not care for everything but subscript variations are highly tolerated. As long as there are integers for indices and a listlist for the third argument, the put( ) will cleanly fill the spreadsheet without any real need to worry about errors.

```
>  put(10,1,MV);

>  put(8,2,MV);

>  put(a,2,MH);
```

Error, put expects its 1st argument, i, to be of type posint, but received a

```
>  put(10,,MH);
```

',' unexpected

Exporting to Excel

As shown, the data collected directly from the spreadsheet are in the form of matrices or lists. This format must be traded for the \t and \n format required to enter Excel directly. An alternative is to put the data in a file using writeline, writedata, print, lprint or printf; then, separately open the file with Excel and, setting the right delimiters, import the data into Excel. Here we undertake an adventure with direct transfer and find it rewarding. Some of the data from above are copied into a listlist.

This is done by typing op(1,) at the [> and cutting and pasting some lines from the spreadsheet into the blank spaces of the function call. Execution makes the data into a listlist. Conversion to a string follows, looking at the name as a symbol. This whole operation is done up front inside the procedure Eparse( ) below. Copying only specific rows and columns is done by highlightening the cells of interest. It may be convenient to use Shift+Click (left mouse button) to end the highlighting.

```
>  MM := MATRIX([[PL1, 7.079, 2.880, 3.722],
>  [WL1, 5.722, 2.808, 3.645], [WL2, 5.480, 2.776, 3.602],
>  [WL3, 5.308, 2.735, 3.546], [N1a, 4.857, 2.710, 3.506],
>  [N1b, 4.945, 2.715, 3.515], [N1c, 5.013, 2.711, 3.510],
>  [N2a, 5.083, 2.381, 3.119]]):

>  op(1,MM); # convert( ,listlist); # MVr5.1
```

$[[PL1, 7.079, 2.880, 3.722], [WL1, 5.722, 2.808, 3.645], [WL2, 5.480, 2.776, 3.602],$
$[WL3, 5.308, 2.735, 3.546], [N1a, 4.857, 2.710, 3.506], [N1b, 4.945, 2.715, 3.515],$
$[N1c, 5.013, 2.711, 3.510], [N2a, 5.083, 2.381, 3.119]]$

```
>  convert(%,string);
```

"[[PL1, 7.079, 2.880, 3.722], [WL1, 5.722, 2.808, 3.645], [WL2, 5.480, 2.776, 3.6\
02], [WL3, 5.308, 2.735, 3.546], [N1a, 4.857, 2.710, 3.506], [N1b, 4.945, \
2.715, 3.515], [N1c, 5.013, 2.711, 3.510], [N2a, 5.083, 2.381, 3.119]]"

The procedure insert replaces one substring in the string with another. The argument list contains: first, the string to be inserted; second, the string; third, the string to be removed. SearchText( ) is a case sensitive search that tells the position of the first character of the string. The procedure selects the characters before the outgoing string ous and, separately, the characters remaining. The new string y is built up of the characters with the incoming substring. The remaining string bit is broken up until SearchText( ) finds no more of the outgoing string ous.

\t or tab and \n or CR are single characters

In this particular case, except for the ends, the commas are replaced with "\t" and the "],[" with a "\n".

```
> insert := proc(ins::string,x::string,ous::string)
> local bit,j,m,y,n;
> j := length(ous);
> bit := x; y := "";
> m:=SearchText(ous,bit);
> while m>0 do;
> n := length(bit);
> substring(bit,1..m-1); y:=cat(y,%,ins);
> substring(bit,m+j..n); bit := %;
> m:=SearchText(ous,bit);
> od;
> y:=cat(y,bit);
> end;
```

$$insert := \mathbf{proc}(ins{::}string,\ x{::}string,\ ous{::}string)$$
$$\mathbf{local}\ bit,\ j,\ m,\ y,\ n;$$
$$j := \text{length}(ous)\,;$$
$$bit := x\,;$$
$$y := \text{""}\,;$$
$$m := \text{SearchText}(ous,\ bit)\,;$$
$$\mathbf{while}\, 0 < m\, \mathbf{do}$$
$$\quad n := \text{length}(bit)\,;$$
$$\quad \text{substring}(bit,\ 1..m-1)\,;$$
$$\quad y := \text{cat}(y,\ \%,\ ins)\,;$$
$$\quad \text{substring}(bit,\ m+j..n)\,;$$
$$\quad bit := \%\,;$$
$$\quad m := \text{SearchText}(ous,\ bit)$$
$$\mathbf{od};$$
$$y := \text{cat}(y,\ bit)$$
$$\mathbf{end}$$

```
>  mh := convert(map(op,MH),string); # test string
```
$$mh := \text{"[WL1, 5.722, 2.808, 3.645]"}$$

After the
commas there are
unseen spaces

```
>  SearchText(", ",mh);
```
The space is visible if Options/Output Display/Maple Notation is used.
$$5$$

```
>  insert("\t",mh,", ");
```
$$\text{"[WL1\t5.722\t2.808\t3.645]"}$$

```
>  mv := convert(map(op,MV),string); # test string
```
$$mv := \text{"[[3.722], [3.645], [3.602], [3.546], [3.506], [3.515], [3.510], [3.119], [3.144]]"}$$

```
>  insert("\n",mv,"], [");
```
$$\text{"[[3.722\n3.645\n3.602\n3.546\n3.506\n3.515\n3.510\n3.119\n3.144]]"}$$

We proceed by ignoring the first [[changing the], [for the \n and the ,
for \t and replacing the last]] with a \n . The individual elements can not be
complex or Excel could not deal with them; so a simple conversion is adequate.

The procedure Eparse includes the extraction of listlist from the MATRIX
form of the Ctrl-c copy from the spreadsheet; Eparse then converts the list to
a string. The montage produces a final string which is converted to a symbol
form and lprint presents it appropriately to the copy buffer.

```
>  Eparse := proc(MatriX)
>  local x,n,y;
>  convert(eval(MatriX),listlist);
>  x := convert(%,string);
>  n := length(x); y:="";
>  if x[1]<>"[" then RETURN fi;
>  if x[2]<>"[" then
>  if x[n]<>"]" then RETURN
>  else
>  insert("\t",x[2..n-1],", ");
>  y:=cat(%,"\n");
>  fi;
>  elif x[n-1..n]<>"]]" then RETURN;
>  else
>  insert("\n",x[3..n-2],"], [");
>  insert("\t", % ,", ");
>  y := cat(%,"\n");
>  fi;
>  fprint(convert(y,symbol)); # use lprint MVr5.1
>  end;
```

$Eparse := \mathbf{proc}(MatriX)$
$\mathbf{local}\, x,\, n,\, y;$
   $\text{convert}(\text{eval}(MatriX),\, listlist)\,;$
   $x := \text{convert}(\%,\, string)\,;$
   $n := \text{length}(x)\,;$
   $y := \text{""}\,;$
   $\mathbf{if}\, x_1 \neq \text{"["} \ \ \mathbf{then}\, RETURN\, \mathbf{fi}\,;$
   $\mathbf{if}\, x_2 \neq \text{"["} \ \ \mathbf{then}$
      $\mathbf{if}\, x_n \neq \text{"]"} \ \ \mathbf{then}\, RETURN\, \mathbf{else}\, \text{insert}(\text{" \t"},\, x_{2..n-1},\, \text{", "})\,;\, y := \text{cat}(\%,\, \text{"\n"})\, \mathbf{fi}$
   $\mathbf{elif}\, x_{n-1..n} \neq \text{"]]"} \ \ \mathbf{then}\, RETURN$
   $\mathbf{else}\, \text{insert}(\text{" \n"},\, x_{3..n-2},\, \text{"],"},\, \text{["})\,;\, \text{insert}(\text{"\t"},\, \%,\, \text{", "})\,;\, y := \text{cat}(\%,\, \text{"\n"})$
   $\mathbf{fi}\,;$
   $\text{fprint}(\text{convert}(y,\, symbol))$
$\mathbf{end}$

```
>  Eparse(MM);

PL1 7.079 2.880 3.722
WL1 5.722 2.808 3.645
WL2 5.480 2.776 3.602
WL3 5.308 2.735 3.546
N1a 4.857 2.710 3.506
N1b 4.945 2.715 3.515
N1c 5.013 2.711 3.510
N2a 5.083 2.381 3.119
```

Highlight this, Ctrl-c copy or use Edit/Copy. Open up an Excel worksheet, Ctrl-v paste or Edit/Paste. The actual process allows a highlight and
copy from the spreadsheet that is pasted directly into the blank field in the
statement below. This makes a mess for a large file; execute the statement
with a CR. Immediately highlight and copy the output; open Excel, paste it
in. Two Undos; Undo Result and Undo Paste should clean up. If not done
immediately, or Undo does not work, don't be suprised.

```
>  Eparse(   ):
```

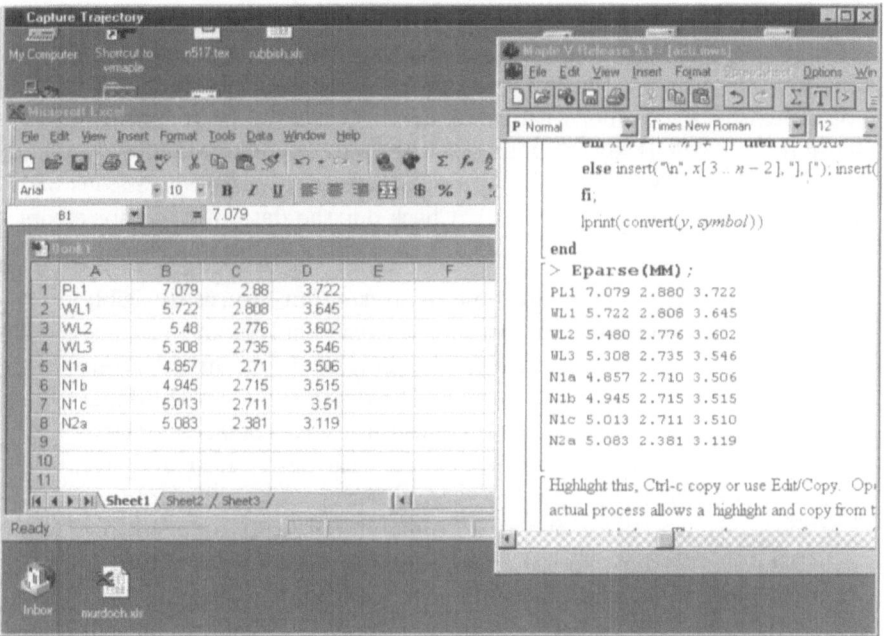

Exercise* 6.6 Automatic Plot

Use the composite statements below to do the plotting directly, with a single procedure. Two matrices are copied from the spreadsheet into the assignment statements; alternatively, the matrices can be copied into the argument list of the subroutine.

```
>  A :=        :
>  B :=        :
>  plotspread := proc(AD,BD) local xl,yl;
>  xl := map(op,AD); # convert( ,listlist);
>  yl := map(op,BD); # Maple V Revision 5.1
>  zip((x,y)->[x,y],xl,yl);
>  plot(%); end:
>  plotspread(A,B);
>  plotspread(   ,   );
```

Exercise* 6.7 Save to a file

Review sections 4.1, 4.2, 4.3 and 5.6.2, page 276, where readline( ) and readdata( ) were used. Reverse the process and and write the data in the above spreadsheet to a file; with writeline( ), writedata( ), or fprintf( ) save the data into an ASCII file in a format with spaces or commas and CR so that Excel can position them correctly. You may want to do some conversions with sscanf( ) or formatting with printf( ). Check out the data in Excel for errors.

> This chapter is more updated and upmarket, but returns to visible analysis, of functions and data. An inverse analysis of functions solves for the required arguments from graphical contours, the conditions for kidney stone formation. Digitization done with Maple and a companion image program unveils the edges of lakes. Smartplots are attractive for playing games with data. Last and not least, spreadsheets in Maple are competitive. They can produce a nice output and use the algorithms, functionality and plotting of Maple. It is noteworthy that programs based on spreadsheets are tending to add-on programming capabilities. Maple has such functionality as a built-in feature, and can be friendly. Indeed, interactive, symbolic patterns in 2D (the spreadsheet) may well spawn future algebras, with 2D synergism.

Acknowledgements for Chapter 6

Stewart Greenhill wrote the Oberon 2 source code for the programs capture, capturef, capturep, capturec, capturew as used in the procedures capture and post. The massage of the image files was completed by Pat Scott. Much of the information in the smartplots and spreadsheet sections comes from either the Maple Help files or Waterloo Maple Inc., through Scott Rabuka and the Frequently Asked Questions of The Maple Reporter.

Bibliography

[1] Forman S. Acton: *Numerical Methods that usually Work*, Harper and Row, New York, 541 pages, (1970).

[2] H. Agler, A. Ahearn, A. Kitchell, N. Lopanik, H. Miller, M. Miller, and D. S. Wethey: Resource Competition in Algae: A Class Project in Mathematical Biology, *MapleTech*, Vol 4, pp78-85, Birkhäuser, (1997).

[3] W. C. Bauldry, and J. R. Fiedler: *Calculus Laboratories with Maple: A Tool, not an Oracle*, Brooks/Cole, Belmont, California, 144 pages, (1991).

[4] J. Bear and A. Verrujt: *Modeling Groundwater Flow and Pollution, with computer programs for sample cases*, Reidel, Dorecht, California, *** pages, (1990).

[5] N. R. Blachman and Michael J. Mossinghoff: *Maple V Quick Reference*, Brooks/Cole, Pacific Grove, California, 522pages, (1991).

[6] G. W. Bluman and J. D. Cole: *Similarity Methods for Differential Equations*, Springer-Verlag, New York, 332pages, (1974).

[7] H. Bouwer: *Groundwater Hydrology*, McGraw-Hill, New York, 480pages, (1978).

[8] B. W. Char, B. W., K. O. Geddes, G. H. Gonnet, B. L. Leong, M. B. Monagan and S. M. Watt: *Maple V Language Reference Manual*, Springer-Verlag, New York, 267pages, (1991).

[9] R. M. Corless: *Essential Maple: An Introduction for Scientific Programmers*, Springer-Verlag, New York, 218pages, (1995).

[10] R. E. Crandall: *Topics in Advanced Scientific Computation*, Springer-Verlag, New York, 340pages, (1996).

[11] W. Ellis Jr. and E. Lodi: *Maple for the Calculus Student: a Tutorial*, Brooks/Cole, Belmont, California, 67pages, (1989).

[12] Wade Ellis, Jr., Eugene W. Johnson, Ed Lodi and Daniel Schwalbe: *Maple V Flight Manual* Tutorials for Calculus, Linear Algebra, and Differential Equations, Brooks/Cole, Pacific Grove, California 183 pages, (1992).

[13] A. Heck: *Introduction to Maple, 2nd Ed.*, Springer-Verlag, New York, 699 pages, (1996).

[14] J. V. Herod: A Model for an HIV Infection *MapleTech*, Fall 1993, pp 72-78, Birkhuser, (1993).

[15] K. M. Heal, M. L. Hansen and K. M. Rickard: *Maple V Learning Guide*, Springer-Verlag, New York, 274pages, (1996).

[16] K. M. Heal, M. L. Hansen and K. M. Rickard: *Maple 6 Learning Guide*, Waterloo Maple Inc., Waterloo, Ontario, 314pages, (2000).

[17] G. Klimek and M. Klimek: *Discovering Curves and Surfaces with Maple*, Springer-Verlag, New York, 217pages, (1997).

[18] P. E. Kloeden and E. Platen: *Numerical Solution of Stochastic Differential Equations*, Springer-Verlag, Heidelberg, 632pages, (1995).

[19] P. E. Kloeden, E. Platen and H Schurz: *Numerical Solution of SDE through Computer Experiments*, Springer-Verlag, Heidelberg, 292pages, (1997).

[20] M. Kofler: *Maple VR4, Einführing und Leitfaden für den Praktiker*, Addison-Wesley, 582pages. (1996).

[21] P. W. Kunkel and G. B. Ralston: *Schaum's Outline Series - Theory and Problems of Biochemistry*, McGraw-Hill, New York, 555pages, (1988).

[22] L. D. Landau and E. M. Lifshiz: *Fluid Mechanics, Course of Theoretical Physics, Volume 6*, Pergamon, Oxford, 536pages, (1959).

[23] M. B. Monagan, K. O. Geddes, K. M. Heal, G. Labahn and S. M. Vorkoetter: *Maple V Release 5 Programming Guide*, Springer, New York, 379pages, (1998).

[24] M. B. Monagan, K. O. Geddes, K. M. Heal, G. Labahn, S. M. Vorkoetter and J. McCarron: *Maple 6 Programming Guide*, Waterloo Maple Inc., Waterloo, Ontario, 585pages, (2000).

[25] W. D. Scott and P. V. Hobbs: The formation of Sulfate in Water Drops *Journal of the Atmospheric Sciences*, Vol 24, pp54-57, (1967).

[26] W. D. Scott T. J. Wrigley and K. M. Webb: A Computer Model of Struvite Solution Chemistry *Talanta*, Vol 38, pp889-895, (1991).

[27] T. Scott R. Pavelle and D. Redfern: Maple in Education *The Maple Technical Newsletter*, Issue 7, pp11-16, (1992).

[28] T. J. Wrigley, W. D. Scott and K. M. Webb: An Improved Model of Struvite Solution Chemistry *Talanta*, Vol 39, pp1597-1603, (1992).

AAArgh->

Build your own Help

Restricted Access

Myhelp

helper

Packaging Help

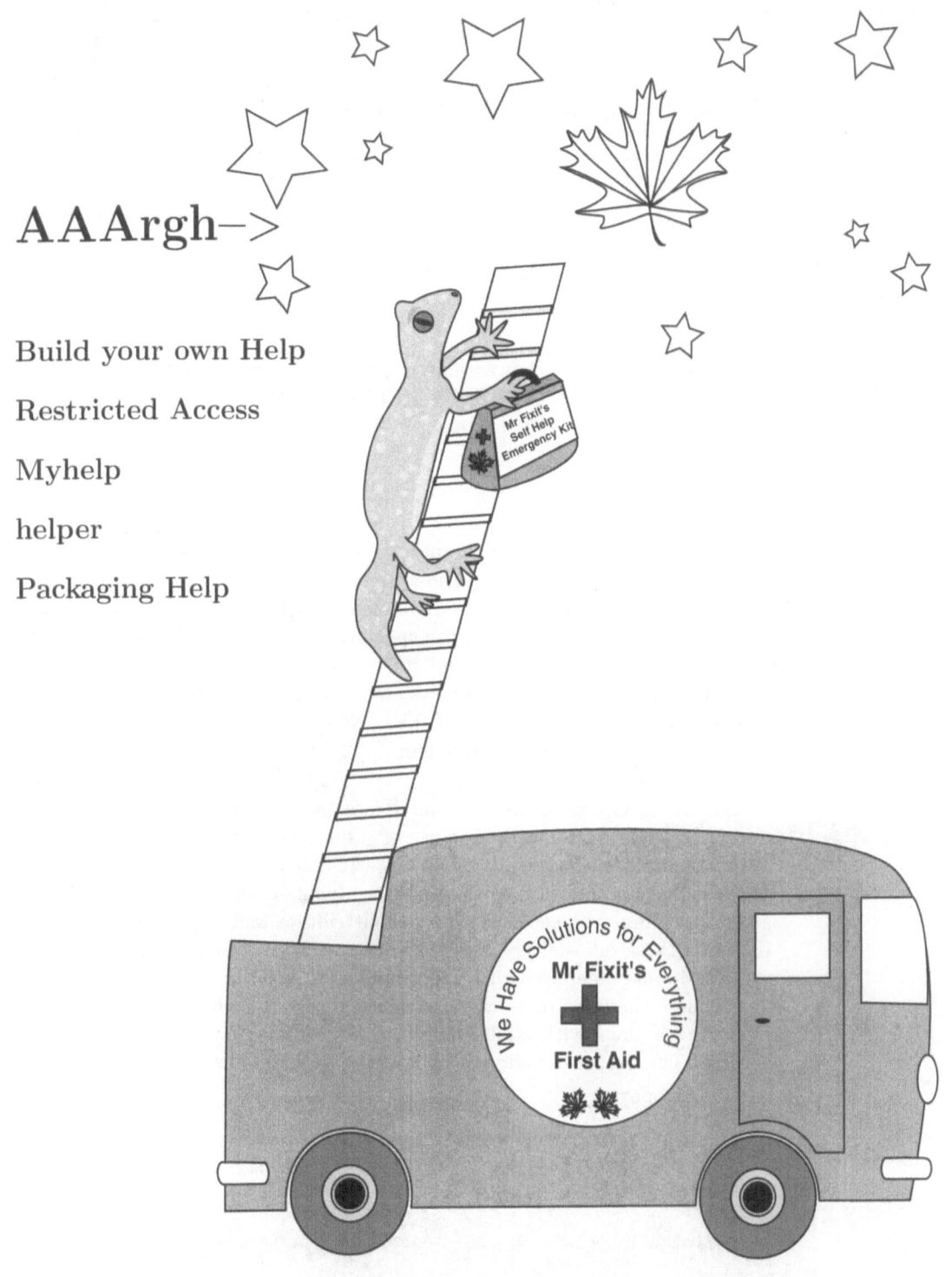

AAArgh->Build your own help

Übersicht

Into the bowels of Maple, to generate a help system. A continuation of the self help of section 2.5, page 97 ; consider making your own help files. Make an environment of your own that you enjoy.

Guidewords:

assume( ), :: , showassumed, ghost variables; Help, TEXT( ), PackHelpfix( ), INTERFACE_HELP( ), Mywith( ), Mywithout( ), Myhelp, helper, init, makehelp

Restricted Access

It is possible to restrict the characteristics of a variable; this can keep you from using a variable or a variable type for other than its intended purpose. Knowing what to expect, Maple can immediately let you know where you went wrong. Two common ways are the specification of variable type at the start of a procedure, type checking, and the use of assume. Type checking appears as a double colon :: with the parameters (arguments) to the procedure. Cross checking that your variables are of the right type is effective in avoiding errors and can help when there is a problem.

Assume

An alternative is to use assume. This comes with is( ) , about( ), additionally( ) and unassume( ). The use of assume will appear from time-to-time in this book; it is effective but powerful; to be avoided by new users of Maple. In the right circumstance (usually unassignment) it can produce 'ghost' 'variables that cannot be accessed, yet have attributes and appear in expressions. In short, these 'ghosts' are like attributes without a name. Nonetheless assume and unassume, used with judgement, can be quite effective. Assume can assign more than one property to a variable as well; as integer,prime. Adding properties is done with additionally or addproperty; not the repeated use of assume. The properties allowed by assume are listed in the help file.

```
>   ?property
>   restart:
>   interface(showassumed=1);# using 2 only alerts
>   int(exp(-a*t),t=0..infinity);
```

Definite integration: Can't determine if the integral is convergent.

Need to know the sign of --> a

Will now try indefinite integration and then take limits.

$$\lim_{t \to \infty} -\frac{e^{(-a\,t)} - 1}{a}$$

```
>  assume(a>0);# a is greater than zero
>  about(a);
```

Originally a, renamed a˜:

   is assumed to be: RealRange(Open(0),infinity)

```
>  int1 := int(exp(-a*t),t=0..infinity);
```

$$int1 := \frac{1}{a˜}$$

Knowing that a was greater than 0 allowed the integration to be completed. Now we explore how a acts as an assumed variable.

```
>  if a>0 then good else bad fi;
```

Error, cannot evaluate boolean

```
>  is(a>0);
```

$$true$$

```
>  if is(a>0) then good else bad fi;
```

The boolean must be true or false

$$good$$

The problem comes when we try to use the variable again.

```
>  a:=evaln(a); # evaluate to a name
```

$$a := a$$

This command is identical to unassign and the use of ' '. Let us see what has happened to the a's we have used. First, a look at the unassigned names for a.

```
>  nops([unames()]);
```

$$1635$$

Searching for a is difficult. Increase the high number in the range until you collect a.

```
>  un := sort([unames()])[1..543]:
>  member(a,un);
```

$$true$$

```
>  sort([unames()])[540..543];
```

$$[\_Y,\ \_Z,\ \_{lnGAMMA0},\ a]$$

```
>  eval(un,1)[543];
```

$$a$$

a is there, and unassigned, as it should be. But what if we use our original function; add an a, supposedly the unevaluated symbol a.

```
>  int2 :=int1 + a;
```

$$int2 := \frac{1}{a^{\tilde{}}} + a$$

> i2:=indets(int2);

$$i2 := \{a, a^{\tilde{}}\}$$

It is clear that the variable a˜ was not unassigned with the evaln statement above. Maple also presumed the value added was the new, unassigned a, the variable used in the evaln statement. Look at the assigned names

> ghosta := op(i2 minus {a});

$$ghosta := a^{\tilde{}}$$

> nops([anames()]);

$$417$$

> an := [anames()]:

The variable isn't assigned

> member(op(i2 minus {a}),[anames()]);

$$false$$

nor is it unassigned

> member(op(i2 minus {a}),[unames()]);

$$false$$

One variable in int1 and int2 doesn't have a recognised name. Proceeding, we can use subs to remove a˜ finally.

> int3:=subs(op(i2 minus {a}) =a,int2);

$$int3 := \frac{1}{a} + a$$

It is clear that you should be careful with assumed variables. This means

- never use assume twice on the same variable in one session.

- restart after using assume.

- remember expressions which have assumed variables, for recovery with indets( ).

The above is a round-about experience, but we have learned that

- Maple carries the last known value of a variable through to the next operation.

"No powerful command is without its dark side"

Overall, the effect is similar to the effect of local variables exported from procedures, **exported local variables**, as mentioned by Corless [9], pgs 139-140. These variables have been called **ghost variables**. It is difficult to simplify expressions with 'ghosts'; they are difficult to use, convert, or evaluate. Assume should be used with care.

Exercise A.1 Ghost Variables with Assume

In a completely new Maple session, change the attributes and assumptions regarding the single variable aa. Use about and unassign to remove it. Try to find it . It may happen that you have, effectively, several different aa symbols. If you substitute for aa, Maple presumes the last aa; some aa's may be hard to find. Heal et al [15], pg 76, suggests using a delayed evaluation that is done before unassignment of a.

```
>  int3 := subs(a='a',int2);
```

Assert

Use of kernelopts and setting of ASSERT=true allows that a message be printed whenever ASSERT is activated. The current state of assert is preset at false.

```
>  kernelopts(ASSERT);
```

$$false$$

First we turn on ASSERT

```
>  kernelopts(ASSERT=true);
```

$$false$$

```
>  a:=1:  ASSERT(a>5,cat('a is small:  ',a));
```

```
Error, assertion failed, a is small: 1
```

This gives a handle on how a is evolving during computations. The last statements could even be made a macro to get an immediate indexMylib response.

Help through Maple's help system

This requires a bit more work. We form the directory Mylib beside the lib directory of Maple or/and in your own working directory, here c:/mplbook/Mylib. Remember that a directory inside of Maple will have limited access. Then, set up a mapleini file with an additional path to this location. Here this is set up to be done with Maple opening, writing and closing the file, so as to avoid problems with file types and extensions. The name of the file is maple.ini in DOS and .mapleini in Unix.

```
>  #f := fopen("maple.ini",WRITE);
```

$$f := 0$$

```
>  #writeline(f,"libname:=
>  ""c:/Program files/Maple V release 5/lib"",
>  ""c:/mplbook/Mylib"";");
```

These lines are purposely commented out to avoid file corruption

```
>  #close(f);
```

There are several points of note here. First, the use of a file descriptor f to take the name of the file in use; it allows a simple name to take the place of the file name; the f:=0 means that this is the 0th file to be opened. A similar effect is obtained by simply assigning the filename, as a string, the name f. Second, is the use of " to make the file name a string; the double values correspond to the use of " within the string, that makes up the line to be put into the file name. Third, and importantly, is the path to the library, which includes a second path to Mylib, separated from the first by a comma. If the first path becomes corrupted, Maple can not find its libraries and will become non-functional. Fourth, notice that the Directory Structure in the path uses the Unix convention, the / ; Maple converts this correctly (parses it) into the appropriate syntax for your computer.

> `restart;`

libname :=" "c:/Program files/Maple V release 5/lib", "c:/mplbook/Mylib"

The proper listing of the path means that Maple has set up its path correctly; Maple can now be used to make a customised maple.hdb file to be included in the your own Maple help system.

Help/Save to Database

Now, start a new worksheet, include whatever help information you like. In DOS, go to the 'Help' menu under 'Save to Database'. Give the worksheet an appropriate name, one you are likely to remember. The name can be a part of a parent directory, as well; perhaps 'Plot/points' where you have made a procedure to make 'largedots'. Add any aliases you think could act as links. Put in the full path to Mylib; here c:/mplbook/Mylib/maple.hbd. Click 'save current'. This creates the maple.hbd file in Mylib.

> `?largedots`

The help file you made should come up in another window. The file is not changeable but can be removed using 'Remove Topic' in the 'Help' menu. Remember that you can copy bits of it at any time, like the built-in Maple Help files.

Don't hesitate to use this facility. It can keep you informed and allows a continuous upgrade of your knowledge, as well as improving Maple's help.

The menu approach to help, as above, can also be handled on the command line with INTERFACE_HELP.

Help with a Package

Here we take on a help file that appears with a package. This uses the somewhat dated technique of writing a series of procedures in a file and putting a large heading at the start of the file. This heading could be put into a worksheet and used as above, with the information going directly into the help system. But such a scheme doesn't allow for old packages, which are many;

they have been written with the old 'header' structure. Also, it doesn't easily allow a package to be included into the help system, as a whole entity, without a lot of effort. Two examples are given, tex and helper. tex is an old package, delivered with the share library with Maple Vr3, written by Yunliang Yu in 1993. The specific details of putting this help on to Maple are given. helper we make here, as a package to make your own help files.

```
>  AAArgh;
```

```
Warning, timeout parameter to ssystem ignored by this operating system
```

"`"

```
>  restart:
```

Updating the file

First we convert the file into a Maple Vr5 format. The file tt is a temporary, intermediate file. The final, converted file is tex, ultimately lodged in the Mylib directory. This could be done outside of Maple in the DOS prompt or on the command line in Unix. Here we work from inside Maple with the system and ssystem commands. The file tex.txt is a copy of the file tex from the Maple share library Maple V Release 3. It is lodged in your directory c:/mplbook/Mylib and needs to be updated. Again, a system command is used, this time the command does everything remotely, with a brief hint of what is going on.

First, let us have a simple read of it.

```
>  restart;
```

$libname :=$ "c:/Program files/Maple V release 5/lib", "c:/mplbook/Mylib"

```
>  readlib(readdata);
```

$$\mathbf{proc}(\mathit{fIdent}) \dots \mathbf{end}$$

```
>  for i from 1 to 3 do:  readline("tex.txt");
>  od;
```

This reads three lines of the file. When evoked repeatedly, it scans the file, 3 lines at a time.

"#\t   Copyright (c) 1993 by Yunliang Yu. All rights reserved."

"#"

"#\tThis package may be freely distributed for non-profit purposes "

We see that, in the least, this is the right file. The system command is used for the update. The editor allows that we correct errors. These commands are set up for IBM compatibles; the commands are different on Unix workstations and Macintosh computers.

```
>  libnme :="c:/mplbook/Mylib";
>  "updt3to4 ".libnme."/tex.txt > ".libnme."/tt";
```

"updt3to4 c:/mplbook/Mylib/tex.txt > c:/mplbook/Mylib/tt"

```
>  "updtsrc ".libnme."/tt > ".libnme."/tex";
```

"updtsrc c:/mplbook/Mylib/tt > c:/mplbook/Mylib/tex"

Using the % is
not the best way
to do this

```
> system(%%);
```
                                        0

```
> system(%%);
```
                                        0

The 0's indicate that the commands were successful. there should be a converted file tex in the directory Mylib, along with a junk file tt and tex.txt file, the old, original file. At some point, you may want to delete the file tt and, perhaps, tex.txt. There were some conversion errors, which should be corrected; the file tex supplied with the CD has been corrected in the hopes that the tex package commands will run properly.

```
> ssystem("cd")[2];
```
                        "C:\\mplbook\n"

```
> ssystem("edit tex")[2];
```

Here, the first use of the ssystem command shows that you are in the current directory mplbook. The tex files are in Mylib, a directory inside of mplbook. This is a little awkward but a use of open in the left, File menu gives a directory tree; clicking on the Mylib directory immediately lists the .txt file in the directory. Erase the .txt from the file name requestor, and you should see the tex file you have just made.

The converted file tex has a structure representative of the general help structure that exists in a varied form in Maple Vr5. Have a look at it in the editor. Other than the up-front comments with # marks, it consists of a series of assignments of names to TEXT structures, strings that are single lines of text that have names like directories. The TEXT structures are the help features of the document. Following the help features, is a series of procedures that allow a massage of LaTeX, or for export of files from Maple in latex format. mtex, particularily, changes the length of the output line in Maple, massaged for import to LaTeX.

Using an ASCII help file

The file is read in as a Maple Text; this automatically assigns the plot structures and reads in the procedures that come with the package. To find out which TEXT structures are read in and which procedures are spawned, we find the assigned names that are of type TEXT and procedure. The use of sets of names allows a subtraction of all the names that are irrelevant to the use of the package tex.

```
> restart;
```
libname := "c:/Program files/Maple V release 5/lib", "c:/mplbook/Mylib"

```
> pbefore := anames(procedure):
```

```
>   Tbefore :=anames(TEXT):

>   read "Mylib/tex";

>   pafter := anames(procedure):

>   anames(TEXT);
```

help/text/mtex, help/text/tex, help/tex/text/usage, help/TeX/text/example,
help/tex/text/example, help/text/TeX, help/TeX/text/usage

```
>   Tafter := 'anames(TEXT)';
```

$$Tafter := \text{anames}(TEXT)$$

The ' ' keeps
down the level of
evaluation

```
>   pnames := op({pafter} minus {pbefore});
```

pnames := tex/minus, tex_tall, tex_sort, tex_split, tex_print, tex_lead, tex, tex_pdt, mtex,
rlms, clct, stex, test, tex/union, tex/SetOps, tex/intersect

```
>   Tnames := op({Tafter} minus {Tbefore});
```

Tnames := help/text/TeX, help/text/tex, help/text/mtex, help/tex/text/usage,
help/TeX/text/usage, help/TeX/text/example, help/tex/text/example

If we want to use the procedures in pnames in another session of Maple, we can save them in Maple's internal format, with a .m extension. We won't do this here, but they could be saved in a library Mylib or in Maple's library with savelib( ), to be read with readlib( ).

```
>   #save(pnames,'tex.m');
```

This command saves the command pnames, from the package tex in the file tex.m in the current directory.

Enlisting the help Information

Utilising the help information requires the manipulation of TEXT structures, to make them easily accessible to the user. First we have a look at one of the TEXT structures.

```
>   'help/text/tex';
```

TEXT(, *mtex* :   *MAPLE* ===> *TeX*,

--,
> *Copyright (c) 1993 by Yunliang Yu. All rights reserved.,*
> *Please send bug reports, comments and ideas to yu@math.duke.edu,*

Main Features :, =============,
> *1). typesetting in AmS − TeX, LaTeX or Plain TeX,*
> *2). automatic line − breaking,*
> *3). multi − level sorting,*
> *4). name and index substitutions.*

Main Deficiencies :,

==================,
> *1). no automatic line − breaking for matrices or quotients,*
> *2). general tables are not implemented yet,*

Procedures Available :,

===================,
> *mtex, stex, tex, rlms, clct, test,*

SEE ALSO : ? *TeX* [*usage*], ? *TeX* [*example*].,

VERSION : $Id : mtex, v 1.22 1993/08/11 13 : 10 : 38 yu Exp yu $,

Repeating, this is the built-in help system, that comes with the package tex. In the file, the TEXT structures have a format that are ASCII text lines, presented as symbols, enclosed in a left or backquote ' ' . Each executed line with a semicolon contains a left hand, assigned variable that has a Directory Structure inside it, as the 'help/text/tex' above; these are symbols in Maple and can be evaluated; they evaluate from the command line to the TEXT Structures. The TEXT Structures are the help file for particular commands.

Displaying the TEXT Structures, or rather simply naming them on an execution line, brings the particular help lines into view when you use the CR.

With Maple Vr5, these help files are no longer viewed automatically. Here we present a method of doing that: first in a simplest way, by printing. In a following subsection we put together a package, helper, to put the information into Maple's help system.

First a test of a TEXT structure. To find the names of the TEXT structures, we use the assigned names command, anames.

```
>   sy := TEXT('this is the first item', 'this is
>   the second item', third);
```

$sy := \text{TEXT}(\textit{this is the first item, this is the second item, third})$

```
>   type(sy,TEXT); whattype(sy);
```

true

function

So we see that a TEXT structure is also a function. Also, note that we have used the ' ' to denote the arguments. These are symbols in Maple Vr5. The argument sequence could just as well be " " or strings. Strings can not be evaluated.

```
>   op(0..3,sy);
```

TEXT, this is the first item, this is the second item, third

```
>   op(1,sy);
```

this is the first item

```
>   whattype(op(1,sy));
```

symbol

```
>   st := TEXT("this is the first item", "this is
>   the second item", "third"):
>   op(0..3,st);
```

TEXT, "this is the first item", "this is the second item", "third"

```
>   whattype(op(1,st));
```

string

The names, except the test names st and sy, were derived from the reading in of the file tex.

```
>   anames(TEXT);
```

st, sy, help/text/mtex, help/text/tex, help/tex/text/usage, help/TeX/text/example,
   help/tex/text/example, help/text/TeX, help/TeX/text/usage

The output looks nicer if we use a variation of the name, in the form of a procedure TexT, to print the output properly.

```
>   tt :=subs(TEXT=TexT,sy);
```

tt := TexT(this is the first item, this is the second item, third)

```
>   TexT :=proc()
>   print(args);
>   end;
```

TexT := **proc**() print(args) **end**

```
>   tt;
```

this is the first item, this is the second item, third

We remove the test names st and sy ; this leaves the other TEXT structures, the help structures associated with the package tex. A special procedure PackHelpfix does this.

```
>  st := 'st':  sy := 'sy':
>  LN :=[anames(TEXT)]; nops(%);
```

$$LN := [help/text/mtex, \, help/text/tex, \, help/tex/text/usage,$$
$$help/TeX/text/example, \, help/tex/text/example, \, help/text/TeX,$$
$$help/TeX/text/usage]$$

7

Note the way the sequence is self made

```
>  PackHelpfix:=proc()
>  local LN,RN,sss,i;
>  LN :=[anames(TEXT)];
>  WLN :=[anames(TEXT)];
>  RN:=subs(TEXT=TexT,eval(LN));
>  sss := NULL;
>  for i from 1 to nops(LN) do;
>  sss := sss,LN[i]=RN[i]; od;
>  assign([sss]);
>  end;
```

$$PackHelpfix := \textbf{proc}()$$
$$\textbf{local } LN, \, RN, \, sss, \, i;$$
$$LN := [\text{anames}(TEXT)];$$
$$RN := \text{subs}(TEXT = TexT, \text{eval}(LN));$$
$$sss := NULL;$$
$$\textbf{for } i \textbf{ to } \text{nops}(LN) \textbf{ do } sss := sss, \, LN_i = RN_i \textbf{ od};$$
$$\text{assign}([sss])$$
$$\textbf{end}$$

Running PackHelpfix( ) does the simple conversion and, with anames, we have a simple help system available, derived from the tex package. Remember, however, that the names are no longer TEXT structures and can not be found so easily; they are print lines.

```
>  anames(TEXT);
```

$$help/text/mtex, \, help/text/tex, \, help/tex/text/usage, \, help/TeX/text/example,$$
$$help/tex/text/example, \, help/text/TeX, \, help/TeX/text/usage$$

```
>  PackHelpfix();
>  anames(TEXT);  # no names are found
>  `help/text/tex`;
```

, *mtex* :   *MAPLE ===> TeX*,

\-,

      Copyright (c) 1993 by Yunliang Yu. All rights reserved.,

    Please send bug reports, comments and ideas to yu@math.duke.edu,

Main Features :, *===============*,

    1). typesetting in AmS − TeX, LaTeX or Plain TeX,

    2). automatic line − breaking,

    3). multi − level sorting,

    4). name and index substitutions.

, *Main Deficiencies* :

==================,

    1). no automatic line − breaking for matrices or quotients,

    2). general tables are not implemented yet,

, *Procedures Available* :

====================,

,   *mtex, stex, tex, rlms, clct, test*

SEE ALSO : ? *TeX* [*usage*], ? *TeX* [*example*].

VERSION : $Id : mtex, v 1.22 1993/08/11 13 : 10 : 38 yu Exp yu $

Updating the Browser

An alternative is to use the Maple Help System, as we did with the personalised help files in the last section. We make and update maple.hdb in Mylib to include the information of tex. This uses a command line version of the menu system. First, we have a simple look at solve.

```
>   INTERFACE_HELP(display,topic=solve);
```

This is a clear entry to the Maple help system; the system only needs updating. As a start, we put the information on 'help/TeX/text/usage' into the help system. Then we have a look at full inclusion, with the package helper.

```
>   ?INTERFACE_HELP
```

It seems that the structures prepared above have disallowed the use of the automatic help. If, in fact, we had put the information in the browser first, then it would not have been necessary to override the work we have done above. In short, if you are happy with the help, as presented above, don't proceed.

```
>   restart;
```

libname := "c:/Program files/Maple V release 5/lib", "c:/mplbook/Mylib"

```
>   read 'c:/mplbook/Mylib/tex';
>   libnme :="c:/mplbook/Mylib";
```
$$libnme := \text{``c:/mplbook/Mylib''}$$
```
>   LN :=[anames(TEXT)]; nops(LN);
```

$$LN := [help/text/mtex, \ help/text/tex, \ help/tex/text/usage,$$
$$help/TeX/text/example, \ help/tex/text/example, \ help/text/TeX,$$
$$help/TeX/text/usage]$$

7

Consider help on "help/text/usage". This has to be a string (not a symbol) to work in the browser.
```
>   sym := 'help/text/usage';
```
$$sym := help/text/usage$$
```
>   str := convert(sym,string);
```
$$str := \text{``help/text/usage''}$$

Below, the procedure ftopic docks off the topic from the end of the help string, to use as information to the browser. The procedure substring follows each letter from the left side, until / is found; the position of the / defines the topic, which is assigned the name tc on evoking the procedure.
```
>   tc := 'tc';
```
$$tc := tc$$
```
>   ftopic:= proc(s) local j,m;
>   for j from 1 to length(s)
>   while (substring(s,-j)<>"/")
>   do; m:=j; od;
>   substring(s,-m..-1);
>   end;
```

$ftopic := \mathbf{proc}(s)$

   $\mathbf{local}\ j,\ m;$

   $\mathbf{for}\ j\ \mathbf{to}\ \text{length}(s)\ \mathbf{while}\ \text{substring}(s,\ -j) \neq \text{``/''}\ \mathbf{do}\ m := j\ \mathbf{od}\ ; \text{substring}(s,\ -m..-1)$

$\mathbf{end}$

```
>   tc := ftopic(str);
```
$$tc := \text{``usage''}$$
```
>   INTERFACE_HELP(insert,
>   topic=tc,text=eval(LN[3]),
>   library=libnme);# adds information into help
>   INTERFACE_HELP(insert,topic=tc,browser=str,
>   library=libnme);# add to Browser
```

Have a look in the Help Menu; under Topic or Full Text Search you should find an item usage. Click on help, topic, and type in usage. A window should appear, with a browser directory at the top. Other access points, help and text, should also work in Full TexT Search.

Deleting from the Help Browser

At this point the workings of help should start to become clear. First, we delete the information we lodged in help, which was just a test. It is important be scrupulously clean and not to build up misinformation.

```
> INTERFACE_HELP(delete,topic=tc,library=libnme);
```

```
> INTERFACE_HELP(delete,browser=str,library=libnme);
```

Finally check that the help browser deleted the message under usage and related topics.

```
> ?usage
```

```
> INTERFACE_HELP(display,topic=tc);
```

```
Could not find any help on "usage"
```

```
> help(usage);
```

```
Could not find any help on "usage"
```

Making a Mywith command

The scene is set to make a with command that loads in a package that contains the TEXT help structures. If this is an older package, it is presumed that it has been updated to Maple V Release 5, using updt3to4 and/or updtsrc and the raw editing has been completed that will make the package work.

Take care with the libraries. The following procedures Mywith( ) and Mywithout( ), require the appending of a second library, which is used to store the help (handbook) file maple.hdb

In Maple 6 use libname; libname:=%,%; CAUTIOUSLY

```
> restart;
```
$libname :=$ "c:/Program files/Maple V release 5/lib", "c:/mplbook/Mylib"
```
> anames(TEXT);
```
```
> PROCnames["tex"];
```
$$PROCnames \text{``tex''}$$
```
> libname[2];
```
$$\text{``c:/mplbook/Mylib''}$$

The global variable PROCnames is defined so that we can acquire the names of procedures that are defined when Mywith is read in. Global variable TEXTnames keeps track of the Directory Structures that are formed so that they may be deleted.

Importantly, the directory path to the file must be presented reliably. The help files are placed in libname[2].

```
> libnme :=libname[2];
```

$$libnme := \text{"c:/mplbook/Mylib"}$$

```
> pg := "tex";
```

$$pg := tex$$

```
> "".libnme."/".pg;
```

$$c:/mplbook/Mylib/tex$$

In Maple 6 put
tex in lib

```
> Mywith := proc(pg)
> local i,j,L,n,m,LI,lbne,s,spg,tc,Tn,fnme,Tb4,Pb4;
> global PROCnames,TEXTnames,ftopic;
> Tb4 := anames(TEXT);
> Pb4 := anames(procedure);
> lbne := libname[2];
> fnme := "".lbne."/".pg;
> spg := "".pg;
> read fnme;
> Tn := {anames(TEXT)} minus {Tb4};
> TEXTnames[spg]:=op(map(convert,Tn,string));
>{anames(procedure)} minus {Pb4};
> PROCnames[spg]:=op(map(convert,%,string));
> L := [op(Tn)];
> n := nops(L);
> for i from 1 to n do;
> LI := L[i];
> s := convert(LI,string);
> for j from 1 to length(s)
> while (substring(s,-j)<>"/") do;
> m:=j; od;
> tc := substring(s,-m..-1);
> INTERFACE_HELP(insert,topic=tc,text=eval(LI),library=lbne);
> # add info into help
> print(%);# necessary to activate INTERFACE
> INTERFACE_HELP(insert,topic=tc,browser=s,library=lbne);
> # add to Browser
> print(%);
> od;
> RETURN(op(Tn));
> end:
```

```
> Mywith("tex");
```

$$help/TeX/text/example, \; help/tex/text/example, \; help/text/TeX,$$
$$help/tex/text/usage, \; help/text/tex, \; help/text/mtex,$$
$$help/TeX/text/usage$$

TeX help also
appears in the
Browser

```
> ?TeX
```

```
> anames(TEXT);
```

help/text/mtex, *help/text/tex*, *help/tex/text/usage*, *help/TeX/text/example*,
   help/tex/text/example, *help/text/TeX*, *help/TeX/text/usage*

```
>  PROCnames["tex"];
```

"tex", "tex_tall", "tex_sort", "tex_split", "tex_print", "tex_lead", "tex/SetOps",
   "tex/union", "tex/minus", "tex_pdt", "mtex", "rlms", "clct", "stex", "test",
   "tex/intersect"

A Mywithout command is also necessary to remove any errors committed,
and keep the help system clean.

```
>  Mywithout := proc(spg)
>  local i,j,L,m,k,P,tc,s,lbne;
>  global PROCnames,TEXTnames,ftopic;
>  lbne := libname[2];
>  L  := [TEXTnames[spg]];
>  m := nops(L);
>  P := [PROCnames[spg]];
>  k := nops(P);
>  print('Deleting the ',m,' TEXTnames');
>  print(op(L));
>  print('Deleting the ',k,' PROCnames');
>  print(PROCnames[spg]);
>  for i to m do;
>  s := L[i];
>  for j from 1 to length(s)
>  while (substring(s,-j)<>"/") do;
>  m:=j; od;
>  tc := substring(s,-m..-1);
>  INTERFACE_HELP(delete,topic=tc,library=lbne);
>  print(%);
>  INTERFACE_HELP(delete,browser=s,library=lbne);
>  print(%);
>  od;

>  unassign(op(map(convert,L,symbol)));

>  unassign(op(map(convert,P,symbol)));

>  unassign(PROCnames,TEXTnames);

>  end:
```

It is clear, here, that not only do the TEXT structures and Procedures
have to be deleted, their names and the references to their names also need to
be deleted.

```
>  anames(TEXT);
```

help/text/mtex, help/text/tex, help/tex/text/usage, help/TeX/text/example,
help/tex/text/example, help/text/TeX, help/TeX/text/usage

```
>  PROCnames["tex"];
```

"tex", "tex_tall", "tex_sort", "tex_split", "tex_print", "tex_lead", "tex/SetOps",
"tex/union", "tex/minus", "tex_pdt", "mtex", "rlms", "clct", "stex",
"test", "tex/intersect"

```
>  Mywithout("tex");
```
Deleting the , 7,   TEXTnames

"help/tex/text/usage", "help/TeX/text/usage", "help/TeX/text/example",
"help/tex/text/example", "help/text/mtex", "help/text/tex",
"help/text/TeX"

Deleting the , 16,   PROCnames

"tex", "tex_tall", "tex_sort", "tex_split", "tex_print", "tex_lead", "tex/SetOps",
"tex/union", "tex/minus", "tex_pdt", "mtex", "rlms", "clct", "stex",
"test", "tex/intersect"

```
During delete of usage - topic not found

During delete of example - topic not found
```

These two topics
had no separate
help files

```
>  PROCnames["tex"];
```
$PROCnames$ "tex"
```
>  anames(TEXT);
>  ?TeX
```
There is no response to the TeX help request. The reference to tex is to
LaTeX in the standard Maple help library.

Myhelp in the Maple Directory

These two commands need to be lodged near the Maple library so as to be
easily available. Again, an additional library path should be placed in your
mapleini file as another companion library. The library Mylib is formed as a
new folder in the directory, as named below.

```
> restart;
```
libname := "c:/Program files/Maple V release 5/lib", "c:/mplbook/Mylib"

```
> libname;
```
"c:/Program files/Maple V release 5/lib", "c:/mplbook/Mylib"

```
> libname := libname,"c:/Program files/Maple V
> release 5/Mylib";
```

libname := "c:/Program files/Maple V release 5/lib", "c:/mplbook/Mylib",
    "c:/Program files/Maple V release 5/Mylib"

```
> ?savelib
> libname := "c:/Program files/Maple V release
> 5/lib", "c:/mplbook/Mylib";
```
libname := "c:/Program files/Maple V release 5/lib", "c:/mplbook/Mylib"

The helper Package

A file was written in standard ASCII text to document the use of package helper. It is available on the CD ROM companion to the book (or can be obtained from the author's web site or the Maple share library). Conversion commands three2four and four2five are included; they work within Maple but there is no guarantee that packages will work after conversion.

```
> restart;
```
libname := "c:/Program files/Maple V release 5/lib", "c:/mplbook/Mylib"

```
> libnme := libname[2];
```
        libnme := "c:/mplbook/Mylib"

```
> read ''.libnme.'/'.helper;
```

helper/infO := TEXT(*helper*,
      active Help and Browser for Maple, Bill Scott, Nov98,
    usage : *Mywith(package)*,   *package is an ASCII file*
  of TEXT Structures and procedures with a setup,
     ?Mywith or use Help menu or backquote,)

helper/With, helper/fix, helper/Mywith, helper/PackHelpfix, helper/three2four,
   helper/four2five, helper/infO

Mywith, TexT, ftopic, Mywithout, PackHelpfix, three2four, four2five

```
>   Mywithout("helper"):
```
$$\textit{There are , 7, \quad TEXT \quad names}$$
$$\textit{Deleting the , 7, \quad TEXTnames}$$

"helper/With", "helper/fix", "helper/Mywith", "helper/PackHelpfix",
    "helper/three2four", "helper/four2five", "helper/infO"

$$\textit{Deleting the , 7, \quad PROCnames}$$

"Mywith", "TexT", "ftopic", "Mywithout", "PackHelpfix", "three2four",
    "four2five"

Exercise A.2 Package Cleanup

Go through the help and the browser topics and see if any of the 'remains' of
the packages tex or helper remain. These would be a result of the various possible efforts during executing the commands, above. Use INTERFACE_HELP
to remove them to make a tidy help system.

A Package as a Table

Maple 6 prefers a
module structure

One way to 'get it all together' when you have tackled a problem, had some
success at definition and/or produced a final solution, is to make a 'package'
containing all the codes but particularly, the documentation. This can be done
effectively as a table of 'working elements' including introductory material,
procedures and help information. The documentation part and saving will
be covered here. This is always an important part of a job or project, but
it is often neglected. The information is stored as a part of the table in
a special procedure init, which contains startup definitions and procedures
and, as necessary, some Directory Structures or a list of Directory Structures.
makehelp or INTERFACE_HELP can put these into Maple's help and browser.

Here we take a hypothetical series of procedures called enlarge that will
make large dots, large letters, and thick lines on graphs. The procedures would
normally be multiple lines; the structures of the procedures are not important
here; they can be 'filled-in' later.

Packaging and Documentating

```
>   restart;
```
libname := "c:/Program files/Maple V release 5/lib", "c:/mplbook/Mylib"

```
>   enlarge :=table();
```

$$enlarge := \text{table}([$$
$$])$$

Use Ctrl+Shift
between lines to
produce a CR

```
>   enlarge[PTS] := proc();   end;
```
$$enlarge_{PTS} := \mathbf{proc}() \ \mathbf{end}$$

```
>   enlarge[letter] := proc();   end;
```
$$enlarge_{letter} := \mathbf{proc}() \ \mathbf{end}$$

```
>   enlarge[Thick]:= proc();   end;
```
$$enlarge_{Thick} := \mathbf{proc}() \ \mathbf{end}$$

```
>   enlarge[init] := TEXT('Package enlarge',
>   'Allows for adjusting the thickness of the lines',
>   'the size of the symbols, and the font size of the',
>   'letters when the size of a plot is made small'):
```

```
>   print(enlarge);
```

$$\text{table}([$$
$$letter = (\mathbf{proc}() \ \mathbf{end})$$
$$PTS = (\mathbf{proc}() \ \mathbf{end})$$
$$Thick = (\mathbf{proc}() \ \mathbf{end})$$
$$init = \text{TEXT}(Package \ enlarge,$$
$$Allows \ for \ adjusting \ the \ thickness \ of \ the \ lines,$$
$$the \ size \ of \ the \ symbols, \ and \ the \ font \ size \ of \ the,$$
$$letters \ when \ the \ size \ of \ a \ plot \ is \ made \ small)$$
$$])$$

Saving into a library

```
>   lbnme:=libname[2];
```
$$lbnme := \text{``c:/mplbook/Mylib''}$$

```
>   sl :="".lbnme."/enlarge.m";
```
$$sl := \text{``c:/mplbook/Mylib/enlarge.m''}$$

```
>   save(enlarge,sl);
```

```
>   restart;
```
$$libname := \text{``c:/Program files/Maple V release 5/lib''}, \text{``c:/mplbook/Mylib''}$$

```
>   with(enlarge);
```
$$[PTS, \ Thick, \ init, \ letter]$$

The with command simply 'scrapes off' the outer shell enlarge and lets you use the procedures and documentation directly; with( ) loads the file enlarge.m from the library c:/mplbook/Mylib and acts as a conduit so that the procedure names can be used conveniently. The procedure enlarge[init] has a special significance, however; it is read first and executed, and places the other procedures in an appropriate, defined environment. Global names are assigned, new types of variables are defined, and Maple's help activated; a general setup for the package is completed.

Saving with a Directory Structure

An augmented way of saving the documentation is with a Directory Structure, or Directory Structures, as companion files. The Directory Structures can be included in Maple's help.

Proecedures are
'filled-in' a little

```
> restart;  enlarge := table():
> enlarge[PTS] := proc(); [[a, b], [c, d]]; end:
> enlarge[letter] := proc(x) 200*x; end:
> enlarge[Thick] := proc() BROAD end:
```

The table structure recognises init as a special table element during initial setup by with( ). Here a banner and a standard list of commands are presented. The init procedure might also assign new variables, define macros or create new types of variables. The command with( ) delivers the short names for the package commands.

The init procedure constructs the foundation for the package. Notice, also, that a table of procedures may contain semicolons ; or multiple statements.

```
> enlarge[init] := proc() local i; global Info;
> Info := 'Package enlarge',
> 'Allows for adjusting the thickness of the lines,',
> 'the size of the symbols, and the font size of the',
> 'letters when the size of a plot is made small';
> for i in Info do print(i) od end;

> enlarge[TexT] := proc() local i; for i in args do
> print(i) od;  end;

> enlarge[Read] := proc() subs(TEXT=TexT,x); %; end;

> 'enlarge/infO' := TEXT(Info);
```

This composes three possible and plausible help scenarios: Firstly, the old TEXT Structure is there as a Directory Structure in 'enlarge/infO'. Secondly, the global variable Info gives the same information directly. Of course, there could be several of these Directory Structures, to give information on different procedures or details for using the package. Lastly, the table entry with init is automatically run when the package is made available with the shortened names, using with( ). Normally this would be saved as a .m file to conserve space but here I have chosen to save the file in ASCII, for editing. The ASCII version also allows that the file be read by any platform or version of Maple.

The TexT procedure is included for convenient formatting of TEXT Structures. Read( ) does the final formatting.

```
>    "c:\\mplbook\\Mylib\\":  # The directory
>    "".%.enlarge:  # "".%."enlarge.m"
>    #save(enlarge,'enlarge/inf0',%)
>    restart;
>    read cat(    ,enlarge):
```

Maple 6 uses ||
for catenation

Fetch directory
from above

```
>    with(enlarge);
```

Package enlarge
Allows for adjusting the thickness of the lines,
the size of the symbols, and the font size of the
letters when the size of a plot is made small
[inf0, PTS, TexT, Thick, init, letter]

Sequence not well
formatted

```
>    Info; # global sequence
```

Package enlarge, Allows for adjusting the thickness of the lines,,
the size of the symbols, and the font size of the,
letters when the size of a plot is made small

```
>    Read('enlarge/inf0';
```

Package enlarge
Allows for adjusting the thickness of the lines,
the size of the symbols, and the font size of the
letters when the size of a plot is made small

```
>    eval(TexT);
```

$\mathbf{proc}()\,\mathbf{local}\,i;\,\mathbf{for}\,i\ \mathbf{in}\,\text{args}\,\mathbf{do}\,\text{print}(i)\ \mathbf{end\,do\,end\,proc}$

```
>    Thick(),PTS(),letter(10); # the package is ready
```

BROAD, $[[a, b], [c, d]]$, 2000

```
>    with(plots):
>    packages(); # the package is available
```

[enlarge, plots]

Exercise* A.3 A Package for Plotting Big

Develop the package above, with the purpose of redefining the size of the points, letters and lines on a given plot. It is possible that these conditions could simply be 'appended' to the present plot( ) command; perhaps a command Plot( ) is in order. Carry the matter through so that a help system is available on call and from the Browser.

All three
arguments to
makehelp( ) need
to be names

Into Maple's help with makehelp

The command makehelp( ) takes the Directory Structure and puts it into the help system. The command takes three arguments: the reference name, the name of the file containing the information (which could be a .mws file), and the directory that will receive the maple.hdb (handbook) file. The maple.hdb files contain and update the data for Maple's help system; there can be several files in quite different locations, perhaps contained with your other workspace material on a given project. To be clear, full paths are included in the file names; some versions of Maple may not respond to partial names. The last argument could be any proper location; we use Maple's lib and makehelp( ) alters the file maple.hdb, the heart of Maple's help system. This produces a readily available help system.

If the location of the maple.hdb file is other than the standard library, libname, the library path for Maple, needs to be appended accordingly. The help preparer makehelp( ) will probably not accept string names, even converted string names. The problem relates to the massaging of backslashes. It is easiest to produce a output form of libname from Maple, copy from the output, and use direct editing to produce a symbol form.

First you bring in makehelp( ) with readlib( ). Then, go into your Maple/lib directory and make another directory Mylib2. The file enlarge.txt is lodged there with writedata( ); enlarge.txt contains the information as an ASCII equivalent, for 'enlarge/infO'; enlarge.txt needn't be permanent because maple.hdb will store the information, once makehelp( ) has done its transfer.

```
>   readlib(makehelp);
```
$$\mathbf{proc}(nm{::}name,\ file{::}name,\ lib{::}name)\ \dots\ \mathbf{end\ proc}$$

Another way to
look at TEXT
Structures

```
>   op('enlarge/infO');
```

Package enlarge, Allows for adjusting the thickness of the lines,,
the size of the symbols, and the font size of the,
letters when the size of a plot is made small

```
>   fn := 'C:\\PROGRAM FILES\\MAPLE 6\\lib\\Mylib2\\enlarge.txt';
```
$$fn := C:\backslash PROGRAM\ FILES\backslash MAPLE\ 6\backslash lib\backslash Mylib2\backslash enlarge.txt$$

A symbol
filename is
compatible below

```
>   F1:=open(fn,WRITE);
```
$$F1 := 0$$
```
>   writedata(F1,[op('enlarge/infO')],string);
>   close(fn);
```

the file
enlarge.txt is
formed

```
>   makehelp(evaln('enlarge/infO'),'a:\\Mylib2\\enlarge.txt',
>   'a:\\Mylib2'); # works from floppy  a:Mylib/maple.hdb
```

You may put the maple.hdb anywhere but you must have the names in symbolic form. A floppy is a bit difficult, however. The Maple/lib used here is a permanent and easily found library position. It should be kept scrupulously clean, however. The last clean up, through INTERFACE_HELP, removes the example information.

```
>  lbnm := 'C:\\PROGRAM FILES\\MAPLE 6\\lib\\Mylib2';
```
$$lbnm := C:\backslash PROGRAM\ FILES\backslash MAPLE\ 6\backslash lib\backslash Mylib2$$
```
>  makehelp(evaln('enlarge/inf0'),fn,lbnm);
>  # works from Maple/lib/maple.hdb
```
$$libname := a:\backslash Mylib2$$
```
>  ?inf0
```

The information appears in the help, with a Topic Search.

As a test, quit Maple. Remove the file enlarge.txt from the directory Maple/lib/Mylib2. Restart Maple. If you do not have appended libraries in your .ini file, restore your library Maple/lib/Mylib2. Retype ?info. The information is retained in the help file maple.hdb. To keep the library clean, purge this example from maple.hdb with INTERFACE_HELP.

```
>  INTERFACE_HELP(delete,topic=evaln('enlarge/inf0'),
>  library=lbnm);
```

Try ?info again. The topic is gone.

> We have looked at the inner workings of Maple, peeked and poked, and put together a self-help system. Help and upgrading are always necessary in any worthwhile software.

Acknowledgements for Appendix AAArgh

Much of the information on the use of packages came from Greg Fee and Michael Monagan of the CECM of Simon Fraser University. The detailed description of init usage comes from Monagan's book [23], pg100.

A Listing of Exercises

A Listing of Exercises

Note that the numbers that are starred * are more difficult, open ended and may be projects.

Improvised Routines

The routines listed below are mostly useful procedures in Maple; a macro, a package, some odd routines and DOS Windows programs are included.

Index Notes

The index entries contain **boldface** page entries, page ranges, as well as tagged pages. The **boldface** pages are more substantial references. Be aware that the page number accuracy is about a page. The tagged pages are for guidance:

i–Maple input, o–Maple output, f–figure, m–marginal note, e–exercise

Index

Back Cover Lookup

An active CDrom version of this book is available. Complementary software is also freely available from the author's web site

http://espc22.murdoch.edu.au/~scott
or maplesoft.com

Production: Druckhaus Beltz, Hemsbach